POLYMER SCIENCE AND TECHNOLOGY
Volume 20

# POLYMER ALLOYS III

Blends, Blocks, Grafts, and
Interpenetrating Networks

# POLYMER SCIENCE AND TECHNOLOGY

*Recent volumes in the series:*

A Continuation Order Plan is available for this series  A continuation order will bring delivery of each new
volume immediately upon publication  Volumes are billed only upon actual shipment  For further information
please contact the publisher

**POLYMER SCIENCE AND TECHNOLOGY**
Volume 20

# POLYMER ALLOYS III

## Blends, Blocks, Grafts, and Interpenetrating Networks

Edited by

## Daniel Klempner
## and Kurt C. Frisch

*Polymer Institute*
*University of Detroit*
*Detroit, Michigan*

Springer Science+Business Media, LLC

Library of Congress Cataloging in Publication Data

Symposium on Polymer Alloys (1981: New York, N.Y. Polymer alloys III.

    (Polymer science and technology; v. 20)
    "Proceedings of a Symposium on Polymer Alloys, sponsored by the Organic
Coating and Plastic Division, held August 23–28, 1981, in New York"—T.p.
verso.
    Includes bibliographical references and index.
    1. Polymers and polymerization—Congresses. I. Klempner, Daniel. II. Frisch,
Kurt Charles, 1918–    . III. American Chemical Society. Division of Organic
Coatings and Plastic Chemistry. IV. Title. V. Series.
QD380.S939 1981                547.7                82-24593
ISBN 978-1-4684-4360-8     ISBN 978-1-4684-4358-5 (eBook)
DOI 10.1007/ 978-1-4684-4358-5

Proceedings of a Symposium on Polymer Alloys, sponsored by the Organic
Coating and Plastic Division, held August 23–28, 1981, in New York

© 1983 Springer Science+Business Media New York
Originally published by Plenum Press, New York in 1983
Softcover reprint of the hardcover 1st edition 1983

PREFACE


On this, the dawning of a new age in high technology,
man is seeking answers to increasingly complex problems.
We are routinely launching reusable vehicles into space,
designing and building computers with seemingly limitless
powers, and developing sophisticated communications systems
using laser technology, fiber optics, holography, etc., all
of which require new and advanced materials.  Polymer alloys
continue to provide new solutions to the materials problems,
and remain an area of ever increasing research.


Polymer alloys are multicomponent macromolecular systems.
The components may be all on the same chain (as in block copolymers), on side chains (as in graft copolymers), or in
different molecules (as in polyblends and interpenetrating
polymer networks).  The variety of morphologies possible
and the synergistic effects on ultimate properties continue
to stimulate research on new polymer alloys.  More and more
studies on synthesis of new alloys, the kinetics and mechanisms of their formation, and their characterization, are
taking place, as well as studies on their processing and
applications.


This book presents the proceedings of the Symposium
on Polymer Alloys, sponsored by the American Chemical Society's
Division of Organic Coatings and Plastics Chemistry held at
the 182nd meeting of the American Chemical Society in New
York, in August, 1981.  The most recent efforts of scientists
and engineers from all over the world in this increasingly
important field are presented in the following pages.

We wish to express our appreciation to the authors who
contributed to this book, as well as to the University of
Detroit for their encouragement.

                                        Daniel Klempner

                                        Kurt C. Frisch

Polymer Institute

University of Detroit

CONTENTS

# MODEL PLASTIC-RUBBER COMPOSITES FROM EMULSION POLYMERS

Maurice Morton, N. K. Agarwal* and M. Cizmecioglu**
Institute of Polymer Science, The University of Akron,
Akron, Ohio 44325, USA

*E. I. du Pont de Nemours & Company, Seaford, Delaware,USA
**Jet Propulsion Laboratories, Pasadena, California, USA

## INTRODUCTION

Two of the most interesting phenomena of polymer science are
the reinforcement of the strength of rubber vulcanizates by
particulate fillers and the toughening (to impact) of plastics by
particulate rubber inclusions. The mechanisms underlying these
phenomena have, however, been difficult to elucidate because of
the complexity of the materials used in their industrial applications. Thus the most common filler used in rubber technology,
carbon black, has a complex structure, making it difficult even
to determine its "particle" size, aside from the question of its
chemical reactions with the rubber. Similarly, the rubber-
toughened plastics, such as high-impact polystyrene, are prepared
by a polymerization process in presence of dissolved rubber,
leading to a precipitation of ill-defined rubber "particles".

The work described herein, which covers a research project in
these laboratories over the past number of years, is concerned
with the preparation and study of well-defined dispersions of
hard, glassy plastics in SBR vulcanizates and rubber particles in
a polystyrene matrix. Both types of composite materials were
prepared by blending latex polymers of known particle size, which,
upon coagulation, became composites containing the desired
particles of plastic "fillers" for the rubber and rubber particles
in the plastic.

## RUBBER VULCANIZATES WITH POLYMERIC FILLERS

In this case, latex blends of SBR (or polybutadiene) and
rigid polymeric fillers, e.g., polystyrene, were used to prepare

1

rubber vulcanizates containing known amounts of hard spherical
fillers of known particle size.[1-3] Vulcanization was carried out
by incorporating either sulfur, etc., or peroxides into the coagulated latex blend and heating at 60-80°C., well below the softening point of the filler.  Furthermore, the presence of any
chemical bonds between the rubber and the filler could be detected
by suitable solvent extraction of the vulcanizate.  These systems
were essentially intended to clarify the mechanism of reinforcement of rubber by particulate fillers, e.g., carbon black, but
using rigid plastics as model fillers.

Effect of Filler Particle Size

     Figure 1 shows replica electron microphotographs of two SBR
vulcanizates filled with polystyrene particles of two different
particle sizes.  The individual filler particles can be clearly
seen, and indicate the absence of aggregates, which generally
occur when dry powders are dispersed in rubber, and which obscure
the effects of particle size.  It is therefore possible, as shown
in Figure 2, to draw some conclusions about the effect of filler
particle size on the tensile strength of these vulcanizates.  It
should be noted, parenthetically, that all of these vulcanizates
were crosslinked to the same extent, for comparative purposes.
From these data, it can be concluded that the fillers extend the
(tensile strength)-temperature plateau, and that the smaller
particles extend it further than the larger ones.  This effect of
particle size on strength is further elucidated in Figure 3,
which shows a single "failure envelope" for SBR reinforced by
polystyrene particles of different average sizes.  It appears that
the smaller particles simply move the tensile strength data
further upward along the same curve, indicating that their effect
is on the viscoelastic response of the network and not on the
crosslink density.  Incidentally, it was also noted[1] that filler
loading had the same effect on the failure envelope, higher loading moving the points to higher strength values on the same
envelope.

Effect of Filler-Rubber Bonds

     The effect of chemical bonds between fillers and the rubber
has never been clearly determined in the case of conventional
fillers, mainly because of the difficulty of establishing unequivocally the presence or absence of such bonds.  In the case of our
model polymeric fillers, the presence of filler-rubber bonds
could be easily determined by examining the solubility of the
polymeric fillers imbedded in the vulcanizate.  For this purpose,
a special polymeric filler was prepared, consisting of a 90/10
wt. percent styrene/butadiene copolymer; and the solubility of the
various fillers, as determined by benzene solvent extraction, is
shown in Table I.  The presence of filler-rubber bonds is clearly

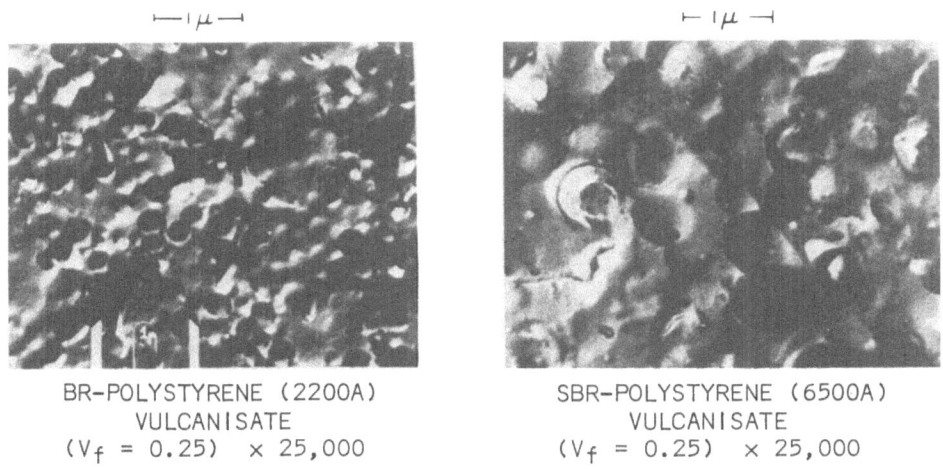

BR-POLYSTYRENE (2200A)
VULCANISATE
($V_f = 0.25$)  × 25,000

SBR-POLYSTYRENE (6500A)
VULCANISATE
($V_f = 0.25$)  × 25,000

Fig. 1  Replica electron microphotographs of polystyrene-filled SBR
        vulcanizates.

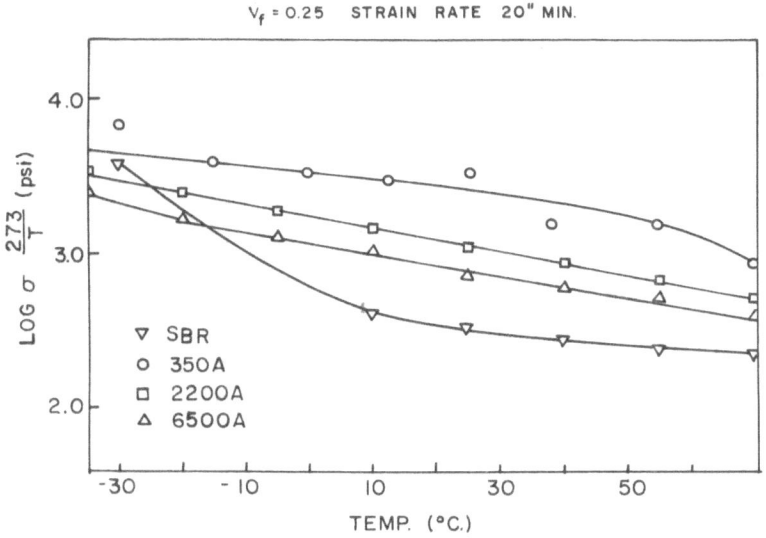

Fig. 2  Effect of polystyrene particle size on tensile strength of
        SBR.

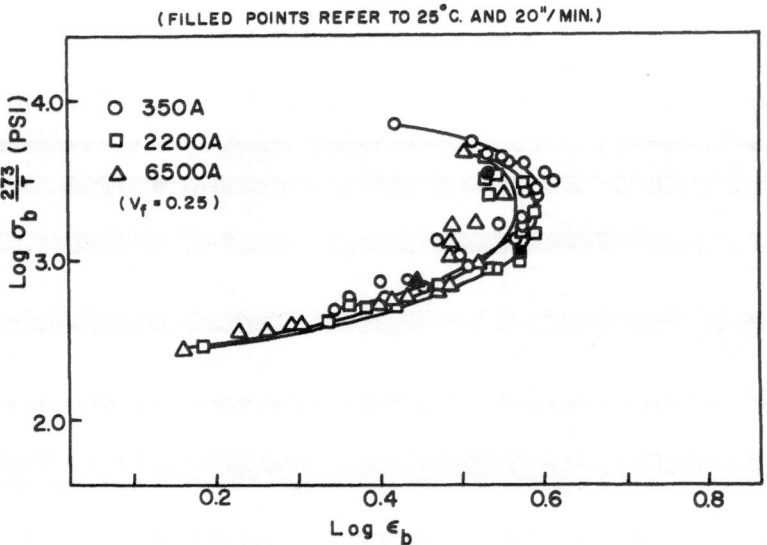

Fig. 3  Effect of polystyrene particle size on failure envelope.

indicated in the case of the SB-10 copolymer.  What effect such
attachments can have on tensile strength is demonstrated in
Figure 4, which shows the somewhat unexpected result that such
chemical bonds actually <u>depress</u> the tensile strength to a small
extent.  It should be noted here, parenthetically, that, from a
knowledge of the molecular weight of the filler polymer, and the
solubility data, it has been estimated[1] that each particle has 50-
100 attachments to the vulcanizate.

Table 1.  Solubility of Filled Vulcanizates[1] (Benzene)

|  | SBR | PS (485A) | | SB10 (535A) | |
| --- | --- | --- | --- | --- | --- |
| Type of Filler | None | Polystyrene | | Styrene-buta-diene (90/10) | |
| $M_v$ of Filler (x10$^{-5}$) | -- | 8.1 | 8.1 | 8.0 | 8.0 |
| $T_g$ of Filler | -- | 107 | 107 | 69 | 69 |
| Vol. % Filler | -- | 15 | 25 | 15 | 25 |
| Wt. % Filler | -- | 16.0 | 26.5 | 16.0 | 26.5 |
| Wt. % Extracted | 3.9 | 16.3 | 28.9 | 5.45 | 8.50 |
| Swelling Ratio | 4.40 | 4.55 | 4.50 | 5.16 | 6.83 |

Effect of Filler Rigidity

    The apparent depression of tensile strength by the introduction of filler-rubber bonds is, however, open to some question, since it should be noted that the presence of the butadiene in the copolymer filler depresses its glass transition temperature ($T_g$) from 107°C to 69°C[1] and hence presumably makes it a "softer" filler. To investigate this possible factor more fully, a series of polymeric fillers was prepared,[2,3] by emulsion polymerization of such monomers as styrene, methyl methacrylate and acenaphthylene, and the flexural modulus of molded bars was measured. The results are shown in Table 2.

    The effect of these three fillers on the tensile strength of SBR vulcanizates is shown in Figure 5. It is obvious at once that the modulus of the fillers <u>does</u> have an effect, the more rigid fillers exhibiting a greater reinforcement of the rubber vulcanizate. Hence the effect of the SB-10 filler in Figure 4 can best be ascribed to its lower modulus rather than to any effect of filler-rubber bonds.

    These effects of particle size and filler modulus on vulcanizate strength raise some interesting questions about possible mechanisms. It has been generally accepted for some time that the strength of elastomers is related to hysteresis effects,[4] the latter acting as an energy-absorbing phenomenon which dissipates the elastic energy which might otherwise be concentrated on propagating a growing crack. Thus the $T_g$ of an elastomer has an influence on its strength. This is illustrated in Figure 6, where the tensile strengths of SBR ($T_g \sim -50$°C) and polybutadiene ($T_g \sim -90$°C) are compared as a function of temperature. The presence of a rigid filler obviously increases the hysteresis of the vulcanizate, and it appears from this work that this increase is a function of the rigidity (or modulus) of the filler, too. Incidentally, in Figure 6, the two fillers indicated are polystyrene (PS) and poly-2,6-dichlorostyrene (DC), which have widely different $T_g$ values (105°C and 165°C, respectively) but have the same modulus of rigidity, and hence a similar effect on the tensile strength of the vulcanizate.

Table 2. Characteristics of Fillers

| Polymeric Filler | Particle Size (A) | $T_g$ (°C) | Flex.Mod. at 25°C (kg cm$^{-2}$ x 10$^{-4}$) |
|---|---|---|---|
| Polyacenaphthylene (PA) | 410 | 240 | 8.0 |
| Polystyrene (PS) | 570 | 108 | 3.2 |
| Poly(methyl methacrylate)(PMMA) | 510 | 110 | 2.4 |

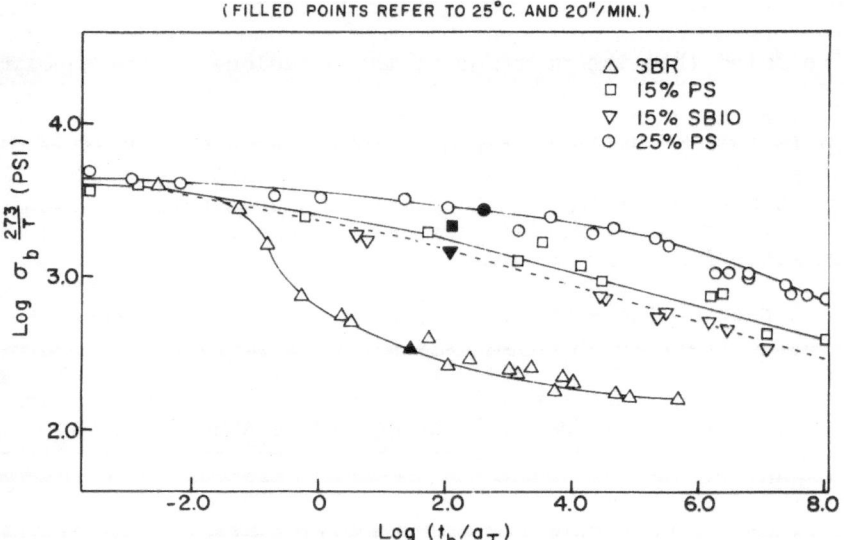

Fig. 4  Effect of fillers on log σ_b vs. log(t_b/a_T).

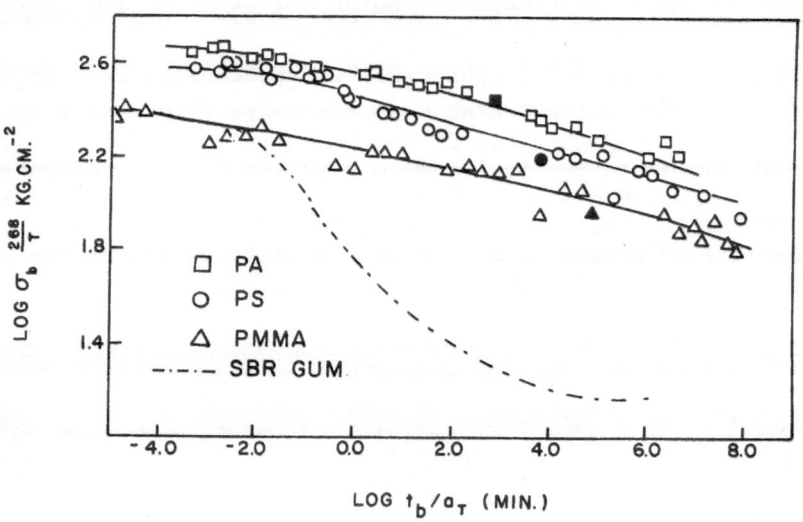

Fig. 5  Effect of filler modulus on tensile strength of SBR.

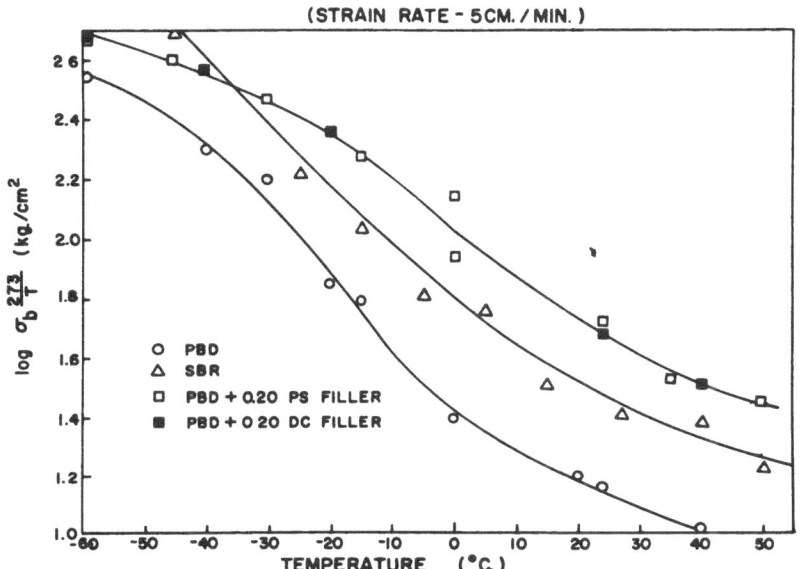

Fig. 6  Tensile strength vs. temperature for SBR and PBD.

Effect of Filler-Rubber Adhesion

In this connection, there is one more parameter that should
be considered, i.e., the degree of contact, or adhesion, between
filler and rubber.  This has previously been explored in some
detail in our laboratories by measurements of the dilation of
filled SBR vulcanizates as a function of strain, using sensitive
density measurements.[3]  Changes in volume during strain were
assumed to be due to vacuole formation at or near the filler-
rubber interface.  In this study the presence of any particulate
matter, other than the filler, was avoided by using peroxide,
rather than sulfur, vulcanization.

Figure 7 shows the effect of strain on the dilation of
polystyrene-filled SBR vulcanizates.  Dilation of the gum vulcanizate was also measured, as a control.  It can be seen that the
fractional change in volume appears to be a linear function of the
strain, at least within the range shown, and that it is directly
dependent on the particle size.  The points designated as "66%S"
and "100%S" refer to polystyrene latices which were blended with
the SBR latex and allowed to dry as a film, rather than being
coagulated in alcohol, as in all other cases.  Hence the soap
emulsifier (sodium oleate) would be expected to remain at the
filler-rubber interface in these cases.  The "66%S" represents the
calculated coverage of the polystyrene particle surface by the
soap, before blending, while, for the "100%S", additional soap was

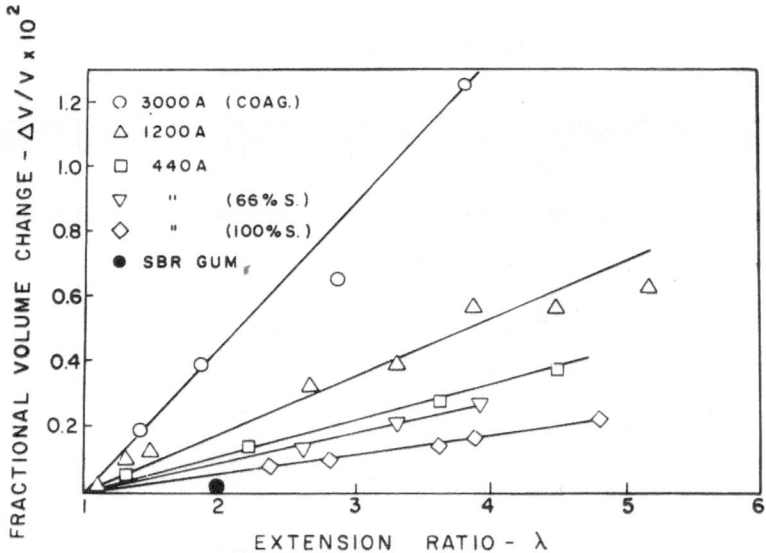

Fig. 7   Effect of PS filler on strain dilation of SBR.

added to the polystyrene latex, for complete coverage, before blending.  It can be seen that the presence of such a layer of soap apparently increases the adhesion at the filler-rubber interface.

Figure 8 shows a plot of the slopes of these lines, i.e., fractional change in volume per unit extension ($\Delta V/V\lambda$), against the filler particle diameter for the three types of filler used, with and without the presence of soap.  (The particle diameters are plotted as D/1.5 since this is equal to the total surface area of the filler per unit volume of rubber, when $V_f = 0.25$.)  It can be seen that this dilation per unit extension is a linear function of particle diameter, and therefore inversely proportional to the filler surface area, as would be expected.  However, what is really interesting is the fact that the different fillers give different slopes, i.e., different dilation values at any given filler particle size.  This must therefore reflect the influence of the filler surface on the filler-rubber "adhesion".  This is also borne out by the fact that the presence of a soap layer increases the adhesion in all cases.

It is probably purely fortuitous that the dilation values in Figure 8 are very similar for the polystyrene filler covered with soap and for the "bare" PMMA filler.  However, it does indicate that the lower dilation values (higher adhesion) in these cases may be related to the higher surface energy of the ester or

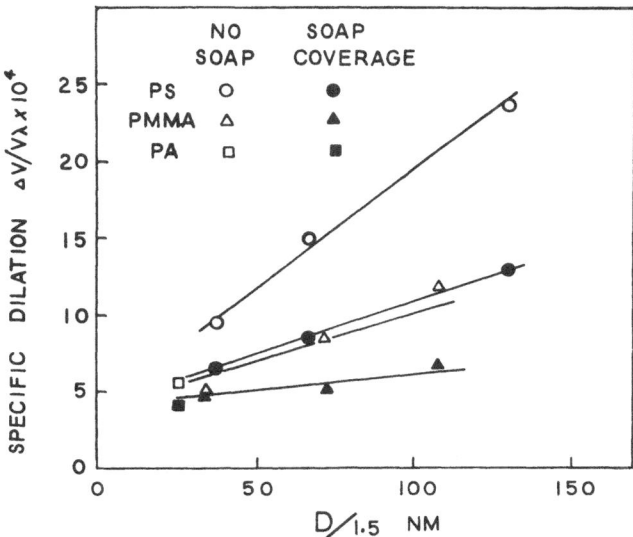

Fig. 8   Volume dilation of filled SBR.

carboxyl groups at the interface.  On the other hand, the presence
of a soap layer on the surface of the PMMA filler increased the adhesion even more. It is entirely possible that the presence of soap
at the filler-rubber interface may influence the extent of chemical
bonding during the vulcanization, but this cannot be established
without further work.

The effect of filler-rubber adhesion on the strength of these
filled vulcanizates is clearly demonstrated in Figures 9 and 10.
Thus figure 9 shows the inter-relation of tear strength, hysteresis
and adhesion in the case of the two fillers, PS and PMMA. It is
clear from these results that both the filler particle size and
nature of the filler surface influence the filler-rubber adhesion,
which also affects the hysteresis and hence the tear strength of
the vulcanizate. Furthermore, the rigidity of the filler also
affects the hysteresis and hence the strength, as shown by the two
separate lines for PS and PMMA fillers. This effect of hysteresis on
strength of rubber corroborates the current hypothesis[4] that it is
the hysteresis during the stretching of rubber which helps to dissi-
pate the elastic energy and to delay the propagation of a growing
crack.

Figure 10 shows a similar relation between tensile strength and
filler-rubber adhesion for these two fillers, again demonstrating the
positive effect of the presence of soap at the filler-rubber interface. Again two lines are obtained, the higher one for PS and the
lower one for the PMMA, just as in the case of Figure 9.

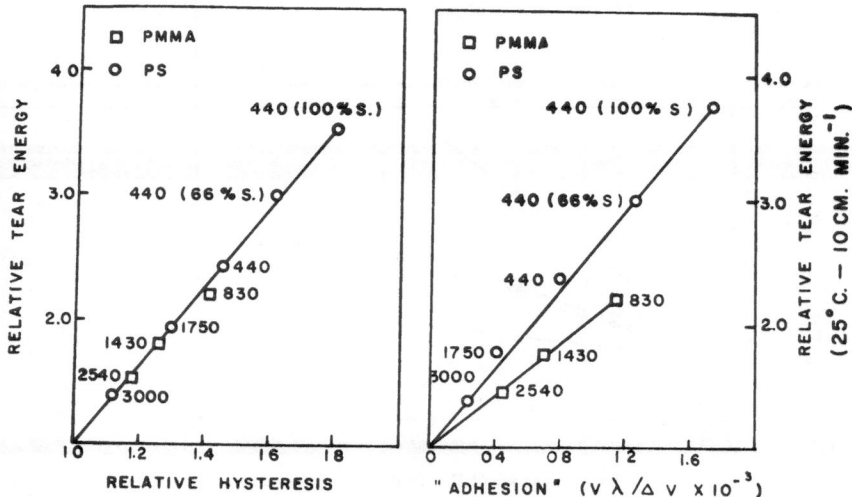

Fig. 9   Relation between "adhesion", hysteresis and tear
         strength of filled SBR.

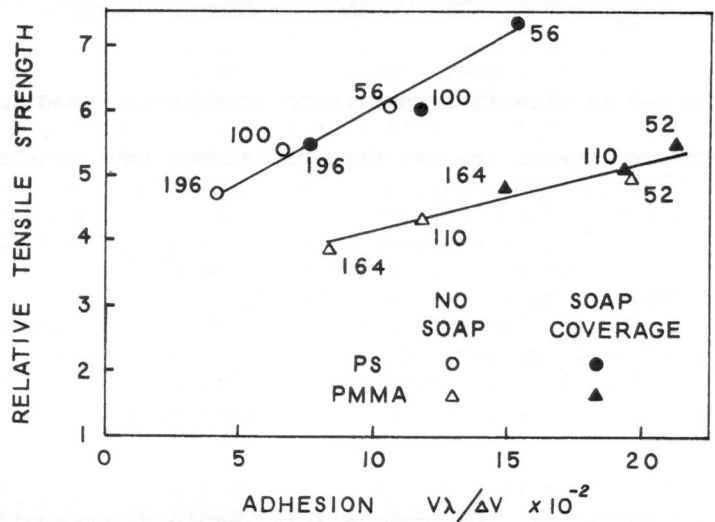

Fig. 10.   Effect of filler-rubber adhesion on tensile strength
           of SBR.

## Effect of Degree of Vulcanization on Filler-Rubber Adhesion

All of the previous results were obtained on vulcanizates having the same degree of vulcanization. More recently, work was carried out on the effect of the extent of vulcanization on the volume dilation, and the results are shown in Figure 11 for a PS-filled SBR vulcanizate. It can be seen that the dilation actually increases linearly with an increase in vulcanization (as measured by the rheometer). Presumably the higher modulus at higher degrees of vulcanization causes increased dilation at any given strain.

This raises some interesting questions about the effect of crosslink density and filler-rubber adhesion on the strength of filled vulcanizates. It is well known that the strength of rubber vulcanizates, with or without filler, goes through a maximum and then decreases with increasing crosslink density. This has been rationalized on the basis of changes in viscoelastic response of the network as well as changes in the network chain orientation vectors.[5] However, since Figure 11 shows that dilation increases monotonically with crosslink density, it is obvious that the extent of vulcanization results in two opposing effects, and the observed increase in strength with increasing crosslink density simply reflects a net difference between the two. In this connection, it should be noted that the tensile strength maximum of the three vulcanizates shown in Figure 11 occurs between 70% and 100% vulcanization.

## Nature of Volume Dilation

Although it was assumed that the volume dilation (density decrease) of the strained vulcanizates was the result of dewetting at the filler-rubber interface, this is not an unambiguous conclusion. It is of interest to note, therefore, that actual observation of the fracture surfaces of these vulcanizates, when frozen in the strained condition, showed the presence of vacuoles around the filler particles, as expected.

## RUBBER-TOUGHENED POLYSTYRENE

The effect of finely-divided rubber inclusions on the impact strength of plastics has been known for some time and represents a well-developed technology.[6] However, the morphology of these heterophase materials is usually quite complex, and has made it difficult to establish unequivocally the influence of the various parameters. Thus, for example, the process of producing high-impact polystyrene (HIPS) is based on the mass polymerization of styrene containing 5-10% of dissolved rubber (polybutadiene or SBR) during the course of which there occurs a phase separation of the rubber as a fine dispersion within the polystyrene. However,

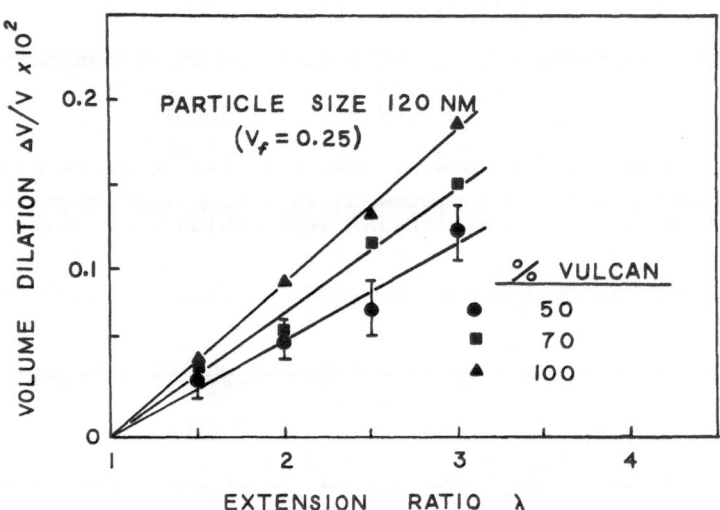

Fig. 11 Effect of degree of vulcanization on volume dilation.

because of the fact that the rubber dispersion is swollen with styrene monomer when it separates, a third phase forms, consisting of polystyrene dispersed within the rubber particles. Furthermore, during the polymerization of the styrene, considerable crosslinking and grafting of the rubber also occurs. Because all of these processes have been difficult to control precisely, it has not been possible to relate the results unequivocally to the final properties of the polystyrene. The concensus[7,8] appears to be that there is an optimum particle size for the rubber dispersion of 1-2 micrometers, and that chemical bonding between the rubber and the polystyrene is desirable.

It occurred to us sometime ago that it might be possible to answer some of these questions by utilizing the same techniques as described above in the case of the model polymeric fillers for rubber reinforcement. This involved the preparation, by emulsion polymerization, of polybutadiene or SBR latices of varying particle sizes, and the blending of suitable proportions of these latices with polystyrene latices prior to coagulation, drying and molding. Furthermore, chemical bonding between the polystyrene and the rubber, and crosslinking of the rubber, could be introduced, and controlled, by inclusion of suitable catalysts, such as peroxides.

In this work polybutadiene latices with average particle sizes ranging from 500A to 4000A were prepared by successive "seeded" polymerization, and a 7000A particle-size latex was obtained from commercial sources. SBR latices were prepared in the same way to give a particle size range of about 800A to 4500A, while a commercial latex of particle size 8650A was also used. Crosslinking of the rubber alone was accomplished by treating the rubber latex with a peroxide catalyst, while grafting of poly-styrene to rubber, together with crosslinking, was attained by incorporating the peroxide into the blended latex prior to coagula-tion, drying and molding.

Some of the significant results of this work are shown in Figures 12-16 and in Table 3. Thus Figure 12 shows a trans-mission electron microphotograph of a thin-microtomed section of molded polystyrene containing a dispersion of SBR of particle size 8650A. This shows that the particles from the rubber latex have been incorporated into the polystyrene without any agglomeration or distortion. Figure 13 shows the effect of the particle size and amount of polybutadiene rubber on the notched Izod impact strength of the polystyrene. As expected both the amount and the larger particle sizes of the rubber gave improved impact resistance. Unfortunately, particle sizes larger than 8650A were not available, so that a maximum in the impact strength was not attained and must await additional work. Figure 14 shows that polybutadiene rubber gave slightly better impact than the SBR, indicating that a more resilient rubber (lower $T_g$) may have better energy-absorbing characteristics in this case.

Figure 15 also shows the expected effect of these rubber inclusions on the flexural modulus of the polystyrene, and again the softer polybutadiene exerts a slightly greater effect than the stiffer SBR. Incidentally, the particle size had no influence on the flexural modulus. A further check on the similarity between these model systems and the commercial varieties is provided by Table 3, which shows the effect of polybutadiene inclusions on the Heat Deflection Temperatures. Thus the value of 93°C for a poly-styrene containing 10% rubber compares favorably with the reported value of 81°C for an analogous commercial HIPS (Styron 495).

Table 3.  Heat Deflection Temperatures for Polystyrene-Polybutadiene Blends

|  | PS | PB* Content of Blends | | |
|---|---|---|---|---|
|  |  | 5% | 10% | 15% |
| Heat Deflection Temp., °C. (Fiber stress = 4.6 Kg.cm$^{-2}$) | 102 | 98 | 96.5 | 94 |
| Heat Deflection Temp., °C. (Fiber stress = 18.6 Kg.cm$^{-2}$) | 97.5 | 95 | 93 | 90 |

*7150A

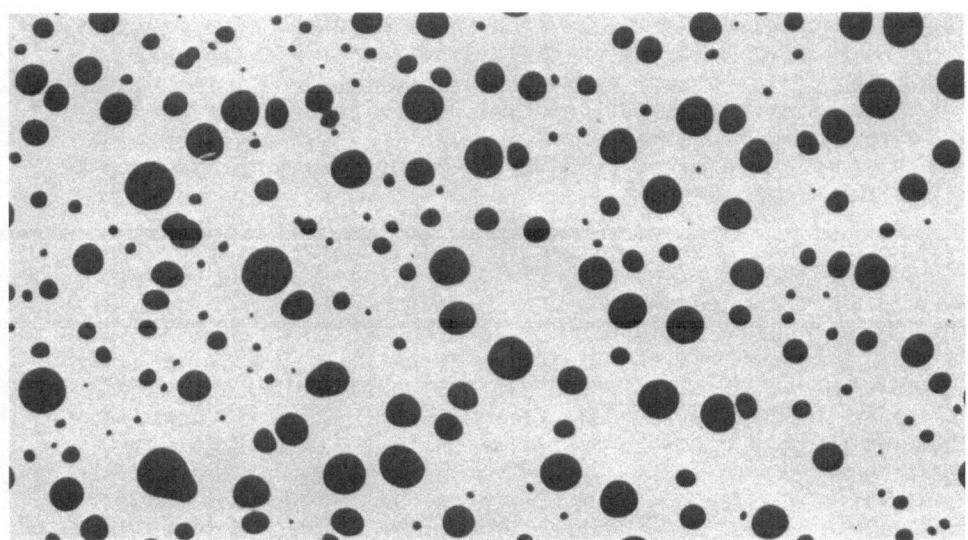

Fig. 12   Transmission electron microphotograph of SBR dispersion
          in polystyrene (8650A).

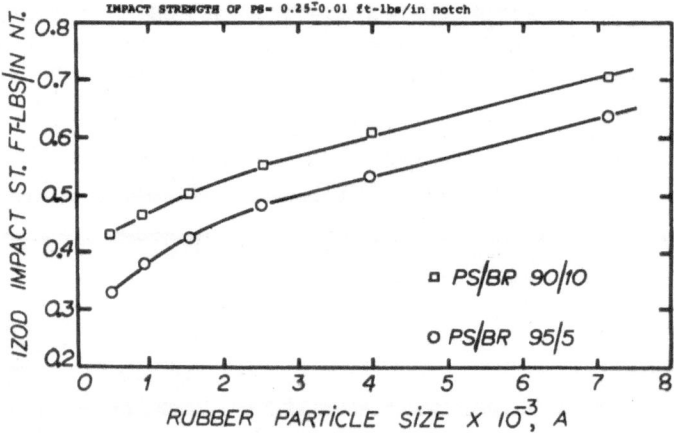

Fig. 13   Effect of polybutadiene particle size on impact strength
          of polystyrene.

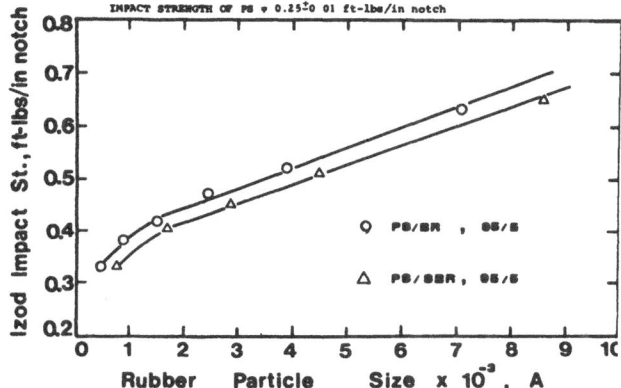

Fig. 14   Effect of polybutadiene and SBR on impact strength of
          polystyrene.

It is interesting to note that the magnitude of the highest
impact strength shown in Figure 13 is substantially lower than the
values usually claimed for commercial HIPS (1.2-1.5 ft. lbs. per
in. notch).   However, this is not surprising, since samples containing larger rubber inclusions could not be prepared at this
time, and also since there was no chemical bonding between the
polystyrene and the rubber.   Figure 16 is of special interest,
therefore, since it shows the effect of both chemical bonding
between polystyrene and rubber, as well as of crosslinking the
rubber.   Thus it can be seen that, at zero grafting between the
two phases (G.I. — "grafting index", i.e., percent by weight of

polystyrene grafted to the crosslinked rubber), the crosslinking
of the rubber leads first to a slight increase in impact strength,
followed by a precipitous decline for the tightly crosslinked
rubber.  However, as the bonding between the polystyrene and the
rubber increases (higher G.I. values), there is a substantial
increase in impact resistance, representing, at its maximum, more
than double the increase attained without plastic-to-rubber
bonding.

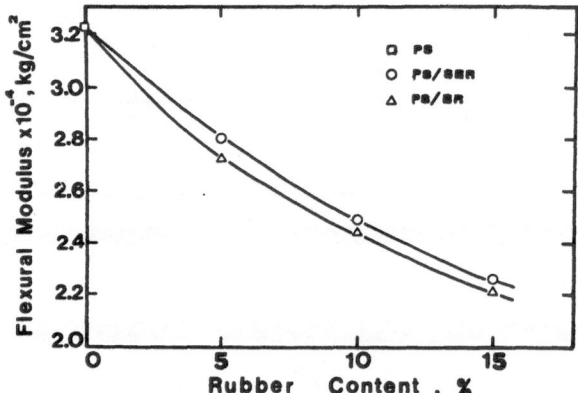

Fig. 15  Effect of SBR and BR on flexural modulus of polystyrene.

        Considering the limitation in attainment of larger particle
sizes of rubber in this work, the maximum value of 1.1-1.2 ft.
lbs. per in. notch reached in Figure 16 is equivalent (or better)
than the usual values exhibited by commercial HIPS. Hence these
results have been successful in providing a model to establish
unequivocally the enhancement of the impact strength by increasing
particle size of the rubber, and by the presence of

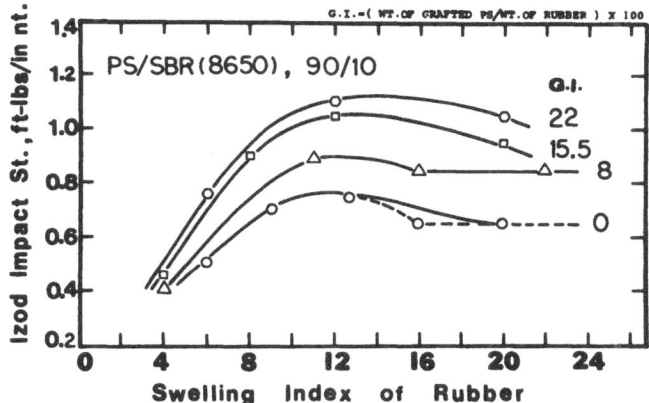

Fig. 16  Effect of grafting and crosslinking on impact strength.

plastic-to-rubber bonds.  These model studies have also brought out what appears to be a previously overlooked result, i.e., that excessive crosslinking of the rubber inclusions actually decreases the impact strength substantially.

REFERENCES

1.  M. Morton, J. C. Healy and R. L. Denecour, Proceedings of the International Rubber Conference 1967, MacLaren and Sons, London, 1968, p. 175; Appl. Polym. Symp., 7, 155 (1968).

2.  M. Morton, J. L. Trout and T. C. Cheng, Proceedings of the International Rubber Conference, 1972, Institution of the Rubber Industry, London, p. G6-1; see also Maurice Morton, Adv. in Chem. Series, No. 99, American Chemical Society, Washington, 1971, p. 490.

3.  M. Morton, R. J. Murphy and T. C. Cheng, Colloques Internationaux du CNRS, No. 231 (1973), p. 3; Adv. in Chem. Series No. 142, American Chemical Society, Washington, 1974, p. 409.

4.  a) L. Mullins, Trans. IRI, 35, 213 (1959).
    b) L. A. Grosch, J.A.C. Harwood and A. R. Payne, J. Appl. Polymer Sci., 12, 889 (1968).

5.  J. C. Halpin and F. J. Bueche, J. Appl. Phys., 35, 36, 3142 (1964).

6.  C. B. Bucknall, "Toughened Plastics," Applied Science Publishers Ltd., London, 1977.

7.  R. F. Boyer and H. Keskkula, Encycl. Polym. Sci. Technol., John Wiley & Sons, Inc., New York, 1970, p. 392.

8.  E. P. Chang and A. Takahashi, Polym. Eng. Sci. 18, 350 (1978).

ON THE CORRELATION OF MECHANICAL PROPERTIES OF HIGH IMPACT POLY-
STYRENE WITH ITS MORPHOLOGY, MOLECULAR-WEIGHT CHARACTERISTICS AND
EXTRUSION CONDITIONS

V.D. Yenalyev, V.I. Melnichenko, O.P. Boukunenko,
A.N. Shelest, N.M. Tchalaya, Y.I. Yegorova, and
N.G. Podosyonova

Donetsk State University
Donetsk, 340055, USSR

Mechanical properties as well as practical ones of high impact polystyrene (HIPS) are defined by a number of parameters,
characterizing its molecular and supermolecular structure: molecular weight (MW) and molecular weight distribution (MWD) of the
polystyrene matrix, quantity, composition and structure of graft
copolymer, dispersed and continuous phases volume ratio, rubber
particles size, their substructure, etc. There are a number of
papers (1-7) in which attempts have been made to study the effect
of the most substantial factors on polymer properties in order
to find their optimal combination. This task concerning HIPS
and ABS plastics is not simple for it is difficult to prepare
polymers with a definite value of one given parameter, keeping
the others constant. In addition, it is now known that some
physical and chemical HIPS characteristics are closely connected
with each other and to clarify the individual effect of one of
them on the polymer properties is a complicated problem. Thus,
the present paper deals with an attempt to establish, by means
of statistical methods, the quantitative correlation between
HIPS elongation at break and impact strength and MW and MWD of
the polystyrene matrix, copolymer and gel quantity, and morphological characteristics (12 parameters in all). With this goal,
two different samples have been obtained from industry and from
the laboratory for investigation. The MW of the polystyrene
matrix was measured by viscosimetry ($K = 1.34 \cdot 10^{-4}$, $\alpha = 0.71$ in
toluene, 25ºC), MWD and were determined by Gel Permeation Chromatography (GPC), and morphology was observed by means of electron microscopy (8). The graft copolymer and gel content were
defined by selective dissolving (9), and the polystyrene graft-

ing degree on polybutadiene was calculated according to the following equation:

$$G_S = \frac{m_g}{K} \frac{m - K}{(100 - K)} 100$$

where $m_g$ is the copolymer and free rubber content in 100g of polymer; m is the total polymer content in 100g of polymerized mass; K is the rubber content in per cent.

The degree of gel crosslinking was determined according to weight percentage of its swelling in benzene. HIPS extrusion was carried out on a twin-screw granulator ZSK-83 "Werner-Pfleiderer" (temperature - 190°C, screw speed - 60 r.p.m.). In the laboratory a one-screw extruder-diameter 32 mm- was used. Sheets with a width of 120 mm and thickness of 4 mm were obtained (temperature 200°C, screw speed - 30 r.p.m.).

From the experimental data we calculated with the help of a computer, the regression equations which related HIPS mechanical properties with polymer molecular and structural characteristics:

$$
\begin{aligned}
E \cdot 10^{-1} = {} & 5.04 - 1.96X_1 - 1.63X_3 - 1.27X_4 + 1.29X_6 \\
& - 2.55X_7 + 0.57X_8 - 0.65X_9 - 1.12X_{10} \\
& + 3.29X_{11} + 0.39X_{12} + 0.009X_1X_2 + 0.90X_1X_3 \\
& - 0.31X_1X_4 - 0.08X_1X_5 - 0.66X_1X_6 + 0.09X_1X_{10} \\
& - 0.004X_2X_5 - 2.00X_3X_4 + 2.35X_3X_6 + 1.41X_4X_6 \\
& - 1.03X_4X_7 + 0.26X_4X_{10} - 0.002X_5X_{10} + \\
& + 2.75X_6X_7 - 0.03X_3^2 + 0.03X_4^2 - 2.29X_6^2
\end{aligned}
\tag{1}
$$

$$
\begin{aligned}
F = {} & 4.53 - 3.51X_1 + 1.99X_2 + 8.29X_3 + 3.49X_4 \\
& - 7.11X_6 + 29.1X_7 - 1.32X_6 + 0.72X_9 \\
& - 9.63X_{10} - 0.049X_{11} + 3.21X_{12} - 0.37X_1X_2 \\
& - 2.63X_1X_3 - 0.20X_2X_5 - 1.52\ X_1X_6 + 1.68X_1X_{10} \\
& - 0.091X_2X_5 + 2.41X_3X_5 - 5.96X_3X_6 - 3.95X_4X_6 \\
& - 6.42X_4X_7 + 1.97X_4X_{10} + 0.165X_5X_{10} \\
& + 0.71X_6X_7 + 1.25X_1^2 + 7.20X_6^2 - 6.68X_7^2 \\
& - 0.02X_9^2
\end{aligned}
\tag{2}
$$

where E is the elongation at break, %; F is the impact strength (notched Izod), kg cm/cm$^2$; $X_1$ - mean size of rubber particles, $\bar{c}$, μ; $X_1$ - rubber particle size polydispersity; $X_3$ - equilibrium

degree of gel swelling, $Q \cdot 10^{-1}$, %; $X_4$ – degree of polystyrene
grafting on polybutadiene, Gs $10^{-2}$, %; $X_5$ – polybutadiene content,
K, %; $X_6$ – volume fraction of dispersed phase, Vf $10^{-1}$, %; $X_7$ –
number-average molecular weight of polystyrene matrix, MN $10^{-5}$;
$X_8$ – weight-average molecular weight of PS matrix, Mw $10^{-5}$; $X_9$ –
Mz $10^{-5}$ of PS matrix; $X_{10}$ – Mw/Mn; $X_{11}$ – Mz/Mn; $X_{12}$ – Mv $10^{-5}$ of
homopolystyrene.

Equations 1 and 2 have both linear square terms and the
effects of double correlation. In both equations the search
was carried out by means of a preliminary analysis of experimental
data, by eliminating negligible effects. It should be mentioned
in particular that in both cases, the coefficient characterizing
the influence of polybutadiene content of HIPS properties was
insufficient, although this may seem contradictory to the existing opinions at first sight. In fact, an increase of polybutadiene content must result in an increase of the volume fraction
of dispersed phase in HIPS, of graft copolymer and of gel quantity. But in this case such regularity is not observed since the
samples considered were prepared with different recipes and different processing conditions. This is very important because using
the periodic method of production, for instance, the dispersed
volume fraction of the graft copolymer at 5-6% rubber content is
equal to or even higher than using a continuous one when its content is 9% or higher. When these results are statistically work-
out, there is a relationship between polybutadiene and HIPS properties which can be noted in terms of morphology parameters and
copolymer characteristics.

According to the coefficients values in equations (1) and (2)
one can judge to some extent the influence of MW and morphological parameters of the polymer being considered on its strength
characteristics. It seems possible to consider the character of
parameters' individual influence on the values E and F. (Table I)

TABLE I

MEAN VALUES OF PHYSICAL AND CHEMICAL CHARACTERISTICS
OF LABORATORY AND INDUSTRIAL HIPS SAMPLES

| PARA-METER | $X_1$ | $X_2$ | $X_3$ | $X_4$ | $X_5$ | $X_6$ | $X_7$ | $X_8$ | $X_9$ | $X_{10}$ | $X_{11}$ | $X_{12}$ |
|---|---|---|---|---|---|---|---|---|---|---|---|---|
| MEAN VALUE | 2.1 | 3.3 | 1.3 | 2.7 | 6.8 | 2.2 | 0.78 | 2.2 | 4.8 | 2.8 | 2.2 | 1.8 |

If the mean values of variables from Table I (excluding $X_4$)
are substituted in equation 1, the particular dependence of

elongation at break on the degree of polystyrene grafting on
polybutadiene may be calculated as:

$$E \cdot 10^{-1} = 7.55 - 1.49X_4 + 0.03X_4^2 \qquad (3)$$

In an analogous manner, a number of other particular dependences
from other variables can be obtained. As is seen from equation 3,
the increase of graft degree decreases the elongation (undesirable
rubber property). The values of other polymer characteristics
depend on the correlation of E and $X_4$. Thus, Fig. 1 shows the
dependence of E on $X_4$ for 4 values of $X_1$. It is seen that the
bigger the rubber particle size, the more noticeable is the fall
of values when the grafting degree is increased. One of the
most important parameters of HIPS mechanical properties is its
impact resistance. With mean values of $X_2 - X_{12}$ (Table I) the
following dependence of F on $X_1$ may be achieved:

$$F = 19.9 - 8.2X_1 + 1.25X_1^2 \qquad (4)$$

Consequently the increase in rubber particles mean size
(up to 5-6 µ) results in a decrease of impact strength; low cross-
linking density and the high degree of grafting reinforce this
dependence (Fig. 2). Detailed analysis of other correlations
allows us to reach the following conclusions:
1. In all cases a small size of rubber particles (0.5 - 5 µ)
and consequently a more developed boundary surface between polystyrene and rubber phases, and more uniform distribution results
in improvement of the strength properties under consideration.
2. The presence of graft copolymer greatly influences the
morphology of HIPS especially at the initial copolymerization
stages, and promotes a higher degree of polystyrene dispersed in
rubber particles.
At the same time the results show that high degree of polystyrene grafting on polybutadiene somewhat lowers the elongation
at break and the impact strength.
3. The increase of crosslinking degree of copolymer results
in a decrease in elongation at break while tensile strength and
impact strength increase.
4. The increase of Mn results in an increase in the elongation at break, impact strength and tensile strength. Practically in all cases, high polydispersity of the matrix leads to a
decrease in strength properties.

Taking into consideration the established correlations, one
can make a prediction of HIPS mechanical properties at varying
MWD characteristics and morphology. A particularly good example
of the utility of these correlations involves the reprocessing of
scrap HIPS. In the repeated extrusion under the influence of

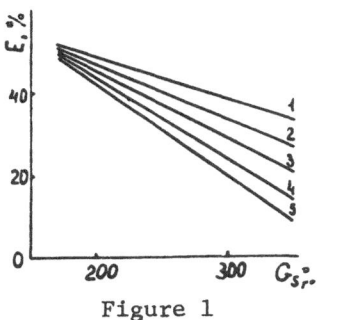

Figure 1

Influence of rummer particles
size ($X_1$) on dependence of elongation at HIPS break vs. the
degrees of polystyrene grafting
on polybutadiene:

1 - 1 μ;   2 - 2 μ;   3 - 3 μ;
4 - 4 μ;   5   5 μ

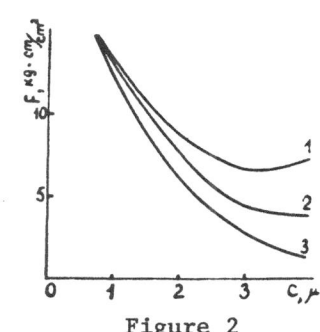

Figure 2

Influence of the degree of gel
swelling ($X_3$ $10^{-1}$) on dependence
of impact strength vs. rubber
particle size:

1 - 1.35%;   2 - 1.70%;   3 - 20.0%

high temperature and strong mechanical treatment, irreversible
changes in molecular weight and structural characteristics in the
polymer may take place.  Table II presents the results of the
effect of multiple extrusion on impact strength F (Izod without
a notch), tensile strength σ, elongation at break E, melt-flow
index I, and structural and molecular weight characteristics of
HIPS specimens.  It is seen that repeated processing considerably
affects the impact strength while the other properties change
very little.  The decrease in impact strength, which is the most
substantial for sample A, may be connected with the change of
structural and MW parametres of the polymer.  Degradation of rubber
phase particles takes place under the repeated influence of high
shearing forces, which is seen in Fig. 3.

     In some particles the interphase with the matrix is broken.
Under repeated processing, these particles are subjected to further degradation and their separate fragments are dispersed in
the matrix.  This in its turn leads to diminishing rubber phase

Table 2.

Influence of Multiple Processing on HIPS Properties.

| Polymer | Extrusion multiple | F, kg·cm/cm² | σ, kg/cm² | E, % | I, g/10min | V̄f, % | C̄, μ | P*, μ | G**, % | Q, % | Mw, 10⁻⁵ | Mw/Mn | Mz/Mw |
|---|---|---|---|---|---|---|---|---|---|---|---|---|---|
| A | 0 | 50.3 | 197 | 33 | 4.2 | 24 | 3.6 | 12.0 | 27.5 | 12.2 | 1.83 | 3.30 | 2.0 |
|   | 1 | 39.1 | 189 | 28 | 4.2 | 20 | 3.2 | 13.7 | 25.7 | 13.2 | – | – | – |
|   | 2 | 24.0 | 194 | 28 | 4.2 | 24 | 3.1 | – | 25.7 | 13.3 | – | – | – |
|   | 3 | 30.2 | 195 | 25 | 4.3 | 19 | 3.2 | 15.3 | 24.4 | 13.8 | – | – | – |
|   | 4 | 20.9 | 196 | 26 | 4.3 | 17 | 2.8 | 13.4 | 24.9 | 14.0 | – | – | – |
|   | 5 | 37.2 | 198 | 28 | 4.3 | 13 | 2.8 | 16.0 | 24.0 | 14.2 | 1.81 | 2.71 | 1.88 |
| B | 0 | 50.8 | 201 | 32 | 4.2 | 15 | 2.3 | 12.6 | 23.2 | 13.0 | 2.01 | 2.66 | 1.88 |
|   | 1 | 51.4 | 203 | 32 | 4.3 | 18 | 3.1 | 13.9 | 23.2 | 13.3 | – | – | – |
|   | 2 | 38.1 | 215 | 34 | 4.3 | 12 | 2.1 | 14.4 | 22.3 | 13.4 | – | – | – |
|   | 3 | 38.6 | 204 | 36 | 4.3 | 15 | 2.7 | 14.8 | 22.2 | 14.9 | – | – | – |
|   | 4 | 38.7 | 205 | 41 | 4.3 | 13 | 2.3 | 18.2 | 20.8 | 15.0 | – | – | – |
|   | 5 | 36.7 | 217 | 37 | 4.4 | 13 | 2.8 | 16.2 | 19.8 | 15.0 | 1.96 | 2.84 | 1.80 |

\* - mean free distance between rubber particles.

\*\* - gel content in g/100 g of polymer.

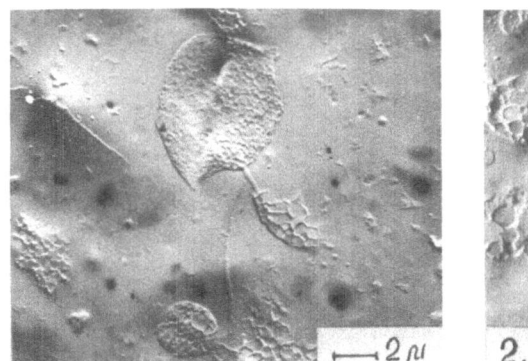 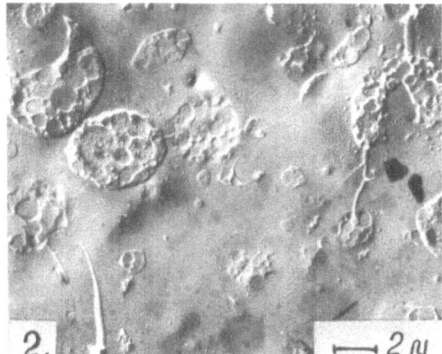

Fig. 3   Influence of three times extrusion on HIPS morphology:   1 - sample A,   2 - sample B.


volume fraction and to increasing distance between particles.
Significant change of MWD parametres and gèl quantity did not
occur.  However, destruction of rubber particles is accompanied
by a scission of the crosslinks, which results in decreased crosslink density.  This is evidenced by an increase in swelling with
increasing reprocessing.

REFERENCES

1.  D. Angier, Rubber Chem. and Techn., $\underline{38}$, 1161 (1965).
2.  K. Fletcher, R.N. Hovard, J. Mann. Chem. and Industry, $\underline{45}$,
    1854 (1965).
3.  G. Cigna, J. Appl. Polym. Sci., $\underline{14}$, 1787 (1979).
4.  B. Baer, J. Appl. Polym. Sci., $\underline{16}$, 1109 (1972).
5.  G. Riess, S. Marti, Y.L. Refregier, M. Schilienger, Polymer
    Alloys, Plenum Press, N.Y., 1977, 10, p. 327.
6.  G. Cigna, S. Mataresse, G.F. Biglione, J. Appl. Polym. Sci.,
    $\underline{20}$, 2285 (1976).
7.  S.G. Turley, H. Keskkula, Polymer, $\underline{21}$, 466 (1980).
8.  G. Bernier, R. Kombour, Macromolecules, $\underline{1}$, 393 (1968).
9.  Minoura J., Y. Mori, M. Imoto, Macromol. Chem., $\underline{24}$, 205 (1957).

# CORRELATION OF MORPHOLOGY, MECHANICAL PROPERTIES AND PROCESSING

# CONDITIONS OF MODIFIED HIGH IMPACT POLYSTYRENE

V.V. Abramov, V.D. Yenalyev, M.S. Akutin,
N.M. Tchalaya, V.I. Melnichenko, and A.N. Shelest

Donetsk State University
Donetsk, 340055, USSR

Due to its great toughness high impact polystyrene (HIPS) is widely applied as a construction material. It is known to be a composite material in which the rubber phase is dispersed in a polystyrene matrix (1). Varying synthesis conditions can change the polymer morphology to some extent, the rubber particles size and volume fraction in particular; this significantly affects HIPS properties (2). It seems possible to improve a number of mechanical properties by introducing a definite quantity of modifying additives of a different nature into the polymer. Triblock styrene and diene copolymers possessing a number of valuable properties (3) can be used as these modifying additions. Thus, this paper deals with the effect of elastomer addition, introduced into the extrusion process on HIPS morphology and some strength properties. To introduce the elastomer, a single screw extruder was used. The morphology of modified HIPS was studied by electron microscopy, impact strength (according to Charpy, without a notch), elongation at break, and tensile strength. The blending conditions of copolymers A and B with HIPS and their effect on mechanical properties are presented in Table I. It is seen that the introduced copolymer greatly affects the impact strength and elongation at break (depending on extrusion conditions). Thus, at 180°C, the increase of shear rate somewhat improves the impact strength of HIPS modified by copolymer A, the MW of which is higher than that of B (characterized by the flow-melt index). When the extrusion temperature is increased this regularity is weakened, and at 200-210°C some decrease in impact strength and elongation at break is observed. With copolymer B (Table I) at a temperature of 180-190°C, the increase of shear rate lowers

TABLE I

PREPARATION CONDITIONS OF MODIFIED HIPS AND THEIR
EFFECT ON MECHANICAL PROPERTIES

| No | Extrusion Temperature °C | Shear rate in dosage zone of a screw sec$^{-1}$ | Mechanical Properties | | | modified copolymer |
|----|----|----|----|----|----|----|
| | | | $F$, $\dfrac{kg\ cm^2}{cm^2}$ | kg/cm$^2$ | Elongation at break, % | |
| 1  | 180 | 1.57  | 26.6 | 177 | 37 | A |
| 2  | 180 | 4.70  | 32   | 182 | 58 | Flow-melt |
| 3  | 190 | 1.57  | 32   | 190 | 43 | index – |
| 4  | 190 | 4.7   | 29.5 | 186 | 48 | 1g/10 min |
| 5  | 190 | 7.85  | 33   | 183 | 41 | at 190°C |
| 6  | 190 | 10.1  | 36   | 181 | 47 | P = 21.6kg |
| 7  | 200 | 1.57  | 35   | 184 | 46 | |
| 8  | 200 | 4.7   | 30   | 180 | 40 | |
| 9  | 200 | 7.85  | 26   | 179 | 38 | |
| 10 | 210 | 1.57  | 27.5 | 187 | 43 | |
| 11 | 210 | 10.1  | 25   | 177 | 42 | |
| 12 | 180 | 1.57  | 30   | 186 | 63 | B |
| 13 | 180 | 4.70  | 25   | 181 | 50 | Flow-melt |
| 14 | 190 | 1.57  | 25   | 182 | 50 | index – |
| 15 | 190 | 4.70  | 20   | 178 | 54 | 13g/10 min |
| 16 | 190 | 10.1  | 16   | 177 | 46 | = 190°C |
| 17 | 190 | 7.85  | 25   | 178 | 44 | P = 21.6kg |
| 18 | 210 | 1.57  | 16   | 130 | 14 | |
| 19 | 210 | 10.1  | 30   | 175 | 16 | |

the polymer impact strength substantially, elongation at break remaining practically unchanged.

The above can be explained based on the morphology of the
modified polymers, this relation to the processing conditions,
and the copolymer properties. The mixing conditions under
consideration don't result in variations of the morphology of
HIPS which was formed, in the synthesis process. In all cases
the dispersed particles size, and volume fraction of the rubber
phase are practically unchanged. This indicates that macromolecules of introduced copolymer do not penetrate into the rubber
particles but form a separate dispersed phase within the polystyrene matrix as is shown in Fig. 1. It is seen that particle
size, and configuration depend significantly on mixing conditions.
Thus, comparing morphology of samples a and b, obtained at 200°C
but with different shear rates, it is not difficult to notice

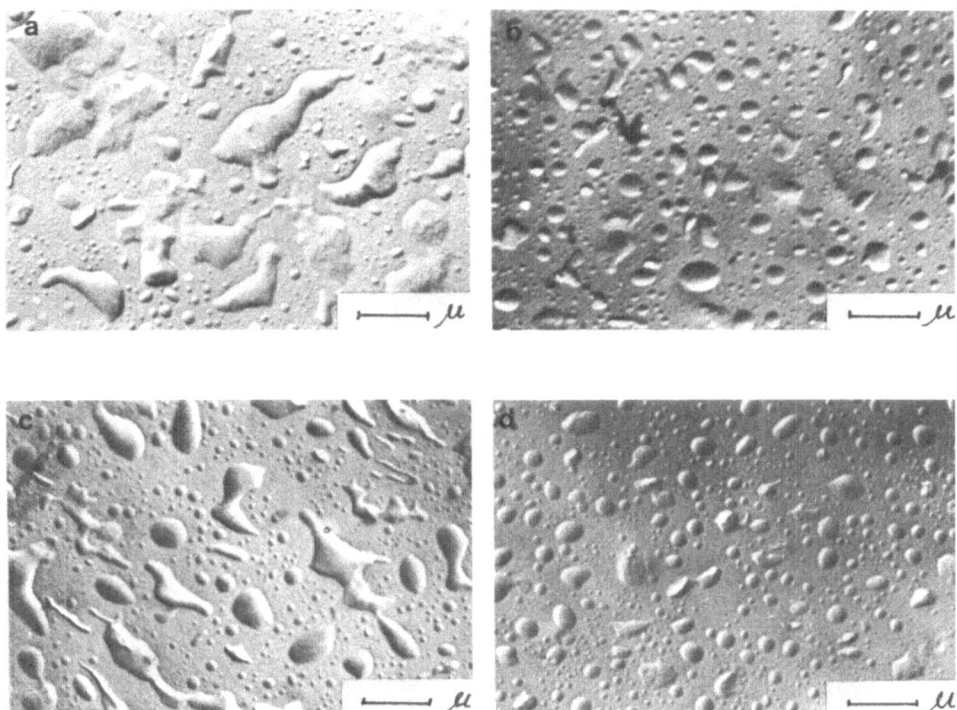

Fig. 1.  Electron micrographs of HIPS, which was modified by co-
         polymer A and B.  Pictures a, b, c, d correspond to samples
         7, 9, 12 and 1 in Table 1 respectively.

that the first one is characterized by a particle size of 0.5 -
1 µ, having an irregular configuration.  Increasing mixing intensity five times resulted in lowering particle size to 0.1 -
0.3 µ.  In addition, the particles are almost spherical.  Similar regularity is observed for samples c and d, obtained under the
same mixing conditions, but with the use of copolymer of different MW.  Thus there is a definite correlation between morphology
of modified polymer and its properties:  the bigger the size of
the particles of introduced copolymer (depending on mixing conditions) the higher are the strength properties of HIPS.  This allows us to state that the introduced copolymer plays a role mainly in increasing the dispersed phase volume thus improving strength properties of HIPS.

    Consequently when choosing mixing parameters for HIPS and
elastomer additions of various composition and structure in the
extrusion process, it is necessary to select the shear rate and
thermal regimes in such a way such that particles of large enough
size to yield an  improvement of mechanical properties are formed.

REFERENCES

1.  G.E. Molau, J. Polym. Sci., 3A, 1267 (1965).
2.  J. Silberberg, C.D. Han, J. Appl. Polym. Sci., 22,
    599 (1978).
3.  J.A. Manson, L.H. Sperling, Polymer Blends and Composites,
    Plenum Press, N.Y., 1976.

INITIATED STYRENE COPOLYMERIZATION WITH POLYBUTADIENE

REACTION CONDITIONS ON THE LENGTH OF POLYSTYRENE GRAFT BRANCHES

V.D. Yenalyev, N.A. Noskova, and O.P. Shmelyova

Donetsk State University

Donetsk, 340055, USSR

The specificity of styrene copolymerization with polybutadiene (PBD) is a heterogeneous polymerization. The creation of
an exhaustive theory of this process must include regulation of
all reactions which take place in both phases and take into account the propagation of graft polystyrene branches as well as
the rubber macromolecule. Results obtained by Mori et al (1)
showing that the length of the graft branches equals the length
of homopolystyrene molecules (PS) - as well as difficulties of
polymer fractionation to homopolystyrene (PS) in both graft copolymers, and selective degradation of copolymer polybutadiene
backbone revealed the fact that the propagation of graft branches
were not investigated. The researchers had identified the graft
branches as free polystyrene using kinetic calculations in a
better case (2), without taking into account polymerization specificity in the rubber phase, i.e. high viscosity of the media
already at the initial stages, large monomer concentration when
the system is near equilibrium (3), rubber and PS incompatibility
in styrene solution, the presence of additional chain transfer,
and unequal initiator distribution in both phases (4). All these
factors may substantially affect reactions taking place in the
rubber phase, even the change of elementary constants. This is
shown by the data presented by investigators (5,6) who showed
that at the initial stages of initiated copolymerization, MW of
the graft PS is higher than that of free PS and this grows with
the increase of PBD concentration whereas the MW of homopolystyrene gets lower. At the same time Kranz and Mortbitzer (7) showed
the regulating action of PBD at the growth of graft chains by using the example of ABS plastics. Our task was to show the dependence of graft PS MW on conditions of obtaining high impact polysty-

rene (HIPS), initiating the process by peroxides, and to compare
it with the MW values of the matrix polystyrene.

Styrene copolymerization with PBD was carried out by the
periodic block method.  Methods of carrying out the experiment
and working out the experimental data were detailed earlier (8).
The method of catalytic oxidation by hydroperoxides in the presence of benzaldehyde (9) was taken as the basis for selective
destruction of the polybutadiene backbone in graft copolymer.
The method was changed by us in the following way:  Chloroben-
zene was substituted for benzene which allowed a lower temperature
to dissolve the polymer and allow oxidizing at 25°C; t-butyl
hydroperoxide was substituted for cumene hydroperoxide; minimal
time of oxidizing – 5 hours, additional benzaldehyde added was
exactly 10 ml/g PBD.

The following regression equations were obtained.  They
relate the molecular weight of graft PS branches in the final
product (Mg) and the ratio of this value to molecular weight
(Mps) of matrix polystyrene (Mg/Mps)) to process parameters of
initiated polymerization, which are as follows:

$$
\begin{aligned}
\text{Mg} \quad 10^{-5} \quad = {} & 2.89 - 1.31X_1 - 0.23X_2 - 0.36X_3 - 0.46X_4 \\
& + 0.16X_1X_2 + 0.27X_1X_3 + 0.28X_1X_4 - 0.20X_1X_5 \\
& - 0.13X_2X_3 + 0.06X_1X_4 - 0.10X_3X_4 - 0.13X_3X_5 \\
& + 0.10X_4X_5 + 0.33X_1^2 + 0.09X_2^2 + 0.32X_3^2 \\
& + 0.04X_5^2 + 0.31X_5
\end{aligned}
\tag{1}
$$

$$
\begin{aligned}
\text{MG/Mps} \quad = {} & 1.28 - 0.03X_3 - 0.1X_4 + 0.04X_5 - 0.03X_1X_2 \\
& - 0.05X_1X_3 - 0.06X_1X_4 + 0.05X_1X_5 - 0.05X_2X_3 \\
& + 0.07X_2X_4 - 0.03X_3X_5 + 0.04X_4X_5 - 0.04X_1^2 \\
& + 0.05X_2^2
\end{aligned}
\tag{2}
$$

where $X_1$ the concentration of benzoyl peroxide (BP)
      $X_2$ the concentration of 5-butyl perbenzoate (TBPB)
      $X_3$ the first stage polymerization temperature ($t_1$)
      $X_4$ the second stage polymerization temperature ($t_2$)
      $X_5$ final first stage monomer conversion (S)

Evaluation of these equations according to Fisher's criterion showed that they adequately described the experimental data.
The presented equations show that influence of each of the variables on investigated parameters is not one value and is de-

fined by a multiple of values of the other process parameters. Figs. 1-3 graphically present the dependence of the parameters obtained by the calculation method on both absolute (curves 1, 2,3) and relative (curves 1', 2', 3') values of MW of graft PS under various copolymerization conditions of styrene with rubber. Fig. 1 shows that the MW of graft polystyrene decreases with increasing t. However, the ratio Mg/Mps increases from 1.12 to 1.37, which indicates much more effect of temperature at this polymerization stage on MW of free polystyrene. The same is observed in Fig. 2 (curves 1 and 3) where in the absence of BP (TBPB under these conditions practically is not decomposed and polymerization can be considered thermal) the ratio Mg/Mps changes until 1 takes place with the increase of reaction duration in conditions of comparatively low temperature of the first stage. This ratio gets even lower under conditions of high radical concentration at the beginning of the process (curve 2 Fig. 2 with low values of S). However in this case the reason for lowering of the graft branch length of homopolystyrene, in our opinion, is different and will be discussed below. Thus, the statements of Riess et al (6) that MW of graft branches is always higher than homopolystyrene MW is not true in all cases. The effect of initiators concentration on the length of free polystyrene is much more than that on graft PS. As a result, the ratio Mg/Mps more often than not grows with the increase of peroxide content and reaches a high value (1.5 - 1.7). Riess (6) explains it by unequal distribution of BP between phases. But taking into consideration the value of the distribution coefficient 0.92 - 0.94 (4), this great difference of molecular weights can not be explained. Moreover temperature $t_2$ influences Mg/Mps in an analogous way. An especially large effect is observed when the content of TBPB - initiator of graft at high stages (curves 1, 2 Fig. 3) - is increased in the system, except for those cases when at the first stage a great number of free radicals is formed (curve 3, Fig. 3). The given facts indicate that the graft PS being formed at the middle and high stages has a much higher degree of polymerization than the matrix PS. With the increase of graft degree, which is promoted by factors under consideration (8), a fraction of graft PS which was formed at this period increases and thus is shown in the increase of the ratio Mg/Mps.

The achieved results allow an understanding of some peculiarities of the graft branches growth at different stages of the process for obtaining high impact polystyrene. Separating the drops of polystyrene phase in the initial homogeneous rubber solution at polymerization causes a sharp reduction of rubber phase volume, which results in a concentration and viscosity increase of at least a factor of ten. The polystyrene phase viscosity increases much less. Consequently, conditions of graft branch growth differ from those of homopolymerization already at

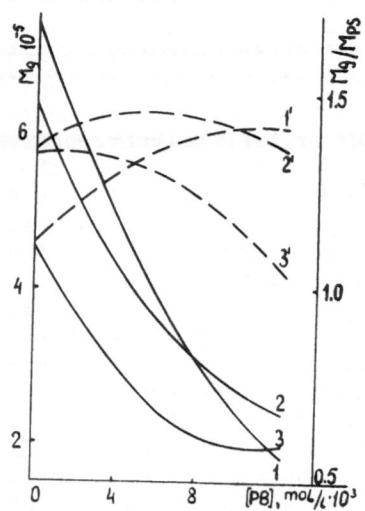

Fig. 1    Average graft polystyrene MW (curves 1,2,3) and ratio
          Mg/Mps (curves 1,2,3) vs. BP concentration:
                    $t_2^0$ = 115°C, S = 45% and

    1,1'/(TBPB) = 0.1    $10^{-2}$mol/1; $t_1$ = 70°C
    2,2'/(TBPB) = 0.1    $10^{-2}$mol/1; $t_1$ = 90°C
    3,3'/(TBPB) = 1.1    $10^{-2}$mol/1; $t_1$ = 90°C

the initial stages.  In the first case two more additional processes will compete, lowering Mg at the expense of chain transfer to PBD and its growth as a result of diminishing value of Kt
caused by the medium viscosity.  If in the thermal process, in
which appearance of a radical in a PBD chain is mainly connected
with transfer reactions, their influence is noticeable - the
graft chain does not exceed much the length of the homopolystyrene one, in the initiated process gel-effect plays a big
role.  It is verified by the already mentioned fact of growth of
graft polystyrene MW with the increase of rubber content in the

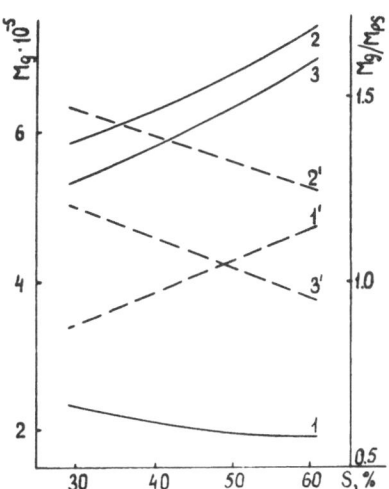

Fig. 2    Average MW of graft polystyrene as a function of the
          first duration (curves 1,2,3) and ratio Mg/Mps (curves
          1', 2', 3') at:

          1.1' (BP) = 1.2  $10^{-2}$mole/1; (TBPB) = 0.9   $10^{-2}$ mole/1;
                $t_1$ = 90$^{\circ}$C; $t_2$ = 115$^{\circ}$C
          2.2' (BP) = 0; (TBPB) = 0.1  $10^{-2}$ mole/1; $t_1$ = 90$^{\circ}$C;
                $t_2$ = 115$^{\circ}$C
          3.3' (BP) = 0; (TBPB) – 1.1  $10^{-2}$ mole/1; $t_1$ = 80$^{\circ}$C;
                $t_2$ = 105$^{\circ}$C

system pointed out by Manarezi (5) and Riess (6). But the fraction of graft polystyrene, formed in the initial period, is so
insufficient that the value of final product MW does not effect
Mg.  The results of the experiment allow us to come to the conclusion that the influence of this copolymer on the rubber phase
structure presupposes conditions of graft chain growth with further polymerization.  Thus with thermal initiation at low temperature at the first stage, insufficient quantity of graft PS with
a high degree of polymerization is formed.  This copolymer will

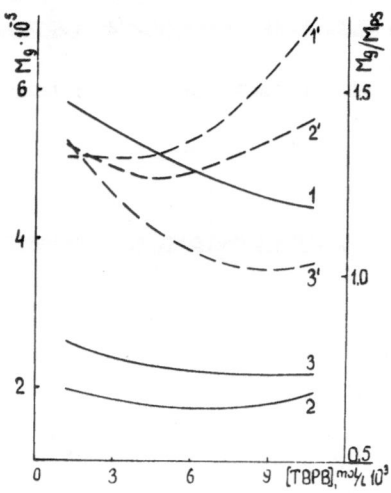

Fig. 3    Average MW of graft polystyrene (curves 1,2,3) and the
          ratio Mg/Mps (curves 1', 2', 3') as a function of con-
          centration TBPB when S = 45% and

          1,1' (BP) = 0; $t_1$ = 80°C; $t_2$ = 125°C
          2,2' (BP) = 1   $10^{-2}$ mole/1; $t_1$ - 80°C; $t_2$ = 125°C
          3,3' (BP) = 1.2   $10^{-2}$ mole/1; $t_1$ = 90°C; $t_2$ = 115°C

be concentrated on the surface of the phase boundary with locali-
zation of polystyrene branches in the matrix.  The rubber phase
will remain homogeneous until high conversions.  The influence
of a higher temperature, $t_2$, on such a system allow formation of
the final product with the mean value Mg close to the matrix MW.
Temperature increase at the first stage or use of a highly effective initiator graft (BP) promotes the growth of graft branch number at the initial stages of the process with simultaneous diminishing of their length.  It allows copolymer distribution between
the surface and the inside of the particles and association of

graft PS into domains, i.e. a heterogeneous system forms as a result of incompatibility with PBD. Since domains act as structural knots, they may fully prevent diffusion of copolymer macromolecules. In this system, as was shown earlier (8), combination termination of radicals becomes difficult. The probability of transfer to rubber is dependent on domain size, which in turn depends on primary length of graft PS. Thus the earlier the polystyrene domains appear in rubber particles and the higher is the MW of graft branches which formed them, the longer will be the growing graft chain under the same conditions with further polymerization. To our mind this explains the growth of Mg/Mps with increase of both $t_1$ and BP concentration at low temperatures of the first stage (curve 1, Fig. 1), TBPB concentration with insufficient BP content, as well as diminishing Mg/Mps with combinations of high PB concentrations and $t_1$ (curve 3, Fig. 3). Some growth of MW of graft PS with increase of TBPB concentration at high value of $t_2$ may be explained by the fact that the methods of defining graft polystyrene branches MW are not perfect because full fractioning of HIPS to homopolystyrene and graft copolymer at high crosslink density is not possible under these conditions and partly preservation of rubber backbone fragments between a pair of graft chains with oxide decomposition of PBD in such a copolymer.

REFERENCES

1. Y. Mori, Y. Minoura, M. Imoto, Macromol. Che., 25, 3-7 (1958).
2. J.P. Fisher, Angew. Makromol. Chem., 33, 35-74 (1973).
3. V.D. Yenalyev, N.A. Noskova, B.V. Kravchenko, Emulsion Polymerization, ACS Symposium Series, 24, 379-396 (1976).
4. W.A. Ludwico, S.L. Rosen, J. Appl. Polym. Sci., 19, 757 (1975).
5. P. Manarezi, V. Passalacqua and F. Pilati, Polymer, 16, No. 7, 520-525 (1975).
6. G. Riess, C. Beslin, J.L. Locatelli, J. Refregier, Polymer Alloys, Plenum Press, N.Y., 1977, 337-346.
7. D. Kranz, L. Morbitzer, K.H. Ott, R. Casper, Agnew. Makromol. Chem., 58-59, 213-226 (1977).
8. V.D. Yenalyev, V.I. Melnichenko, N.A. Noskova and others. Polymer Alloys II, Plenum Press, N.Y., 1980, 59-68.
9. P. Hubin-Eschger, Angew. Makromol. Chem, 26, 107-115 (1972).

# ON THE INFLUENCE OF THE INITIAL STAGES OF STYRENE COPOLYMERIZATION WITH POLYBUTADIENE ON HIGH IMPACT POLYSTYRENE MORPHOLOGY

V.D. Yenalyev, N.A. Noskova, V.I. Melnichenko,
O.P. Bovkunenko, and V.M. Bulatova

Donetsk State University
Donetsk, 340055, USSR

As has already been mentioned in previous publications, one of the main factors influencing the industrial properties of high impact polystyrene (HIPS) is the morphology of this composite material i.e. the rubber volume fraction (Vf) and mean size of dispersed particles (c). For instance, references 1 and 2 deal with the correlation between rubber phase volume and strength properties of HIPS. The conclusion was made that impact resistance is increased with the growth of Vf, whereas the tensile strength decreases. One should bear in mind that the increase in volume in mechanical mixtures may only be possible at the expense of introducing additional polybutadiene (PBD). In copolymerization the same effect is achieved due to polystyrene occlusions. Introducing free and graft polystyrene (PS) into rubber particles proves to have the same toughening effect (3-5). That is why so much attention is paid to studying the influence of graft copolymer on formation of HIPS morphology. As had previously been pointed out by us, a role is given to copolymer which was formed at the initial stages of the process. The growth in the number of graft branches at this period leads not only to better dispersion of the rubber (6), but also stimulates the appearance of occlusions and in this way promotes an increase in rubber phase volume. Formation of copolymer with a great number of short graft branches results in developing particles of all character structure (7). The further increase of polystyrene branch number with the simultaneous decrease in their length produces a change of morphology (8). In this case formation of particles of honey-comb structure is not observed. On the other hand as has already been mentioned (9) the absolute absence of graft polystyrene branches in rubber particles at the stage of prepolymerization under HIPS synthesis

by means of a continuous procedure promotes the retention of rubber in solution (homogeneous) until high conversion values.  In
this process the particles with a small amount of polystyrene are
formed and the fraction volume of dispersed phase is diminished
twice as fast compared to periodic conditions.  One of the ways
of improving HIPS morphology may be changing of the continuous
procedure for obtaining it, i.e. introduction of additional preliminary polymerization of polybutadiene solution before adding
to the continuous prepolymerizer (10).  The purpose of this investigation is to study the effect of preliminary polymerization
conditions on HIPS morphology.

Prepolymerization was carried out in a cylindric apparatus
of continuous action with constant stirring (screw stirrer).
Monomer conversion (S) in the apparatus was maintained at 32 – 35%
at the $t^o = 120^oC$.  The following polymerization was carried out
in soldered glass ampules using a stepped temperature scheme
(150–190$^o$C).  Polybutadiene rubber SKD-PS (90% cis-links) in
the quantity of 6% by weight was used.  Under the same conditions
polymer was obtained by the periodic method for comparison; HIPS
morphology was studied by electron microscopy (1).

Table 1 presents the conditions for preliminary polymerization
of polybutadiene solution and the results of calculations of the
HIPS morphology parameters according to the electron micrographs.
As is seen from the table, a change of conditions of polymerization
resulted in a material with a different mean size of dispersed
phase particles.  The latter, as is known (2,7), influences
greatly the volume fraction of the rubber phase.  This is shown
once again by the examples 1a and 1b in Table 1 obtained under
continuous prepolymerization at different stirring speeds.  In
this case the value of Vf can not be characterized rationally by
using either method.  In order to appreciate adequately the efficiency of the additional stage, we introduced the coefficient $K_1$,
calculated as the ratio of Vf to t he total volume of rubber phase
with commensurable size of dispersed particles which was obtained
by the periodic method under analogous conditions (initiated ther-
mally).  The coefficient $K_2$ is calculated as the ratio of Vf of
HIPS to the same denominator; HIPS was obtained by the periodic
method with introduction of the additional stage of prepolymerization under conditions given in Table 1.  Visual representation
of the morphology is shown by electron micrographs presented in
Fig. 1.  Comparing the microstructure of the samples a and b
(Fig. 1) one can see that the shift to the continuous method of
prepolymerization leads to diminishing fraction volume of the
rubber phase (K = 0.55).  Particles size polydispersity is significantly increased:  there are particles of 0.2 μ to 8–10 μ with
various degree of filling by polystyrene occlusions.  At the same

time when comparing thermal and benzoyl peroxide (PB) initiated processes, the latter proves to be more efficient as is seen from the data given in Table 1 ($K_1$ = 0.55 and 0.68 with samples 1 and 9). Preliminary polymerization (until monomer conversion of 5-7%) before addition of rubber solution into the continuous action apparatus changes the polymer morphology (a, d, Fig. 1), - i.e. polydispersity is diminished and the particles begin to look more alike, all of which increases the volume fraction of rubber phase.

Analysis of the regime effect on $K_1$ (Table 1) shows that both monomer conversion in the range of 6-11% and temperature increase at the initial stage to $140^{\circ}C$ (samples 2, 3, 8 Table 1) change $K_1$ negligibly (0.71 - 0.79). More effective is the preliminary polymerization of highly concentrated solution of polybutadiene (for the given series 9-12%) at the experimental temperature or lower followed by diluting with monomer to achieve the necessary rubber concentration (samples 4,7) or using additions such as initiator, chain transfer agent, low molecular weight homopolystyrene (samples 5, 6, 10, 11, 12) or preliminary polymerization with high content of PBD. These exert a positive influence even on the periodic thermal process ($K_2$ = 1.4, sample 7). In the initiated polymerization such an effect was not observed (compared value $K_2$ with the samples 9 and 11). The results are in good agreement with the following mechanism for formation of rubber particles - inner structure. It is known that addition of rubber solution to the continuous action prepolymerizer results in its dispersion with a simultaneous volume decrease as a result of monomer redistribution. If PBD solution is not subjected to preliminary polymerization the particles will be in a homogeneous system of high concentration.

The presence of graft copolymer in the dosing solution promotes the appearance of occlusions by means of associations of incompatible polymer with PBD graft branches, homopolystyrene moleclues and monomer ones. Low molecular fractions of free PS can diffuse into domains, thereby increasing their size (10). When mixing such a system with prepolymer, the graft copolymer results in easier formation of rubber dispersion. Domains within particles which were being formed increase the rubber phase volume also because of additional styrene which is retained in the particles by polystyrene branches and promote the occlusion growth with further polymerization. Thus a big role in HIPS morphology is played by both the number of graft branches which are formed at the preliminary stage of polymerization, their length, and the ratio of the number of graft branches to the homopolystyrene macromolecules. Thermal polymerization of diluted PBD solution leads to formation of a small number of PS graft branches which are close in size to macromolecules of homopolystyrene. That is why

TABLE I

CONDITIONS OF CARRYING OUT PRELIMINARY POLYMERIZATION STAGES AND MORPHOLOGICAL PARAMETERS OF THE HIPS

| No. of Experiment | Recipes | | Conditions of Preliminary Polymerization | | | Parameters | | | | |
|---|---|---|---|---|---|---|---|---|---|---|
| | BP, % | mercaptane % | t°C | S, %. | Additional Conditions | Vf, % | C | $K_1$ | $K_2$ | Gs, % |
| 1a | | | | – | | 14.3 | 2.20 | 0.55 | | 226 |
| b | | | | – | | 7.9 | 1.01 | 0.52 | | 219 |
| 2 | | | 120 | 7.4 | | 11.7 | 1.33 | 0.71 | | 227 |
| 3 | | | 120 | 10.7 | | 18.4 | 2.33 | 0.77 | | 227 |
| 4 | | | 120 | 6.8 | diluted | 18.3 | 1.50 | 0.96 | | |
| 5 | | 0.01 | 120 | 9.1 | | 26.5 | 2.71 | 0.98 | | 234 |
| 6 | | | 120 | 6.8 | added 1% PS MW = 0.7 $10^5$ | 15.6 | 1.30 | 0.92 | | |
| 7 | | | 110 | 5.5 | diluted | 15.7 | 1.0 | 0.98 | 1.4 | 233 |
| 8 | | | 140 | 6.2 | | 13.1 | 1.19 | 0.79 | | 216 |
| 9 | 0.02 | | | – | | 17.0 | 2.8 | 0.68 | 1.7 | 222 |
| 10 | 0.02 | | 100 | 4.5 | | 24.4 | 2.7 | 0.98 | | |
| 11 | 0.02 | | 100 | 5.2 | diluted | 14.0 | 0.96 | 0.93 | 1.64 | 259 |
| 12 | 0.02 | 0.01 | 100 | 6.1 | * | 21.3 | 1.87 | 0.99 | 1.46 | 291 |

*Mercaptan was introduced after finishing the preliminary stage of polymerization initiated by BP.

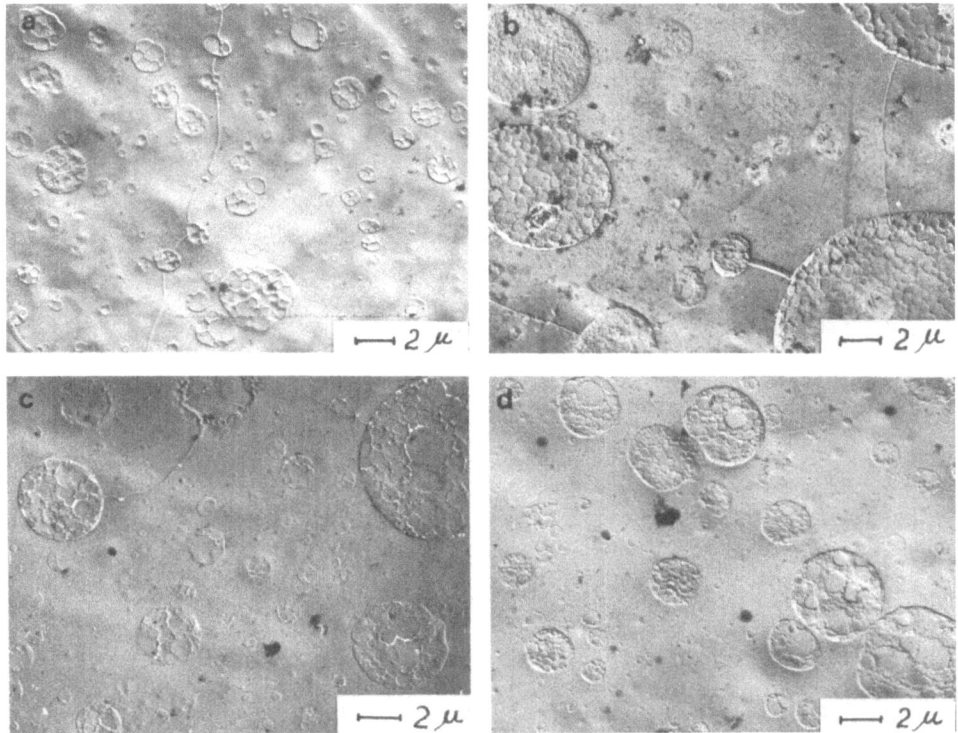

Fig. 1 HIPS morphology:
    a - obtained by periodic styrene copolymerization
    with PBD at thermal initiation,
    b,c,d - samples Ia, 7 and 12 Table 1 respectively.

the morphology change which is being observed at such a regime
of preliminary polymerization is comparatively small.  High concentration of PBD at this stage promotes formation of longer graft
branches (11, 12).  This results in bigger accumulation of graft
PS in the system at equal polymer content after the preliminary
stage is over.  Analogous results are obtained when using, at
this stage, a highly effective grafting initiator - BP- benzoyl
peroxide.  The use of BP in the continuous method without prelimi-
mary polymerization is not effective because of its low temperature
of decomposition.  Comparing to a thermal process, BP increases
both the copolymer quantity and the relative graft branches length.
When introducing the chain transfer agent into the reaction mass
mixture or adding a little quantity of low molecular weight PS,
the same result is obtained as a result of increasing the ratio
of MW of graft polystyrene to matrix molecular weight due to lowering the latter.

Thus, it is seen that any method of producing enrichment of
the initial solution of polybutadiene by graft and low molecular
weight polystyrene results in sufficient improvement of HIPS
morphology obtained by the continuous block method.

REFERENCES

1.  Yenalyev, V.D., Noskova, N.A., Kravchenko, B.V., Emulsion
    Polymerization ACS Symposium Series, 24, 379-396 (1967).
2.  Turley, S.G., Keskkula, H. Polymer, 21, No. 4, 466 (1980).
3.  Wagner, E.R., Robeson, L.M., Rubb. Chem. Technol., 43, 1129
    (1970).
4.  Baer, M., J. Appl. Polymer Sci., 16, 1109 (1972).
5.  Cigna, G., Appl. Polymer Sci., 14, 1781 (1970).
6.  Molay, E.G., J. Polym. Sci., A-3, 1267 (1965).
7.  Yenalyev, V.D., Noskova, N.A., Melnichenko, V.I., Zhuravel,
    Y.N., Bulatova, V.M., Polymer Alloys II, Plenum Press, N.Y.,
    1980, pp. 69-77.
8.  Keskkula, H., Plast. and Rubber Mater. and Appl., 4, No. 2,
    66 (1979).
9.  Yenalyev, V.D., Noskova, N.A., Kravchenko, B.V., ACS Polymer
    Preprints, 37, No. 1, 639 (1977).

10. Echte A., Angew Makromol. Chemie, 58/59, 175 (1977).
11. Manaresi, P., Passalacqua, V. and Pilati, F. Polymer, 16,
    520 (1975).
12. Riess, G., Beslin, C., Lacatelli, Y.S., Refregier, Y.S.,
    Polymer Alloys, Plenum Press, No. 4, 1977, p. 337-346.

CHARACTERIZATION OF POLYMER BLENDS, BLOCK COPOLYMERS,
AND GRAFT COPOLYMERS BY FRACTIONATION PROCEDURES
USING DEMIXING SOLVENTS

Rainer Kuhn

Central Research and Development
Bayer AG
5090 Leverkusen, W. Germany

INTRODUCTION

In polymer analysis the scientific and technical importance
of polymeric multi-component systems has increased to such an extent that problems whose solution depends on the separation of
chemically different polymers are arising more and more frequently.
Thus, where e.g. graft copolymers are concerned, such separation
techniques are important as a means of determining the relationship
between synthesis parameters and grafting efficiency.

The purpose of this paper is to describe a new method for the
fractionation of polymers of differing chemical composition, which,
as will be shown later, has several advantages over the familiar
separation methods. The term we use for this method is "fractionation with demixing solvents".

FUNDAMENTALS

In the simplest case binary solvent mixtures which are miscible
at high temperatures, but have a miscibility gap at low temperatures,
are used for this method of fractionation with demixing solvents.

The principle of the new fractionation method - fractionation
with demixing solvents - as applied in the simplest case, i.e. that
of polymer blends consisting of two chemically different polymers,
can be summarized as follows (Fig. 1):

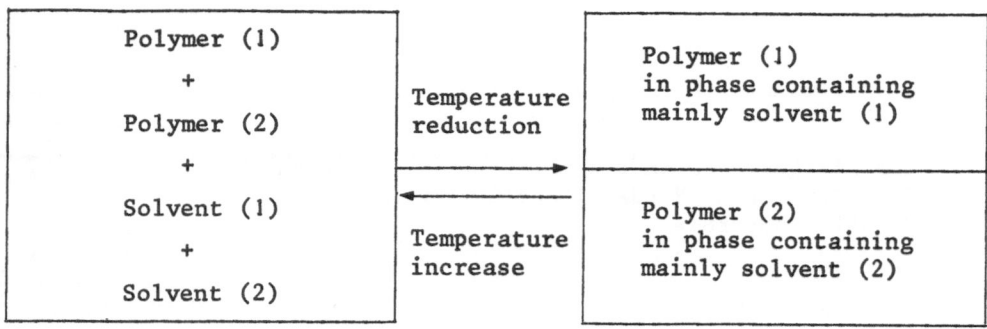

Fig. 1.  Very simple case of fractionation of a polymer blend via the
         demixing of a homogeneous solution into homogeneous phases,
         if solvent (1) has a lower density than solvent (2).


     Two chemically different polymers forming a homogeneous solution
in a binary solvent mixture are separated by lowering the temperature
in the manner indicated. One of the polymers enters the upper phase,
while the other enters the lower phase, this migration generally being
quantitative in both cases. As a rule, the homogeneous solution is
fractionated into solutions which are likewise homogeneous. This process can be explained very simply according to the solubility parameters of polymers and solvents and approximate prediction of the
fractionation conditions is therefore possible[1,2].

     It is also possible, of course, to describe the new fractiona-
tion method with the aid of the concept of the second virial coefficient, which is combined with the Flory Huggins parameter[1].


BINARY DEMIXING SOLVENT PAIRS

     Several binary demixing solvent systems which, at certain
ratios of the constituents, separate reversibly into two phases
when the temperature is reduced are listed in Table 1. Table 1  also
gives, for a volume ratio of the constituents of 1:1, the volume
fraction of the lower phase at 25 $^\circ$C and the critical temperature $T_c$
above which the system becomes homogeneous. It also gives the density
differences and solubility parameters $\delta_{SOLV}$ (cal$^{0.5}$ cm$^{-1.5}$) of the
individual constituents.


EXPERIMENTAL

     The blends of the chemically different polymers, together with
the graft polymers, were dissolved at temperatures above those

Table 1. Selection of binary systems SOLV(1)/SOLV(2) which demix when the temperature is reduced. Solubility parameters $\delta$ SOLV(1) of solvents (1). Differences between the solubility parameters $\delta$ and densities $\rho$ of the solvent pairs. Critical temperature $T_c$ below which, at the volume ratio 1:1, demixing occurs. Corresponding volume fraction of lower phase at 25 °C.

| SOLV(1) | SOLV(2) | $\delta_{SOLV(1)}$ | $\delta_{SOLV(1)} - \delta_{SOLV(2)}$ | $\rho_{SOLV(1)} - \rho_{SOLV(2)}$ | $T_c$/°C | Volume fraction of lower phase |
|---|---|---|---|---|---|---|
| N,N-dimethyl formamide | Cyclopentane | 12.1 | 3.4 | 0.171 | 34.5 | 0.77 |
| N,N-dimethyl formamide | Cyclohexane | 12.1 | 3.9 | 0.171 | 47.5 | 0.64 |
| N,N-dimethyl formamide | Methyl cyclohexane | 12.1 | 4.3 | 0.180 | 52.5 | 0.62 |
| N,N-dimethyl formamide | Heptane | 12.1 | 4.7 | 0.266 | 71 | 0.57 |
| N,N-dimethyl formamide | Decalin | 12.1 | - | 0.066 | 69 | - |
| Methanol | Cyclohexane | 14.5 | 6.3 | 0.013 | 41 | 0.76 |
| Methanol | Methyl cyclohexane | 14.5 | 6.7 | 0.022 | 42 | 0.75 |
| Methanol | Heptane | 14.5 | 7.1 | 0.107 | 49 | 0.70 |
| N,N-dimethyl acetamide | Heptane | 11.1 | 3.7 | 0.254 | 38 | 0.63 |
| N,N-dimethyl acetamide | Isooctane | 11.1 | | | 40 | 0.64 |
| Acetic anhydride | Cyclohexane | 10.3 | 2.1 | 0.30 | 52 | 0.58 |
| Acetic anhydride | Methyl cyclohexane | 10.3 | 2.5 | 0.31 | 56 | 0.57 |
| Acetic anhydride | Heptane | 10.3 | 2.9 | 0.396 | 70 | 0.54 |
| Propylene carbonate | Perchloroethylene | 13.3 | 4.0 | -0.435 | 96 | 0.39 |
| Propylene carbonate | Carbon tetrachloride | 13.3 | 4.7 | -0.392 | 33 | 0.08 |
| Ethylene carbonate | Tetralin | 14.7 | | 0.35 | 97 | 0.58 |
| Ethylene carbonate | Xylene | 14.7 | 5.9 | 0.46 | 70 | 0.60 |
| Ethylene carbonate | n-propanol | 14.7 | 3.2 | 0.52 | 36 | 0.61 |
| Ethylene carbonate | Cyclohexanol | 14.7 | 3.3 | 0.38 | 57.5 | 0.54 |
| Ethylene carbonate | Ethyl benzene | 14.7 | 5.9 | 0.45 | 44.5 | 0.76 |
| Butyrolactone | Decalin | 12.6 | | 0.250 | 129 | 0.52 |

of the solubility curve in solvent mixtures of suitable composition, with the result that all the systems were clear solutions before the demixing fractionation of the low-molecular components. Crystalline polymers, e.g. polyethylene, were dissolved in a dilute solution at temperatures above their crystallization temperatures. The materials were dissolved under refluxing. In general the polymer concentrations at high temperatures were so low (lower than 1 % by weight; in most cases 1 g polymer and 500 ml solvent mixture) that before the separation the solution was homogeneous and not – in consequence of the incompatibility of the chemically different polymers – turbid. The separation from homogeneous solution by temperature reduction was carried out in graduated temperature-controllable cylindrical separating funnels. It was found that, while the cooling was in progress, stirring, followed by slight vibration of the separating funnels, were conducive to rapid and quantitative fractionation. The vibrations prevent small droplets of the counterphase remaining on the wall within the other phase.

The polymers were commercial products or laboratory preparations. The copolymers, except for the block and graft copolymers were chemically uniform. In the majority of cases where PS was treated no dependence of the fractionation effect on the molecular weight was seen under the chosen fractionation conditions within the weight average molecular weight range $M_w$ of approximately $10^4$ to $3 \cdot 10^6$. It is therefore unnecessary to state the exact molecular structures of the numerous polymer blend preparations, whose $M_w$ values were generally between about $10^4$ and $2 \cdot 10^6$. Before each fractionation of a polymer blend a blank test with only one polymer was carried out to make sure that the entire polymer (> 99 % by weight) would be fractionated either into the upper or into the lower phase. In the case of the polymer blend and graft or block copolymer fractionations the determination of the chemical composition of the polymers in the individual phases was supplemented by elemental analysis (e.g. in the case of PVC), or by NMR or IR spectroscopy (e.g. in the case of polymers containing styrene). So in the case of polymer blends it was certain that – within the accuracy of the measurement procedure and the detection limit – the entire quantity of a polymer species (> 99 % by weight) had been fractionated either into the upper or into the lower phase.

Depending on the system and on the molecular weight of the polymer, a fractionation takes ten minutes to several days. In most cases the fractionation was complete after one hour in as far as pure polymer blends, and not block or graft copolymers, were concerned.

It will be seen later that even the binary solvent system methyl cyclohexane (MCH)/N,N-dimethyl formamide (DMF) is suitable for the separation of many chemically different polymers. In this case the fact that the densities of the constituents differ greatly is con-

ducive to rapid phase separation. At constant temperature the composition of the phases and thus the fractionation efficiency are independent of the volume fraction of the solvent pairs[1,5].

RESULTS FOR POLYMER BLENDS

Table 2 gives a few examples of the fractionation efficiencies
of polymer blends obtainable with binary demixing solvent systems[3].
If a polymer is partly crystalline at the separation temperature, it
accumulates in a small intermediate layer between the upper phase and
the lower phase, which is of higher density. The same occurs if, e.g.
a rubber is crosslinked. In cases of this kind the described fractionation method enables three different polymers to be separated,
provided that one of the polymers accumulates in the upper phase and
remains in solution after the temperature has fallen below the separation point.

By a simple combinatorial process the number of polymer systems
separable with demixing solvent mixtures listed in the tables can be
increased many times.

RESULTS FOR GRAFT COPOLYMERS

In the fractionation of complex mixtures consisting, for
example, of block and graft copolymers which form homogeneous solutions at elevated temperatures it is often impossible, if only one
of the two polymer components is genuinely dissolved, to obtain clear
solutions of the polymers in the two phases by reducing the temperature. Thus, where graft or block copolymers are concerned it is possible, with binary demixing solvents, to separate, e.g., the substrate unaffected by grafting, which appears in the upper, homogeneous phase, the ungrafted secondary polymer, which enters the lower,
homogeneous phase, and the genuine graft polymer, which enters an
intermediate, nonhomogeneous, and turbid layer, the separation being
almost quantitative and achieved in a single fractionation step.
The intermediate phase may be either within the upper or lower phase
or within both (see Fig. 2).

For example, when graft copolymers consisting of styrene and
polydimethyl siloxane have been dissolved in the demixing solvents
dimethyl formamide and methyl cyclohexane, a reduction in the temperature to 25 °C causes the polystyrene unaffected by grafting to
appear in the lower phase (containing a high proportion of DMF), the
ungrafted polydimethyl siloxane to appear in the upper phase (containing a high proportion of MCH), and the genuine graft copolymer
to appear in a turbid intermediate layer existing within the upper
phase.  The separation is almost quantitative and achieved in a

Table 2.

| Polymer fractionated into upper phase | Polymer fractionated into lower phase | Volume fractions of the solvents | Separation temperature in °C |
|---|---|---|---|
| cis 1,4-polybutadiene | Polystyrene | 0.3 DMF, 0.7 CH | 25 |
| cis 1,4-polybutadiene | " | 0.4 DMF, 0.6 MCH | 25 |
| Polyisobutylene | " | 0.24 DMF, 0.76 MCH | 45 |
| Polypentenamer | " | 0.2 DMF, 0.8 MCH | 40 |
| Poly-4-methylpentene | " | 0.25 DMF, 0.75 MCH | 25 |
| Polyethylene, high density | " | 0.25 DMF, 0.75 MCH | 25 |
| Polyethylene, low density | " | 0.24 DMF, 0.76 MCH | 25 |
| Polydimethyl siloxane | " | 0.4 DMF, 0.6 MCH | 25 |
| Polypropylene | " | 0.4 DMF, 0.6 Decalin | 45 |
| Polypropylene oxide | " | 0.5 DMF, 0.5 MCH | 25 |
| Polyethylhexyl acrylate | " | 0.4 DMF, 0.6 MCH | 25 |
| Ethylene (53)/propylene (47) copolymer | " | 0.2 DMF, 0.8 MCH | 25 |
| Ethylene (50)/propylene (43)/ dicyclopentadiene (7) copolymer | " | 0.2 DMF, 0.8 MCH | 25 |
| Ethylene (91.5)/vinyl acetate copolymer | " | 0.3 DMF, 0.7 MCH | 25 |
| Ethylene (60)/vinyl acetate copolymer | " | 0.4 DMF, 0.6 MCH | 25 |
| Polydimethyl siloxane | Polymethyl methacrylate | 0.50 DMF, 0.5 MCH | 25 |
| Polypropylene oxide | Poly-caprolactone | 0.5 DMF, 0.5 MCH | 25 |
| Polydimethyl siloxane | Polyurethane | 0.4 DMF, 0.6 MCH | 25 |
| Polydimethyl siloxane | Polybutyl acrylate | 0.4 DMF, 0.6 n-heptane | 25 |
| Polybutadiene | Polybutyl acrylate | 0.4 DMF, 0.6 n-heptane | 25 |
| Polydimethyl siloxane | Bisphenol-A-polycarbonate | 0.5 DMF, 0.5 MCH | 25 |
| Polydimethyl siloxane | Poly-α-methyl styrene | 0.4 DMF, 0.6 MCH | 25 |
| Polyisobutylene | " | 0.32 DMF, 0.68 MCH | 40 |
| cis 1,4-polybutadiene | " | 0.34 DMF, 0.68 MCH | 40 |

| Polymer 1 | Polymer 2 | Solvent 1 | Solvent 2 | T (°C) |
|---|---|---|---|---|
| Polydimethyl siloxane | Poly-4-chlorostyrene | 0.4 DMF | 0.6 MCH | 25 |
| Polydimethyl siloxane | Polytrifluoromethyl styrene | 0.3 DMF | 0.7 MCH | 25 |
| cis 1,4-polybutadiene | Polyvinyl chloride | 0.4 DMF | 0.6 MCH | 25 |
| Polydimethyl siloxane | Polyvinyl chloride | 0.4 DMF | 0.6 MCH | 25 |
| Polyethylene, low density | Polyvinyl chloride | 0.25 DMF | 0.75 MCH | 25 |
| Chlorinated polyethylene (chlorine content <36 % weight) | Polyvinyl chloride | 0.4 DMF | 0.6 MCH | 25 |
| Polydimethyl siloxane | Polyvinyl chloride | 0.2 DMF | 0.8 MCH | 25 |
| Polydimethyl siloxane | Polyethylene oxide | 0.28 $CH_3OH$ | 0.72 heptane | 25 |
| Polyisobutylene | " | 0.2 DMF | 0.8 MCH | 25 |
| Polyethylene, low density | " | 0.2 DMF | 0.8 MCH | 25 |
| Polydimethyl siloxane | Polyvinyl acetate | 0.5 DMF | 0.5 MCH | 25 |
| Polydimethyl siloxane | Polyvinyl pyridine | 0.6 DMF | 0.4 MCH | 25 |
| Polydimethyl siloxane | Ethylene (30)/vinyl acetate (70) copolymer | 0.48 $CH_3OH$ | 0.52 n-heptane | 25 |
| Polyisobutylene | Styrene (72)/butyl acrylate (28) copolymer | 0.2 DMF | 0.8 MCH | 25 |
| Ethylene (53)/propylene (47) copolymer | Styrene (72)/butyl acrylate (28) copolymer | 0.2 DMF | 0.8 MCH | 25 |
| Polypropylene oxide | Styrene (75)/acrylonitrile (25) copolymer | 0.5 DMF | 0.5 MCH | 25 |
| cis 1,4-polybutadiene | Styrene (75)/acrylonitrile (25) copolymer | 0.43 DMF | 0.57 MCH | 25 |
| Polypropylene | Styrene (75)/acrylonitrile (25) copolymer | 0.4 DMF | 0.6 Dekalin | 45 |
| Polydimethyl siloxane | Maleic anhydride (50)/isobutylene (50) copolymer | 0.4 DMF | 0.6 MCH | 25 |
| Polydimethyl siloxane | Acrylonitrile (70)/vinyl acetate (20)/styrene (10) terpolymer | 0.55 DMF | 0.45 MCH | 25 |

DMF = N,N-dimethyl formamide
CH = cyclohexane
MCH = methyl cyclohexane

homogeneous
solution

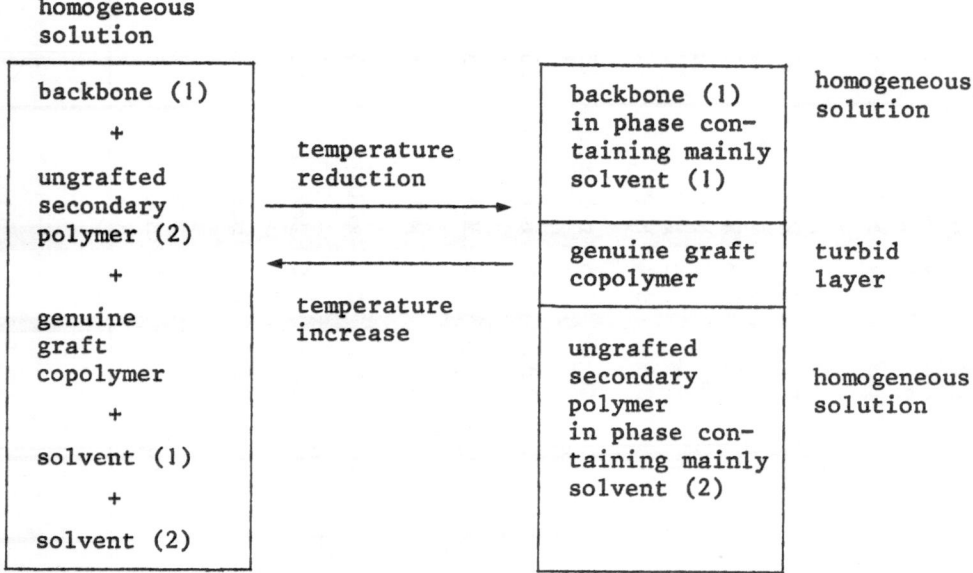

Fig. 2.   Fractionation of graft copolymers with demixing solvents

single step. This separation method is applicable to numerous graft copolymer systems[3], especially in cases where other fractionation methods, e.g. fractional precipitation, do not give satisfactory results. The separation method is even successful when the polymers to be separated are partly crosslinked. In the case of graft systems with crosslinked substrate, for example, one can determine the graft efficiency and proportion of uncrosslinked substrate very simply, and in a single fractionation step, even when the solubility parameters of the polymers are very similar. Occlusion or inclusion effects of the sort which occur in extraction are not observed, as the fractionation is performed at a high temperature and the crosslinked substrate is in the swollen state. Thus, when graft copolymers of styrene on crosslinked polydimethyl siloxane are separated with the demixing system DMF/MCH (volume fractions 0.4/0.6), polystyrene unaffected by grafting enters the lower, DMF-rich phase, crosslinked, grafted, or ungrafted polydimethyl siloxane appears in a small intermediate phase, and uncrosslinked polydimethyl siloxane appears in the upper phase. Experiments have shown that this separation method is applicable to numerous graft copolymer systems having crosslinked substrates.

MULTIPLE DEMIXING SOLVENT SYSTEMS

The demixing liquids may also be multiple, e.g. ternary or quarternary mixing, since the solubility parameter of the homogeneous

phase and those of the conjugated phases formed when the temperature
is reduced may be influenced not only by the temperature but also by
a third low-molecular constituent, or further additional ones, which
is or are differently distributed among the phases formed by the temperature drop, whose composition is thus altered.

By adding a third (or additional) solvent it is also possible
in many cases to confer a miscibility gap on miscible solvent systems which do not demix when the temperature is reduced or, in the
opposite manner, to increase the mutual compatibility of the constituents of binary systems that are not completely miscible at any
temperature to such an extent that the system acquires an upper critical demixing point.

From binary miscible and binary unmiscible solvent systems, each
of which has a solvent in common it is generally possible to form
demixing ternary solvent systems whose composition varies within
wide limits.

Thus the binary system γ-butyrolactone/perchloroethylene is
homogeneous, whereas at many ratios γ-butyrolactone/cyclohexane are
not miscible, even at high temperatures. But the ternary solvent
system butyrolactone/perchloroethylene/cyclohexane demixing at
many ratios. With such ternary systems many polymers whose solubili-
ty parameters differ by less than $1 \cdot 10^3 \ J^{0.5} \ m^{-1.5}$ (corresponding to
less than approximately 0.5 $cal^{0.5} \ cm^{-1.5}$) can be separated from one
another quantitatively (see Table 3).


CROSS FRACTIONATION

More complex polymer mixtures can only be separated into their
individual constituents if the fractions obtained in a first separation step are fractionated again with a different demixing system.
This is known as cross fractionation.

X-fold cross fractionation with demixing solvents enables polymer blends consisting of x+1 heterogeneous polymers to be separated
from one another quantitatively so that the individual constituents
are obtained, as the following example shows. In the case of a polymer blend consisting of polystyrene, SAN copolymer, and polybutadiene the DMF/MCH system enables one – as already indicated by a combinatorial analysis of the examples given in Table 2 – to start by
separating the polystyrene and SAN copolymer into the DMF-rich phase.
The polystyrene can then be separated from the SAN copolymer with the
butyrolactone/cyclohexane/perchloroethylene system just referred to.
It has been found that such separation techniques are applicable to
many polymer systems[3]. By means of a simple combinatorial analysis
one can determine from the tables what ternary polymer systems can
be separated with what demixing solvents.

Table 3.

| Polymer fractionated into upper phase | Polymer fractionated into lower phase | Volume fractions of the solvents | Separation temp. in °C |
|---|---|---|---|
| Polymethyl methacrylate | Polybutyl acrylate | 0.6 PCE, 0.2 CH, 0.2 butyrolactone | 25 |
| Polymethyl methacrylate | Polystyrene | 0.6 PCE, 0.2 CH, 0.2 butyrolactone | 25 |
| Polyvinyl chloride | Polystyrene | 0.6 PCE, 0.2 CH, 0.2 butyrolactone | 25 |
| Acrylonitrile co-polymer (acrylo-nitrile content ≳ 10 % weight) | Polystyrene | 0.6 PCE, 0.2 CH, 0.2 butyrolactone | 25 |
| Polystyrene | Polyacrylonitrile | 0.4 Tetralin, 0.2 CH, 0.4 propylene carbonate | |
| cis 1,4-poly-butadiene | Butadiene (62)/ acrylonitrile (38) copolymer | 0.36 DMF, 0.56 MCH, 0.07 THF | 25 |

PCE = perchloroethylene      THF = tetrahydrofuran
CH  = cyclohexane            DMF and MCH see Table 2.

With graft copolymers consisting of polymer systems whose constituents differ little in respect to their solubility parameters,
fractionation with demixing solvents often results merely in the
quantitative removal of one individual, e.g. of the secondary, but
ungrafted polymer. In such cases cross fractionation often permits
additional separation of the system into substrate unaffected by
grafting and genuine graft copolymer[3].

COGRAFT COPOLYMER ANALYSIS

Let us now consider briefly the analysis of co-graft copolymers,
that is to say graft copolymers for whose grafting reactions chemically different substrates are used[4]. In the simplest case, in which a
single monomer is grafted onto two chemically different substrates,
the system contains six different fundamental types of polymer:
ungrafted substrate (1), ungrafted substrate (2), ungrafted secondary

polymer (3), genuine graft copolymer from polymer (3) on substrate (1)or
(2), and, possibly, substrate (2) coupled with substrate (1) via graft
chain bridges of polymer (3) in consequence of transfer reactions or
recombination. With such systems it is possible, for example, to separate secondary but ungrafted polymer (3) from the other polymers
quantitatively by fractionation with demixing solvents, as indicated
by experiments with homopolymer blends and NMR-spectroscopic investigation of the corresponding fractions. Thus, in a very simple manner,
the entire graft efficiency on substrates (1) and (2) of the polymer
formed in the graft reaction (3) is calculated directly from the
quantities of the fractions obtained in the first fractionation step
and their overall composition.

It has been found that with a different demixing solvent system
ungrafted substrate (2), secondary but ungrafted polymer (3), and
polymer (3) grafted onto substrate (2), for example, can be separated
from the other polymers.

In this way the graft efficiencies obtained on the individual
substrates, and also the proportion of substrates (1) and (2) which
have been coupled via graft chain bridges, can be calculated very
easily from the quantities and chemical compositions of the fractions
obtained in the second fractionation step and the results of the
first fractionation step[3].

With co-graft copolymers having co-graft substrates, suitable
fractionations with demixing solvents therefore give the graft efficiency of the secondary polymer on the individual substrates and the
degree of coupling of the chemically different substrate molecules
via graft chain bridges. This fractionation principle is not, of
course, universally applicable to co-graft copolymers. The fractionation of graft copolymers with co-substrates naturally depends very
much on the polymer systems and demixing solvent systems used. In
some cases cross fractionation with demixing solvents enables the
substrates coupled via graft chain bridges to be separated from the
other polymers. Depending on the co-substrate system and grafting
conditions, efficiencies of 0 to 80 % by weight have so far been
determined for the coupling of chemically different substrates via
graft chain bridges[3].

MULTISTEP FRACTIONATION PROCEDURES

Thus far only single-step separation procedures and cross fractionation, i.e. the use of single-step separation procedures in
succession, have been considered. But multi-step fractionation procedures using demixing solvents can be developed also. Their purpose
is to reveal the chemical distributions of chemically heterogeneous
copolymers, block copolymers, or graft copolymers, that is to say of
substances consisting of many chemically different molecule types. In
order to determine the chemical distributions of binary copolymers,

block copolymers, or graft copolymers by fractionation with demixing solvents and subsequent analysis of the fractions to reveal their chemical composition, the following two conditions must be met.

a) It must be possible for homopolymers from the chemically different monomer units to be fractionated from a homogeneous solution into the mutually opposite phases when the temperature is reduced.

b) The chemical compositions of the demixing solvents must be sufficiently modifiable – e.g. in consequence of a reduction of the temperature or the addition of a third solvent – to ensure that, in the case of true block and copolymers containing no homopolymeric matter, the polymers are distributed among the two phases in accordance with their chemical composition in such a way that the entire polymer may be found either in the upper or in the lower phase. For this purpose it is not necessary to obtain homogeneous solutions after the demixing of the low molecular solvents, provided that, for example, only one of the two chemically different sequences is genuinely dissolved in the individual phases.

In principle three fractionation methods using demixing solvents can then be used to determine the chemical distribution, namely the summative, successive, and semi-summative or successive-summative methods[5]. As an example, successive fractionation with demixing solvents will be described (Fig. 3).

For this puspose the polymer is dissolved in the demixing solvent system, the temperature is reduced for the first fractionation step, in which the phase which should contain only a small percentage of the polymer, e.g. the lower one, is separated off; the polymer contained in this phase is isolated and its quantity and chemical composition are determined. For the second fractionation step an addition of a solvent mixture whose quantity and chemical composition are roughly equal to those of the low-molecular phase removed at the first step is made to the remaining phase, the temperature is raised to make the solution homogeneous, and the solution is fractionated at a temperature different from that used for the first step, with the result that the phase from which the first fraction was obtained again contains only a small percentage of the polymer.

Some of the results obtained in the fractionation of styrene-butadiene block copolymers are presented in Fig. 4.

It is apparent that the block copolymer, which, according to GPC determinations, has a molecular non-uniformity of about 0.01, still has a relatively broad chemical distribution.

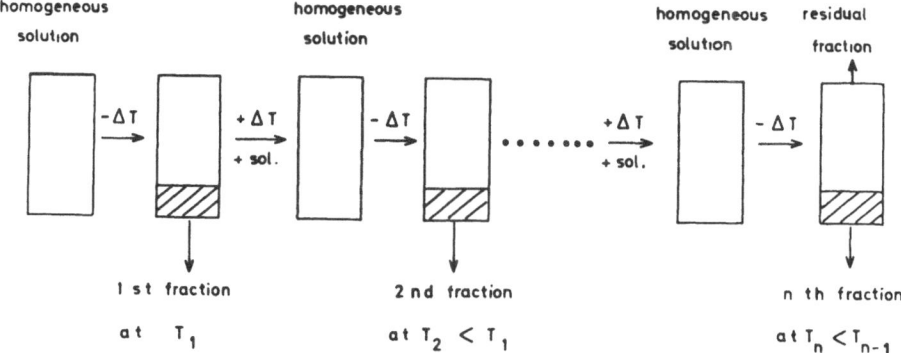

Fig. 3.   Procedure used for successive fractionation with demixing
          solvents.


Thus, whatever the ratio, the chemical distribution of styrene-
butadiene block copolymers can be determined by fractionation with
demixing solvents.

By a very simple head-tail fractionation procedure, according
to which, for example, one first removes the lower phase with a
high polystyrene content at a high temperature and the upper phase
with a high polybutadiene content at a low temperature, macromolecules whose chemical composition differs by only about $\pm$ 1-2 % are
obtained in the lower phase at a low temperature.

In principle the chemical distributions of all chemically non-
uniform polymers can be determined by fractionation with demixing
solvents if the corresponding homopolymers are separable into opposite phases.


CONCLUSION

It has been seen that the described fractionation method using
demixing solvents can be employed to determine the chemical composition of polymer blends and the chemical distributions of copolymers, especially of graft and block copolymers, with very simple
laboratory equipment and that it can therefore supplement the
familiar analytical methods, such as ultracentrifugation in the
density gradient[6], thin layer chromatography[7], and, above all, gel

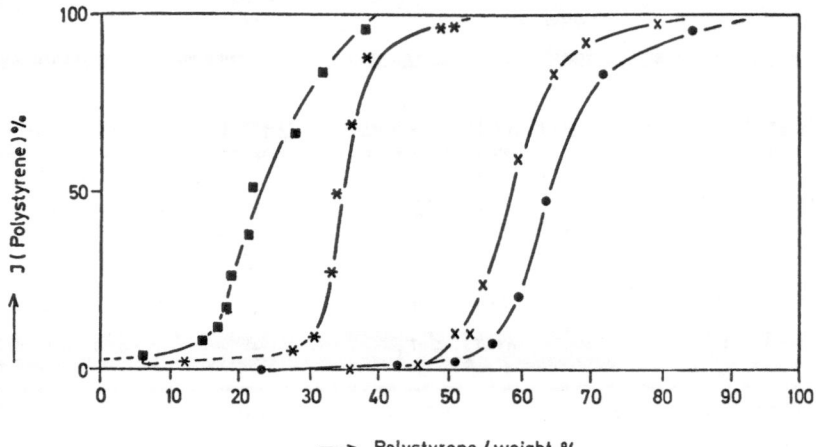

Fig. 4.  Evaluation of the chemical distributions of two block copolymers of styrene and butadiene with different chemical
compositions by preparative fractionation using demixing
solvents[5].

Average styrene content: • 64 weight %, x 60 weight %, ✳ 35
weight %, ■ 23 weight %. The fractionation was carried out
with following demixing solvent systems: • dimethyl formamide (DMF)/cyclohexane; x DMF/cyclohexane; ✳ DMF/methyl
cyclohexane, ■ DMF/heptane.

permeation chromatography in conjunction with turbidity titration[8].
The ability of fractionation with demixing solvents to separate substances efficiently in accordance with their chemical compositions
can be attributed to the fact that in the phase separation of the
low molecular components, which is caused by the reduction of the
temperature, the chemical compositions of the two phases change continuously in a "diametrically opposed" manner. The separation efficiency is additionally favoured by the preferential solvatation of
the chemically different macromolecules and also by the density differences of the phases.


REFERENCES

1. R. Kuhn, Makromolekulare Chemie 177, 1525 (1976).
2. R. Kuhn, Offenlegungsschrift Nr. 2528213 (1975).
3. R. Kuhn, unpublished Bayer AG Reports (1973-1981).
4. H. Alberts, unpublished Bayer AG Reports (1978-1981).
5. R. Kuhn, Makromolekulare Chemie 181, 725 (1980).
6. C. J. Stacy, J. Appl. Polym. Sci. 21, 2231 (1977)
7. H. Inagaki, T. Kotaka, T. Min. Pure Appl. Chem. 46, 61 (1976).
8. M. Hoffmann, H. Urban, Makromol. Chemie 178, 2683 (1977).

# A CRITICAL ASSESSMENT OF THE APPLICATION OF FT-IR SPECTROSCOPY TO THE STUDY OF CRYSTALLINE/COMPATIBLE POLYMER BLENDS

M.M. Coleman, D.F. Varnell and J.P. Runt

Polymer Science Section
The Department of Materials Science and Engineering
The Pennsylvania State University
University Park, PA 16802

## INTRODUCTION

For those of us interested in polymer vibrational spectroscopy, the advent of computer-assisted instruments, especially Fourier transform infrared (FT-IR) spectrometers, served to rekindle our interest in applying the technique to complex multicomponent polymer systems such as blends and alloys. FT-IR spectrometers, are inherently more sensitive and accurate than conventional dispersive instruments but if viewed critically, the major advantages of the newer instruments are a direct consequence of the dedicated minicomputer. The use of signal averaging to enhance signal to noise ratio and the ability to manipulate the spectral data by techniques such as spectral subtraction and addition; least squares fitting of spectra; curve resolving and factor analysis has markedly increased our ability to characterize complex multicomponent polymer systems.

Polymer blends may be subdivided into a number of different categories. Probably the most important distinction is between the so-called incompatible and compatible blends. Incompatible polymer blends, which represent the large majority of all blend systems, (a consequence of the relatively small change in entropy upon macromolecular mixing), have gained wide commercial significance. There are many examples of useful incompatible polymer blends but perhaps the most widely recognized are high impact polystyrene (HIPS) and the acrylonitrile-butadiene-styrene (ABS) materials. Compatible polymer blends are in the minority but they nonetheless do exist. A good example of an industrially significant compatible blend is General Electric's Noryl polymer which is a blend of polystyrene and poly(2,6-dimethyl-1,4-phenylene oxide).

The term 'compatible blend' is not well defined.  We prefer to
use this term to describe mixing of the component polymers on a
molecular level.  However, polymer blends which exist in separate
phases but in very small domains and thus appear transparent to the
eye, have also been loosely termed 'compatible'.  From thermodynamic
considerations a compatible polymer blend infers a system where the
free energy of mixing is zero or negative.  Assuming that the
entropic contribution is minimal, the enthalpy of mixing must be
negative or very close to zero.  Although somewhat simplistic, the
enthalpy of mixing may be discussed in terms of the solubility
parameter concept.  If the solubility parameters of the two component polymers in the blend are very similar then one would predict miscibility.  However, the underlying theory of solubility
parameters is really only applicable to non-polar molecules where we
only have to consider Van der Waals forces between the components.
When the possibility of dipolar interactions and/or hydrogen bonding is introduced the situation becomes more complex.  An intermolecular interaction of this type will affect both the enthalpy
and entropy of mixing.  This is difficult to quantify (although
progress has been made throuth the introduction of the theoretical
Flory-Prigogine equation of state (1)), but in simple terms one
can readily envisage these intermolecular interactions contributing
a negative factor to enthalpy and a positive factor to entropy.  The
important question is just how much of an effect do the interactions
have on the compatibilization of specific polymer blends.  We will
return to this subject later in the text.

Further complexity is introduced when one studies a subdivision
of compatible blend systems in which one or both of the components
is capable of crystallization; the so-called crystalline/compatible
blends.  Again, in simple terms, the blend in the solid state may
be viewed as a multicomponent system in which crystallites of the
crystallizable polymer are embedded in a compatible amorphous matrix.
Two excellent examples of such systems are blends of poly($\epsilon$-capro-
lactone) with poly (vinyl chloride)(2) and poly(bis-phenol A-carbonate) (3).

In this paper, which is basically a review of our FT-IR studies
of crystalline/compatible polymer blends, we will critically discuss
the application of infrared spectroscopy to the study of these complex blend systems.  Specifically, we will pose the question, "What
information can we obtain from FT-IR spectroscopy?"

DISCUSSION

The first question that comes to mind when considering the
application of infrared spectroscopy to the study of polymer blends
is "can infrared spectroscopy be used to differentiate between compatible and incompatible blends?"  The answer is not as straight-

forward as the question posed.  In certain cases, which we will
discuss in greater detail below, compatibility or incompatibility
can only be inferred.  Let us consider a simple example of two
polymers A and B which are completely miscible on the molecular
level.  Additionally, let us further assume that a definite favorable chemical interaction (dipolar or hydrogen bonding) exists
between the two components.  Given that this interaction is strong
enough to (a) significantly perturb some of the normal vibrations
of the polymer chain of either or both components (F matrix), and/or
(b) cause significant changes in chain conformation (G matrix),
then it is possible to observe spectral features in the infrared
spectrum of the blend that can be correlated with compatibility.
These spectral features can be rather subtle and certainly the
advent of FT-IR spectrometers, with their inherent accuracy and
facility to perform spectral manipulations, has been a major
benefit to the study of multicomponent systems.  The first polymer
blend system that we studied (4) was poly(methylmethacrylate)(PMMA)-
poly(vinylidene fluoride) (PVDF).  We reasoned that if the two
polymers were completely incompatible, which implies phase separation, then in infrared spectral terms, PMMA would not recognize
that PVDF exists in the blend and vice versa.  Accordingly, the
infrared spectrum of the blend should be identical to that of a
weighted addition of the spectra of the two pure components.
Alternatively, using difference spectroscopy, it should be possible
to subtract the spectrum of one of the pure components (e.g. PMMA)
from the spectrum of the blend yielding the spectrum of the other
pure component (e.g. PVDF).   On the other hand, if the two polymers
are intimately mixed and there are favorable chemical interactions
and/or conformational changes occurring between the two component
polymers, then the infrared spectrum of the blend will be significantly different from that of the weighted addition of the spectra
of the pure components.  Similarly,  successful subtraction of the
spectrum of a pure component (e.g. PMMA) from that of the blend
will not be possible without evidence of band distortions and
negative absorbances (a consequence of frequency shifts and band
broadening).  It must be emphasized at this stage that considerable
care should be exercised to ensure that spectral subtraction and
addition is performed in an absorbance range where the Beer-Lambert
law is obeyed (5).  Otherwise, band distortions will occur which
have nothing to do with interactions or changes in conformation.
Another factor that should be mentioned is the possibility of dispersion effects which result from variations in the refractive
index of the samples used (6).  In essence, our preliminary studies
of the PMMA-PVDF blends served to substantiate that there were
indeed spectral differences in the spectra of the compatible and
incompatible samples that could be interpreted on the basis of
specific chemical interactions and/or changes in chain conformation.

    Subsequent studies of poly(vinylchloride) (PVC) blends (7,8)
with the polyesters poly($\varepsilon$-caprolactone) (PCL) and poly($\beta$-propio-

lactone) (PPL) are also informative. Blends of PCL and PVC are
compatible in the amorphous state over the entire composition range
(9). From our FT-IR studies of this system we observed that the
carbonyl band of PCL was sensitive to the presence of PVC in the
blends. Figure 1 shows the infrared spectra in the carbonyl region
(1675-1775 cm$^{-1}$) of PCL and blends of PCL and PVC obtained in the
amorphous state at 75°C (above the melting point of PCL). It is
apparent that the carbonyl band shifts to lower frequency and
broadens as a function of PVC concentration. We interpreted this
result as evidence for an interaction involving the carbonyl bond
of PCL with most probably the methine hydrogen of PVC and suggested
that this favorable interaction could contribute to the compatibilization of this blend system. The next system that we studied we
did so for the wrong reason. It was postulated that if the interaction occurring in the PCL-PVC system contributes to the compatibilization of this blend surely the same interaction should exist
in the PPL-PVC system. Additionally, as PPL has a shorter translational repeating unit than PCL, then for a given blend composition
the number of favorable interactions should be greater than in the
case of PCL. However, much to our chagrin, PPL-PVC blends were
found to be incompatible over the entire range of compositions in
the amorphous state! Nevertheless, the result was useful in that
we could check whether or not the frequency shifts and band
broadening of the carbonyl band observed for the PCL-PVC were truly
indicative of interactions between the components. Figure 2 shows
the infrared spectra in the carbonyl region of PPL and blends of
PPL-PVC in the amorphous state recorded at 80°C (above the melting
point of PPL). It is immediately apparent that there are no
analogous frequency shifts or broadening of the carbonyl band of
PPL as a function of PVC concentration in the blends. This result
also served to allay our fears that dispersive effects might contribute significantly to the frequency shifts observed in the PCL-
PVC blends.

More recent studies of other compatible PVC-polyester blends
(10), specifically poly(δ-valerolactone) (PVL) and poly(α-methyl-
α-n-propyl-β-propiolactone) (MPPL) also gave results consistent
with the presence of specific interactions involving the carbonyl
bond of the polyester. Thus we can infer compatibility in polyester-
PVC blends on the basis of observed frequency shifts and band
broadening of the carbonyl band. Why the PPL-PVC blend system is
incompatible remains a dilemma. We have attempted to employ the
solubility parameter concept to rationalize this apparent anomaly
but frankly, there are problems with this simple approach. The
linear polyesters PCL and PVL have calculated solubility parameters
close to that of PVC. On the other hand the calculated solubility
parameter of PPL differs from that of PVC by about 1 (cal/cm$^3$)$^{1/2}$
which is outside the range where macromolecular mixing would be
predicted. However, the calculated solubility parameter of MPPL
also differs from PVC by the same amount and this blend system is

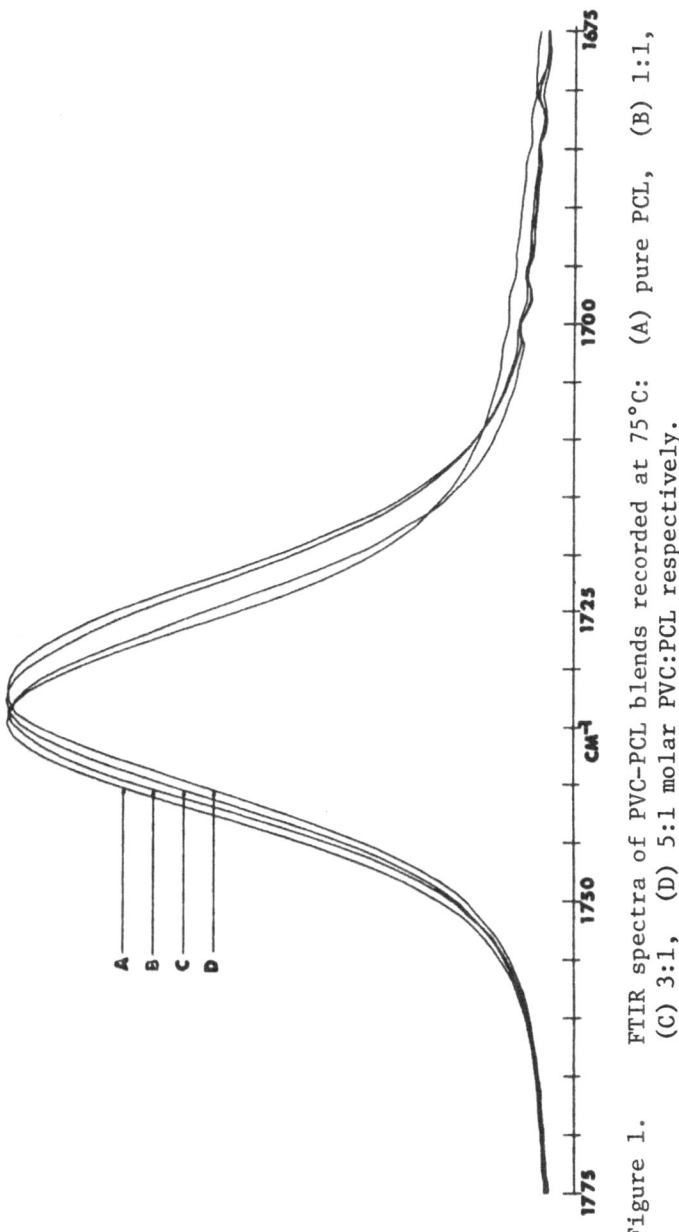

Figure 1.    FTIR spectra of PVC-PCL blends recorded at 75°C:   (A) pure PCL,   (B) 1:1,
(C) 3:1,   (D) 5:1 molar PVC:PCL respectively.

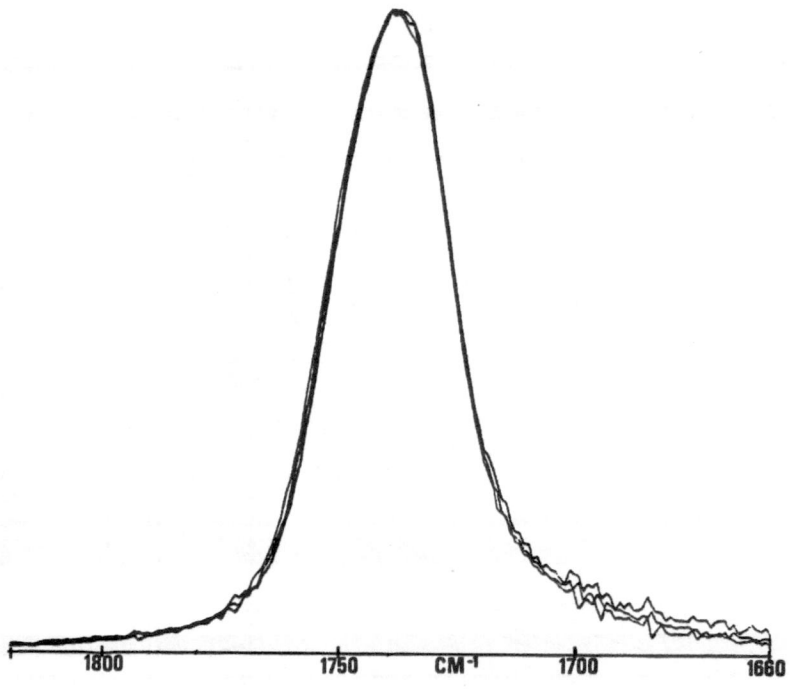

Figure 2.       FTIR spectra for PPL, PPL-PVC (55:45 molar) and PPL-
                PVC (20:80 molar) recorded at 80°C.

compatible.  It is evident that other factors have to be considere:
if we are to be able to predict compatibility.

     Before leaving the subject of infrared spectroscopy and its
ability to differentiate between compatible and incompatible
systems it is important to recognize that in some cases it will be
difficult, if not impossible, to determine whether or not a
particular blend is miscible.  Consider for example two non-polar
polymers that have very similar solubility parameters and are thus
miscible.  The perturbation to the spectra of the two components
will be correspondingly very small and it is unlikely to be detected by infrared spectroscopy.  It is our opinion that the technique will be restricted to those systems where there are relatively
strong dipolar interactions or hydrogen type bonding.

     As mentioned in the Introduction, crystalline/compatible polymer blends are complex materials where one or both of the components crystallize in a compatible amorphous matrix.  An important question that should be addressed is, "can infrared spectroscopy be used to follow changes in the degree of crystallinity of a crystallizable polymer in the various blends?"  Frankly, polymer spectroscopists have been rather loose in their terminology when they talk about 'crystalline bands'.  The infrared spectrum of an amorphous polymer is generally characterized by relatively broad absorbances which reflect the large number of conformations of the polymer chain in the amorphous state (exceptions to this 'rule of thumb' occur when the polymer chain contains rigid functionalities and the normal modes associated with these groups do not couple with the backbone modes; e.g. certain modes of the phenyl group in polystyrene or polycarbonate).  Upon crystallization, the most striking change observed in the infrared spectrum is the sharpening of many of the spectral bands.  In fact, the spectra of most semi-crystalline polymers may be viewed as a composite of two contributions; one attributable to the amorphous material and the other to the polymer in its preferred conformation.  Rarely do we observe spectral features that can be interpreted on the basis of three dimensional order (the classic exception being polyethylene; a polymer with a relatively small translational repeat unit).  Accordingly, it is more correct in most cases to refer to spectral bands associated with the polymer in its preferred chain conformation rather than to employ the term 'crystalline band'.  It follows that infrared spectroscopy cannot be used as an absolute method to measure the degree of crystallinity of a semi-crystalline polymer.  Calibration methods have been employed using X-ray or density results as standards (11).  However, even in these cases the results must be viewed with caution.  It is entirely feasible that a reasonable amount of the polymer units exist in the preferred conformation in the amorphous state and this could vary with sample preparation.

The spectrum of the preferred conformation of many polymers, particularly those with relatively large translational repeating units, may be obtained by difference spectroscopy.  This may be achieved by subtracting the spectrum of an amorphous polymer from that of the spectrum of the polymer in its semi-crystalline state. We have successfully employed this technique to a variety of polymers including trans-1,4-polychloroprene (12), trans-1,4-poly-isoprene (13) and PCL (7).  Similar results may be obtained for poly(bis-phenol A-carbonate) (PC).  Of particular interest to our studies is the fact that in the amorphous state the carbonyl band of PC absorbs at 1775 $cm^{-1}$ while the analogous band associated with the preferred conformation occurs at 1768 $cm^{-1}$.  This is illustrated in figure 3.  Note also that, as anticipated, the width at half height of the band assigned to the carbonyl band of the preferred conformation is considerably narrower than that of the amorphous band.  Given the reservations mentioned above, we can nonetheless correlate the presence of the 1768 $cm^{-1}$ in PC with development of crystallinity (to reiterate, the basic assumption being that in the crystalline phase the polymer will be in its preferred conformation). The separation of the carbonyl stretching frequencies associated with the amorphous and preferred conformations is particularly useful and not just restricted to PC. The polyesters mentioned previously, (PCL, PPL, PVL, and MPPL (7,8,10)) also exhibit analogous frequency differences. The carbonyl stretching frequencies of PCL, for example, occur at 1724 and 1737 cm$^{-1}$ for the preferred

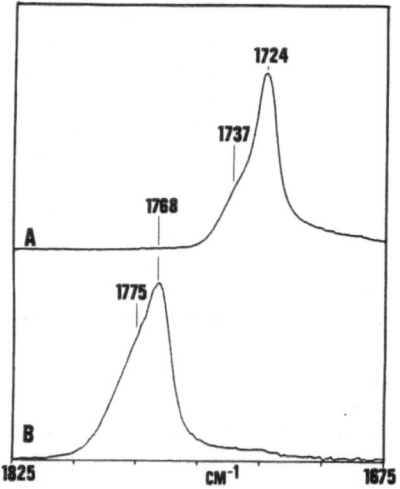

Figure 3.     FTIR spectra in the carbonyl stretching region of (A) PCL and (B) PC.

and amorphous conformations respectively. It is also fortuitous that the carbonyl stretching frequencies of PC and PCL are well separated as illustrated in figure 3. Thus we are now in a position to employ infrared spectroscopy to study the complex crystalline/ compatible blend system; PC-PCL.

Polymer blends containing PC and PCL are the most complex that we have studied to date by infrared spectroscopy (14,15). Both

polymers are capable of crystallization although they have widely
different crystalline melting points, (approximately 230°C and 70°C,
respectively).  There is a correspondingly large difference in the
glass transition temperatures of PC and PCL (approximately 149°C
and -71°C, respectively).  Blends of PC and PCL have been reported
to be compatible in the amorphous state (3) and we thought it important to first of all determine whether we could detect the
presence of specific chemical interactions in this system similar to
those observed in the PCL-PVC blends (see figure 1).  Figure 4
shows the carbonyl stretching frequency of PCL (A) and blends of
PCL and PC containing 50 and 80 mole percent of PC (denoted (B) and
(C) respectively).  These spectra were recorded at 75°C, above the
crystalline melting point of PCL.  It is immediately apparent that
we are observing the familar shift to lower frequency with increasing
concentration of PC indicative of an interaction involving the carbonyl band of PCL.

In two previous publications (14,15), we have presented detailed results of FT-IR studies of the crystalline/compatible PC-
PCL blend system.  Not only is the state of order of the two
polymeric components in the blend a function of composition and
temperature but it is also markedly dependent upon the method of
film preparation.  FT-IR spectroscopy was employed to study the
variation in crystallinity of both components in the blend from
films cast from different solvents and at different evaporation
rates.  The effect of exposure to acetone vapor of the blend films
was also studied.  In addition, stepwise temperature studies were
performed on PC-PCL blends.  Polymer and solvent induced crystallization of PC was readily observed and the results were considered
in terms of the glass transition temperature of the amorphous phase.
The interested reader is referred to these publications for details.
In this paper, we will conclude by considering one example which
demonstrates how the state of order of both components in the blend
can be readily determined.  Figure 5 shows the infrared spectra in
the carbonyl stretching region of a 20:80 mole percent PC-PCL blend
cast from methylene chloride (rapid evaporation) at ambient temperature and heated in a step-wise fashion in 5° intervals to 60°C.
It should be noted that the carbonyl stretching frequencies of PC
and PCL are presented separately and both are expanded to full scale.
The spectra denoted A thru D were obtained at temperatures of
35, 50, 55 and 60°C respectively.  At 35°C the PC component of the
blend is in the amorphous state as indicated by the relatively broad
band centered at 1775 cm$^{-1}$.  In contrast, the PCL component is
semi-crystalline as suggested by the two spectral contributions at
1724 (attributable to PCL in its preferred conformation) and at
1737 cm$^{-1}$ (attributable to PCL in the amorphous state).  When the
temperature is raised to 50°C, the PC component remains in the
amorphous state but there is a small but definite change in the
nature of the PCL component.  The relative amount of the amorphous
PCL material has increased.  Between 50 and 55°C the PCL component

becomes increasingly more amorphous and concurrently, the PC component crystallizes to some extent as seen by the spectral contribution at 1768 cm$^{-1}$.  We interpret this phenomenon in terms of polymer-induced crystallization.  In other words, the additional amorphous PCL material, which arises from the partial melting of PCL crystallites, mixes with the existing amorphous phase forming

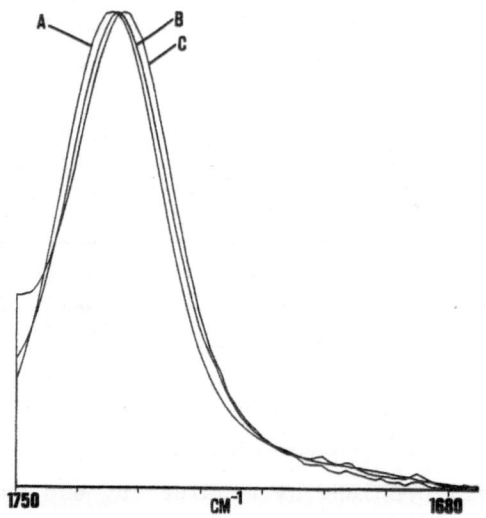

Figure 4.    FTIR spectra obtained at 75°C.  (A) pure PCL. (B) and (C) Blends of PC-PCL containing 50 and 80 mole percent PC respectively.

a compatible blend with an effective glass transition temperature close to or below the experimental temperature.  Thus there is sufficient mobility to induce a portion of the PC to crystallize. Between 55° and 60°C, the PCL completely melts and is obviously amorphous and we can definitely see that the PC component is semi-crystalline.

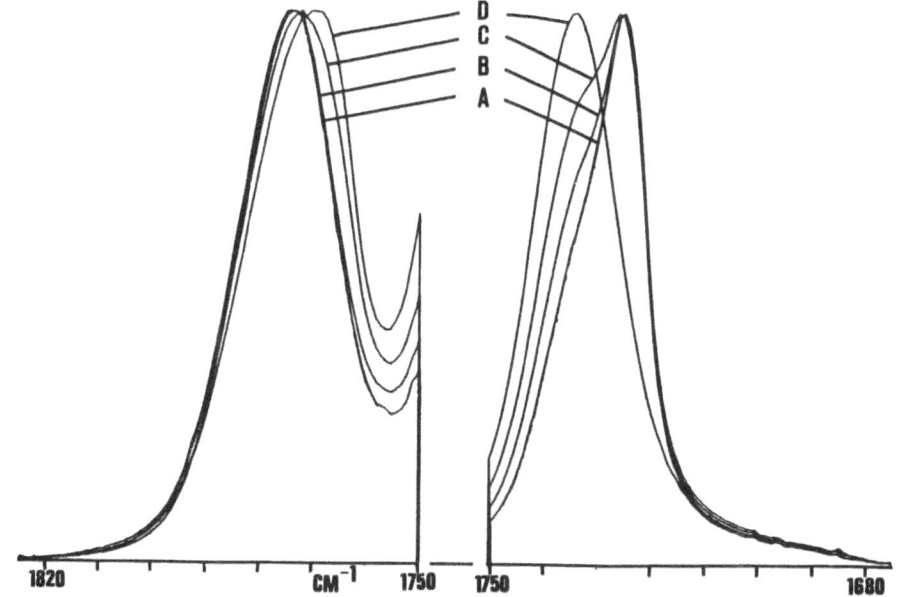

Figure 5.   FT-IR spectra in the range 1675–1825 cm$^{-1}$ for a PC–PCL
20:80 molar blend recorded at:(A) 35$^{\circ}$C,

(A)  35$^{\circ}$C,
(B)  50$\circ$C,
(c)  55$^{\circ}$C, and
(D)  60$\circ$C.

In summary, we believe that FT-IR spectroscopy has considerable potential for studying complex multicomponent systems such as crystalline/compatible polymer blends.  However, it must be recognized that there will be many systems which will be difficult, if not impossible, to study with this technique.  On the other hand, if we carefully select the polymer blend systems for study with due consideration of the type of information that can be obtained, FT-IR spectroscopy will take its place in the arsenal of experimental techniques needed to elucidate the structure of polymer blends.

ACKNOWLEDGEMENTS

The authors wish to express their appreciation for the financial support of the National Science Foundation, Grants DMR-7910841, CPE-8008060 Polymers Program.

REFERENCES

1.    P.J. Flory, J. Am. Chem. Soc. 87, 1833 (1965).
2.    C.J. Ong and F.P. Price, J. Polym. Sci. Polym. Symp. 63, 45 (1978).
3.    C.A. Cruz, D.R. Paul and J.W. Barlow, J. Appl. Polym. Sci. 23, 589 (1979).
4.    M.M. Coleman, J. Zarian, D.F. Varnell and P.C. Painter, J. Polym. Sci. Polym. Lett. Ed. 15, 745 (1977).
5.    M.M. Coleman and P.C. Painter, J. Macromol. Sci. Rev. Macromol. Chem. C16, 197 (1977).
6.    D.L. Allara, Appl. Spectrosc. 33, 358 (1979).
7.    M.M. Coleman and J. Zarian, J. Polym. Sci. Polym. Phys. Ed. 17, 1837 (1979).
8.    M.M. Coleman and D.F. Varnell, J. Polym. Sci. Polym. Phys. Ed. 18, 1403 (1980).
9.    J.V. Koleske and R.D. Lundberg, J. Polym. Sci. A-2, 7, 795 (1969).
10.   D.F. Varnell and M.M. Coleman, Polymer 22, 1324 (1981).
11.   T. Okada and L. Mandelkern, J. Polym. Sci. A-2, 5, 239 (1967).
12.   M.M. Coleman, P.C. Painter, D.L. Tabb and J.L. Koenig, J. Polym. Sci. Polym. Lett. Ed. 12, 577 (1974).
13.   R.J. Petcavich and M.M. Coleman, J. Polym. Sci. Polym. Phys. Ed. 18, 2097 (1980).
14.   M.M. Coleman, D.F. Varnell and J. Runt Contemporary Topics in Polymer Science, (W.J. Bailey, Ed.) Plenum Press, Vol. 4, In press.
15.   D.F. Varnell, J. Runt and M.M. Coleman, Macromolecules, In press.

# FRACTURE TOUGHNESS OF POLY(METHYL METHACRYLATE) BLENDS

R.P. Kusy and W.F. Simmons

Dental Research Center
University of North Carolina
Chapel Hill, NC   27514

## ABSTRACT

A series of acrylic blends was prepared from three feedstocks having number average molecular weights ($\bar{M}_n$) of $4 \times 10^3$ (L), $3 \times 10^4$ (M), and $1 \times 10^5$ (H). After single-edge-notched tension bars were machined from molded sheets, the fracture surface energy ($\gamma$) was determined using the Brown and Srawley equation. When the mean $\gamma$ was plotted against the blend composition, results showed a different dependence for each binary combination: H/M suggested a linear relationship, while H/L and M/L appeared concave downward and upward, respectively. Further reduction of this data to a molecular weight basis permitted analysis via three general approaches: that a property, $P \propto \bar{M}_n^{-1}$, that $P \propto \log \bar{M}_w$, and that $P \propto \log \bar{x}_n$ (where $\bar{x}_n = \bar{M}_n/M_o$). The first two relationships gave straight line segments for each binary combination and provided interesting parallels with several previous investigators--Sookne, Flory, Nielsen, and Berry. In the second case, moreover, the intersection of the H/L and M/L data lines suggested a common origin, which corresponded to $\gamma = 525$ erg/cm$^2$ at $\bar{M}_w = 26,000$ (the entanglement molecular weight, $M_o\epsilon$?). Of the three models, however, the last ($P \propto \log \bar{x}_n$) was most informative. Like the earlier single component materials, the binary blends followed a sigmoidal relationship; however, the effective molecular weight necessary for plastic deformation (x) had decreased. For these blends this shift corresponded to an increase in the strain energy release rate ($G_{I_c} = 2\gamma$) for a given $\bar{x}_n$. These observations are consistent with Manson's fatigue test results on acrylic blends and provide confirmation that the fracture morphology is dependent upon $\gamma$ and not $\bar{x}_n$.

INTRODUCTION

In earlier work the fracture surface energy of poly(methyl methacrylate) (PMMA) was measured as a function of number average molecular weight ($\overline{M}_n$) from $1.5 \times 10^3$ to $5.9 \times 10^5$.[1-3] Results showed that the energy associated with the fracture process could be described by a series of functions, the uppermost of which had a sigmoidal shape (cf ref. 4, too). In that region alone, the model considered the fracture surface energy to be composed of three components, of which the energy of plastic deformation ($\gamma_2$) was predominant. By estimating the energy attributable to the rupture of covalent bonds ($\gamma_1$) and the parting of surfaces ($\gamma_0$), the entanglement degree of polymerization ($\epsilon$), the critical degree of polymerization necessary for plastic deformation (x), and the plastic deformation fracture energy of the "infinite" molecular weight polymer ($\gamma_c$), either the total fracture surface energy ($\gamma$) or the strain energy release rate ($G_{I_c}$) could be determined.[3]

In the present effort the fracture toughness of PMMA blends is reported as a function of composition and molecular weight. Results show that, for a given $\overline{M}_n$, a binary blend composed of a low molecular weight addition can increase $\gamma$ over that of the single-component polymer alone. When the enhanced chain mobility is taken into account via x, the model represents these binaries, too.


EXPERIMENTAL

Binary blends were prepared from three feedstocks: H, $\overline{M}_n$ = $1.22 \times 10^5$; M, $\overline{M}_n$ = $3.21 \times 10^4$; and L, $\overline{M}_n$ = $3.90 \times 10^3$. The first two were commercial products (cf footnotes, Table 1), whereas the third was prepared by solution polymerization of methyl methacrylate at 60°C in toluene.[5,6] After dissolution in acetone and shock precipitation in 10 volumes of a hexane solvent (-20°C), the molecular weight distribution of these starting materials (cf Fig. 1) plus control samples 100S (cf Table 1) was determined by GPC in tetrahydrofuran at room temperature. Overall twenty blends, whose percent compositions are designated as H/M, H/L, and M/L, were formulated in 20g batches using the appropriate solvent/non-solvent system.[3,7] After stepwise drying under vacuum to constant weight, the glass transition temperatures were measured to confirm that no phase separation had occurred.[8] Subsequently each blend was molded into two sheets, 35 X 60 X 3 mm, either in a conventional split die mold or in an open tin foil mold. Sheets were sawed into six 35 mm long slabs and notched using a razor blade/vise-type fixture. Following an annealing treatment at 80°C for $10\frac{1}{2}$ hrs to relieve any residual stresses that might be present, the single-edge-notched specimens were pulled to failure in an Instron machine at a crosshead separation rate of 0.1 cm min$^{-1}$. If the samples were strong enough to withstand the gripping pressure

Table 1. Composition, Molecular Weight, and Fracture
Toughness of Poly(methyl methacrylate) Blends

| Designation | $\bar{M}_n$ | $\bar{M}_w$ | $\bar{\gamma}(erg/cm^2)$ | $\log_{10}\bar{\gamma}$ |
|---|---|---|---|---|
| 100H[a] | $1.22 \times 10^5$ | $3.64 \times 10^5$ | $3.61 \pm 0.81 \times 10^5$ (11) | 5.56 |
| 95H/5M | $(1.07 \times 10^5)$ | $(3.49 \times 10^5)$ | $2.94 \pm 1.03 \times 10^5$ (11) | 5.47 |
| 90H/10M | $(9.53 \times 10^4)$ | $(3.34 \times 10^5)$ | $2.63 \pm 0.89 \times 10^5$ (12) | 5.42 |
| 50H/50M | $(5.08 \times 10^4)$ | $(2.14 \times 10^5)$ | $2.93 \pm 1.38 \times 10^5$ (12) | 5.47 |
| 10H/90M | $(3.46 \times 10^4)$ | $(9.44 \times 10^4)$ | $2.50 \pm 0.81 \times 10^5$ (10) | 5.40 |
| 5H/95M | $(3.33 \times 10^4)$ | $(7.94 \times 10^4)$ | $2.48 \pm 0.92 \times 10^5$ (10) | 5.39 |
| 100M[b] | $3.21 \times 10^4$ | $6.44 \times 10^4$ | $2.42 \pm 0.76 \times 10^5$ (10) | 5.38 |
| | | | | |
| 100H[a] | $1.22 \times 10^5$ | $3.64 \times 10^5$ | $3.74 \pm 0.84 \times 10^5$ (11) | 5.57 |
| 97.5H/2.5L | $(6.94 \times 10^4)$ | $(3.55 \times 10^5)$ | $4.96 \pm 1.47 \times 10^5$ (10) | 5.70 |
| | | | $3.50 \pm 1.30 \times 10^5$ (10) | 5.54 |
| 95H/5L | $(4.85 \times 10^4)$ | $(3.46 \times 10^5)$ | $4.96 \pm 1.47 \times 10^5$ (10) | 5.70 |
| | | | $4.40 \pm 0.72 \times 10^5$ (10) | 5.64 |
| 92.5H/7.5L | $(3.73 \times 10^4)$ | $(3.37 \times 10^5)$ | $3.70 \pm 3.07 \times 10^5$ (12) | 5.57 |
| | | | $3.37 \pm 0.53 \times 10^5$ ( 6) | 5.53 |
| 90H/10L | $(3.03 \times 10^4)$ | $(3.28 \times 10^5)$ | $5.06 \pm 1.93 \times 10^5$ (11) | 5.70 |
| | | | $4.44 \pm 0.64 \times 10^5$ ( 9) | 5.65 |
| 85H/15L | $(2.20 \times 10^4)$ | $(3.10 \times 10^5)$ | $3.28 \pm 1.07 \times 10^5$ ( 4) | 5.52 |
| 80H/20L | $(1.73 \times 10^4)$ | $(2.92 \times 10^5)$ | $3.48 \pm 1.17 \times 10^5$ (11) | 5.54 |
| 70H/30L | $(1.21 \times 10^4)$ | $(2.56 \times 10^5)$ | $1.85 \pm 0.80 \times 10^5$ ( 5) | 5.27 |
| 60H/40L | $(9.30 \times 10^3)$ | $(2.20 \times 10^5)$ | $1.91 \pm 0.23 \times 10^5$ (10) | 5.28 |
| 50H/50L | $(7.56 \times 10^3)$ | $(1.85 \times 10^5)$ | $9.35 \pm 3.55 \times 10^4$ ( 5) | 4.97 |
| 100L | $3.90 \times 10^3$ | $5.03 \times 10^3$ | $2.50 \pm 1.30 \times 10^2$ ( 4)[e] | 2.40 |
| | | | | |
| 100M[b] | $3.21 \times 10^4$ | $6.44 \times 10^4$ | $2.37 \pm 0.74 \times 10^5$ (10) | 5.37 |
| 95M/5L | $(2.36 \times 10^4)$ | $(6.14 \times 10^4)$ | $1.99 \pm 0.86 \times 10^5$ (10) | 5.30 |
| 90M/10L | $(1.86 \times 10^4)$ | $(5.85 \times 10^4)$ | $1.07 \pm 0.48 \times 10^5$ ( 7) | 5.03 |
| 80M/20L | $(1.31 \times 10^4)$ | $(5.25 \times 10^4)$ | $4.64 \pm 2.63 \times 10^4$ ( 7) | 4.67 |
| 70M/30L | $(1.01 \times 10^4)$ | $(4.66 \times 10^4)$ | $3.55 \pm 2.52 \times 10^4$ (11) | 4.55 |
| 60M/40L | $(8.25 \times 10^3)$ | $(4.07 \times 10^4)$ | $6.71 \pm 5.52 \times 10^3$ (10) | 3.83 |
| 50M/50L | $(6.96 \times 10^3)$ | $(3.47 \times 10^4)$ | $5.22 \pm 5.21 \times 10^3$ ( 8) | 3.72 |
| 100L | $3.90 \times 10^3$ | $5.03 \times 10^3$ | $2.50 \pm 1.30 \times 10^2$ ( 4)[e] | 2.40 |
| | | | | |
| C-0[c] | $5.87 \times 10^5$ | $1.11 \times 10^6$ | $5.04 \pm 1.15 \times 10^5$ ( 7) | 5.70 |
| C-10 | $5.40 \times 10^4$ | $1.02 \times 10^5$ | $2.61 \pm 0.37 \times 10^5$ ( 5) | 5.42 |
| C-19 | $3.56 \times 10^4$ | $6.72 \times 10^4$ | $1.72 \pm 0.37 \times 10^5$ ( 5) | 5.24 |
| C-31 | $2.25 \times 10^4$ | $4.26 \times 10^4$ | $3.30 \pm 1.51 \times 10^4$ ( 4) | 4.52 |
| C-61 | $1.34 \times 10^4$ | $2.53 \times 10^4$ | $2.08 \pm 1.08 \times 10^3$ ( 5) | 3.32 |
| | | | | |
| 100S[d] | $2.96 \times 10^4$ | $6.15 \times 10^4$ | | |

[a]Aldrich Chemical Company, Inc., Milwaukee, Wis. (cat. #18,226-5,
[n] = 1.3 for $\bar{M}_w \approx 5 \times 10^5$).
[b]Polysciences, Inc., Warrington, Pa. (cat. #4554, lot #1888-125,
[n] = 0.2 for $\bar{M}_w \approx 5 \times 10^4$).
[c]Rohm and Haas Co., Philadelphia, Pa. (Plexiglas G).
[d]Aldrich Chemical Company, Inc., Milwaukee, Wis. (cat. #18,225-7,
lot #03, $\bar{M}_n = 3.32 \times 10^4$ and $\bar{M}_w = 6.06 \times 10^4$).
[e]Abstracted from Table 8, ref. 3.

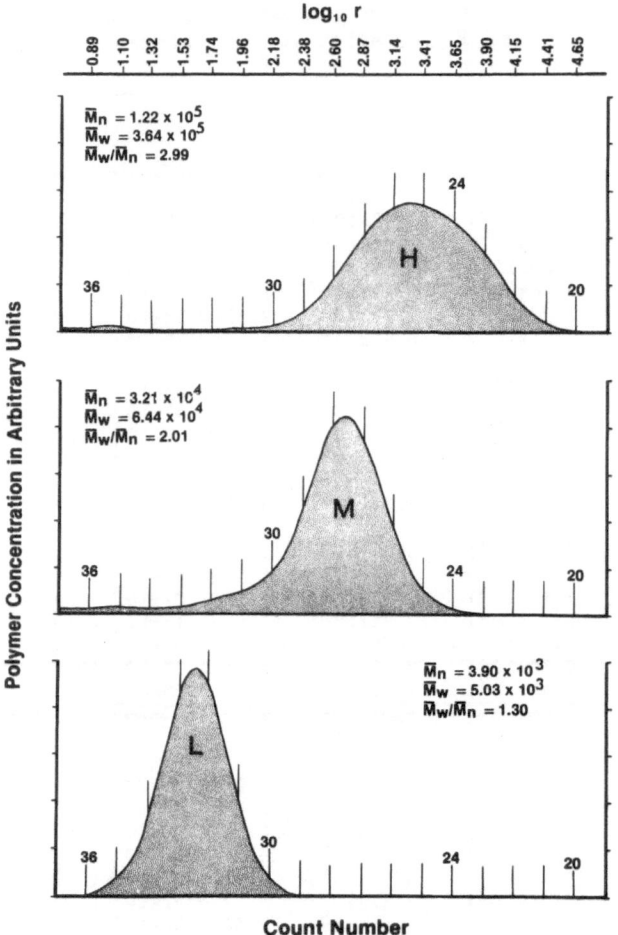

Fig. 1.    GPC scans of the three PMMA feedstocks studied:  H, $\bar{M}_n$ = 1.22X10$^5$; M, $\bar{M}_n$ = 3.21X10$^4$; and L, $\bar{M}_n$ = 3.90X10$^3$.  The corresponding r-th mer is computed for each count number.

requirements the grip separation, L, equalled 13 mm; otherwise, lap joints were fashioned, e.g., for 50H/50L and for 90M/10L to 50M/50L (cf Fig. 2).  By measuring the force at failure (f), the modulus of elasticity (E), and the geometric parameters (c, t, and w; cf Fig. 2a), the fracture surface energy ($\gamma$ = $G_{I_c}$/2) was evaluated:[9]

$$\gamma = (f^2/2Et^2w) \; [7.59(c/w)-32(c/w)^2+117(c/w)^3]. \tag{1}$$

As a control five single component batches were prepared by the
radiolysis of an unmodified PMMA, Plexiglas G (C-series, cf Table
1).[10]  For these samples, $\bar{M}_v$ was determined via limiting viscosity
measurements ($\bar{M}_v \approx \bar{M}_w$),[11] and $\bar{M}_n$ was evaluated from the known
degradation behavior of PMMA.[12]  In contrast the molecular weights of
the binary mixtures, set-off by parentheses in Table 1, were computed
from their base component materials (1 and 2) according to these
relationships,[13]

$$\bar{M}_n = [(w/\bar{M}_n)_1 + (w/\bar{M}_n)_2]^{-1} \text{ and} \tag{2}$$

$$\bar{M}_w = (w \cdot \bar{M}_w)_1 + (w \cdot \bar{M}_w)_2 \tag{3}$$

where for PMMA the average degree of polymerization; $\bar{x}_n = \bar{M}_n/10^2$
($M_o = 100$).  Thereafter all computations were made using a Wang 2200
PCS-II computer, and each fracture surface was evaluated using a
Zeiss Universal microscope.

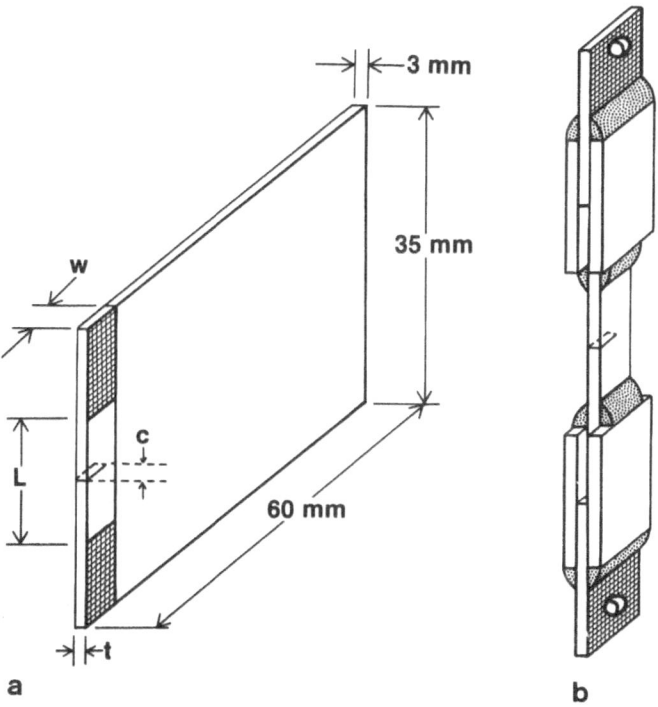

Fig. 2.   Geometry of single-edge-notched tension specimens:  (a),
          original cast sheet showing sample layout; and (b), indirect
          gripping configuration for the more brittle blends ($\gamma < 10^5$
          erg/cm$^2$).  Gripping surfaces and epoxy fillets are indicated
          by hatched and stippled regions, respectively.

RESULTS

Table 1, column 4, lists each mean $\gamma$ measurement ($\bar{\gamma}$) and its standard deviation for the number of specimens indicated in parentheses. These values are based, in part, upon the linear regression line of E versus $10^5/\bar{M}_n$ in which the intercept equalled

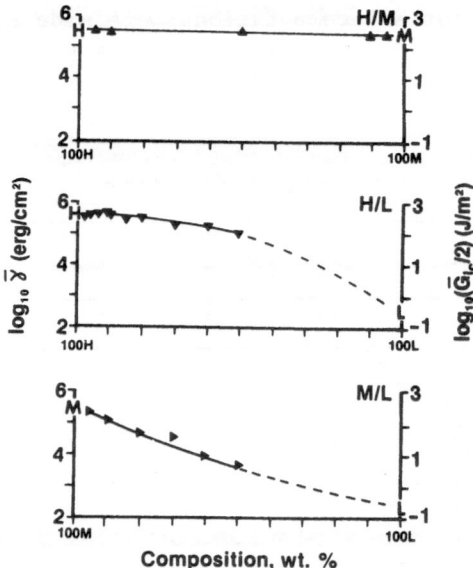

Fig. 3.   Influence of composition on the fracture surface energy ($\bar{\gamma}$) of PMMA blends:  H/M (▲), H/L (▼), and M/L (▶).  Data for each feedstock is lettered accordingly.

$4.42 \times 10^5$ psi and the slope was nearly zero (-0.0170).[14] As was observed previously for single component polymers, $\gamma$ became less precise not only as the blends became more brittle but also as the specimen geometry became smaller. Nevertheless when the $\bar{\gamma}$ values were plotted against composition for each binary (cf Fig. 3: H/M

($\blacktriangle$), H/L ($\blacktriangledown$), and M/L ($\blacktriangleright$)), the accuracy of the mean was interpreted to be, once again, a satisfactory indicator.[2,3] Although $\bar{\gamma}$ generally decreased as more low molecular weight component was added, no fundamental relationships were obvious. To further define these results, each data point was expressed as the mean molecular weight of its components. Several years ago Fox and Flory established that heterogeneous compositions could be treated as the average of their chain length.[15,16] Consequently equations 2 and 3 were used to derive Fig. 4,5,7 and 10.

DISCUSSION

First Theoretical Approach:  That a Property, $P \propto \bar{M}_n^{-1}$

Nearly forty years ago the tensile strength (UTS) of cellulose acetate fractions was shown to be dependent upon $\bar{M}_n$, and the UTS of its blends was shown to be dependent upon the weight average of its components.[17,18] Two years later Flory[19] reconciled these two statements of Sookne and Harris when he showed that

$$UTS = a_0 + a_1 \cdot M^{-1} \tag{4}$$

for $a_2$, $a_3$, ..., $a_j$ = 0 in which $M = \bar{M}_n$ for heterogeneous polymers. Such a form has been used to describe not only tensile or flexural strength,[13,17,18,20-23] but also the modulus,[13,21,22] fold endurance,[17,18] or elongation[17,18,21,22] for rubber,[13,20] polyethylene,[23] polystyrene,[21-23] and PMMA.[23] Although a similar relationship was found between $\bar{\gamma}$ and $\bar{M}_v^{-1}$ in the high molecular weight region,[2,24] measurements over a range four times greater suggested something other than a linear relationship.[25] When the log $\bar{\gamma}$ was plotted against $\bar{M}_n^{-1}$ (Fig. 4), three straight line segments were seen. Now the detrimental effects of L in the M/L binary ($\blacktriangleright$) and the enhanced $\bar{\gamma}$ values of small L component additions to H ($\blacktriangledown$) were obvious. Whether or not these straight lines have any physical significance was not clear, however, since a similar analysis of the data using the Griffith criterion[26] prompted similar conclusions but from curvilinear lines.[14]

Second Theoretical Approach:  $P \propto \log \bar{M}_w$

The second approach considers the log $\bar{\gamma}$ versus log $\bar{M}_w$ (cf Fig. 5). Here the theoretical model alluded to earlier is plotted (dashed line) assuming the most probable distribution (k = 1) in which $\gamma_0 + \gamma_1$ = 522 erg/cm$^2$, $\varepsilon$ = 80, x = 900, and $\gamma_c$ = 5.0X10$^5$ erg/cm$^2$ or,[3]

$$\gamma = 522 + 5.0X10^5[\exp(-900/\bar{x}_n)] \ [1+(900/\bar{x}_n)] \tag{5}$$

In addition the base components (H, M, and L) as well as the

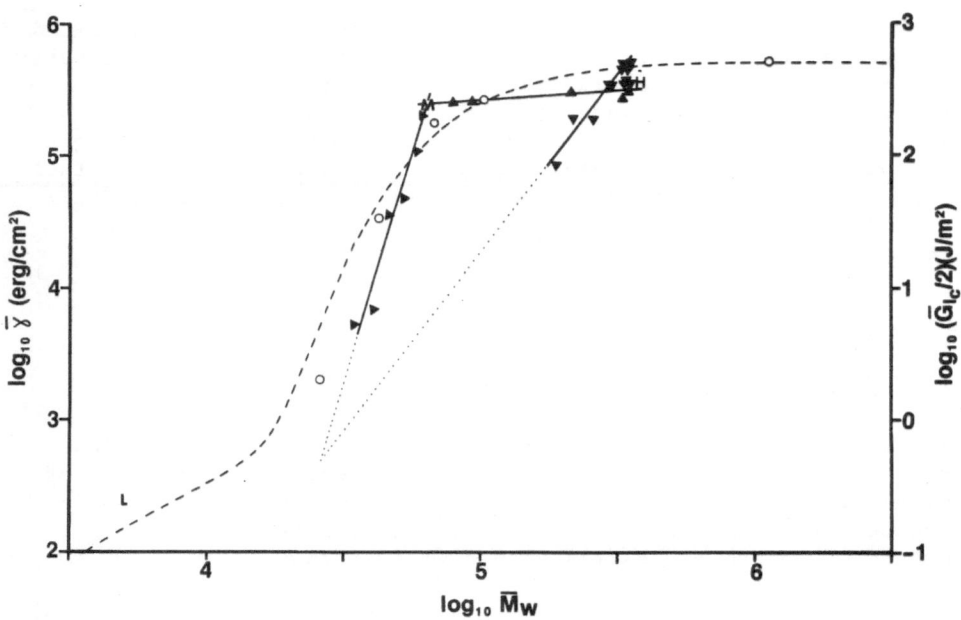

Fig. 5.   Logarithmic dependence of $\bar{\gamma}$ on $\bar{M}_W$.  For notation, cf Fig. 3;
          in addition note that the irradiated controls are identified
          by open symbols (o).  The dashed curve denotes the theoreti-
          cal $\gamma$ function for these PMMA controls (x = 900, $\gamma_c$ =
          $5.0 \times 10^5$ erg/cm$^2$; cf equation 5) while, in order of
          increasing slope, the three solid lines represent the $\gamma$
          function for H/M, H/L, and M/L blends.


structure-property relationship assumes the chord between the primary
components.  In the present work these ratios are 31.3, 8.23, and
3.80 for the H/L, M/L, and H/M binaries, respectively; hence H/L
departs most from the pure component line.

     An interesting application of the foregoing experimental
procedure is suggested by the extrapolation of the log $\bar{\gamma}$ versus log
$\bar{M}_W$ lines of the L component blends (dotted lines).  Over the $\bar{\gamma}$ region
studied these data sets do not approach the value for $\bar{\gamma}$ at 100 L.
Now by convention the first two regions of the $\gamma$ function have as a
common boundary the entanglement degree of polymerization, $\varepsilon$, the
reasoning being that significant plastic deformation can be achieved
only after entanglements occur and that below this value deformation
is limited to the rupture of covalent bonds and the parting of
surfaces.  Since the energy associated with the latter two processes
is usually small relative to the first, a blend comprised of any $\bar{M}_W$

in the upper plateau region (e.g., B,C,D, or E) against any $\bar{M}_w$ in the
low $\gamma$ region (e.g., A) will behave as though $\epsilon$ were the lower
boundary (cf schematic, Fig. 6). If this reasoning is correct, then
a series of blends may be tested and a family of tangents may be
drawn, the extrapolated tails of which will intersect at the
entanglement molecular weight. In Fig. 5 these $(\gamma, \bar{M}_w)$ coordinates
are 525 erg/cm$^2$ and 26,000 for an $\bar{M}_n \approx 13,700$. While this value of
$\epsilon$ = 137 is well within the range of results reported to date, it is
outside of the range of preferred values of $\bar{x}_n$ = 37-106.[27-29] In a
recent comparison of molecular weights which depend upon entanglements, Turner[30] reported that the critical weight average molecular
weight above which zero shear viscosity values rose markedly equalled
27,500. Although no special significance was attached to the value

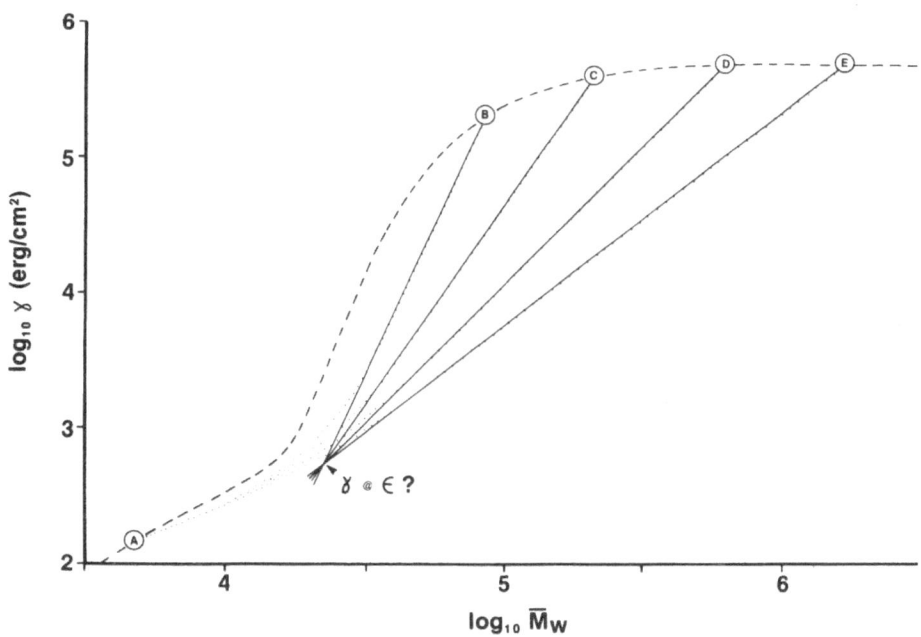

Fig. 6.  Schematic drawing illustrating a mechanical test methodology
         by which the entanglement molecular weight ($\epsilon$) may be
         determined. The dashed curve denotes the theoretical $\gamma$
         function for the base components, while the dotted lines
         represent the hypothetical $\gamma$ versus $\bar{M}_w$ results for four
         blends--B/A, C/A, D/A, and E/A. When tangents are extended
         from the upper traces of these latter curves, their
         intersection estimates $\epsilon$ for the classical Griffith solid
         (= $\gamma_0$ + $\gamma_1$).

itself, Berry did observe that both his fracture surface energy results (25°C)[24] and Vincent's flexure strength data (-196°C)[23] extrapolated to zero at $\overline{M}_v$ and $\overline{M}_n$ of 25,000, respectively. Note that the corresponding $\gamma = 525$ erg/cm$^2$ substantiates the expectations of several authors in which $\gamma$ of the classical Griffith solid (= $\gamma_0 + \gamma_1$) is assigned a value of 400-750 erg/cm$^2$ [1-3,31] (cf refs. 4 and 32, too). Certainly a more detailed analysis is warranted.

## Third Approach: $P \propto \log \overline{x}_n$

The last and most informative approach is detailed in Fig. 7. Once again the fracture toughness model is sketched-in along with the control sample set, and the PMMA blend data is presented on this log-log plot of $\overline{\gamma}$ versus $\overline{x}_n$. Results showed that the three blend data sets (two of which, H/M (▲) and M/L (▶), might be interpreted as complementary) did not superpose the homologous polymer series (H,M,L, and open circles). Noting that the Schulz-type of distribution, implicit in equation 5, was at least a first-order approxima-

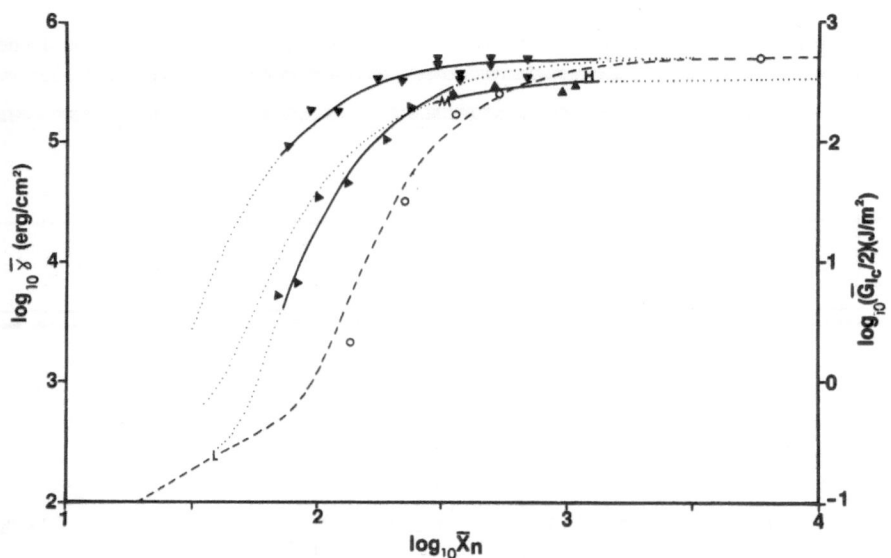

Fig. 7. Logarithmic dependence of $\overline{\gamma}$ on $\overline{x}_n$. For notation, cf Fig. 5. Again, the dashed curve denotes the theoretical $\gamma$ function for PMMA base components (x = 900, $\gamma_c$ = 5.0X10$^5$ erg/cm$^2$; cf equation 5), while the three solid curves represent the best theoretical $\gamma$ function for H/M (x = 368, $\gamma_{c_2}$ = 3.3X10$^5$ erg/cm$^2$), H/L (x = 235, $\gamma_c$ = 5.0X10$^5$ erg/cm$^2$), and M/L (x = 500, $\gamma_c$ = 5.0X10$^5$ erg/cm$^2$) blends.

tion for the log-normal distributions of the base components (cf Fig. 1.6.5 in ref. 33 and Fig. 8), the hypothesis was entertained that this $\gamma$ function might still suffice if it could be translated somehow, say via x. Accordingly, by changing x = 900 to x = 235 (H/L) and x = 500 (M/L) while maintaining $\gamma_c$ = 5.0X10$^5$ erg/cm$^2$, excellent agreement was obtained over the compositional range investigated, i.e., from (50/50) to (100/0) (cf Fig. 7). In the M/L binary, moreover, the extrapolated curve passed through the low

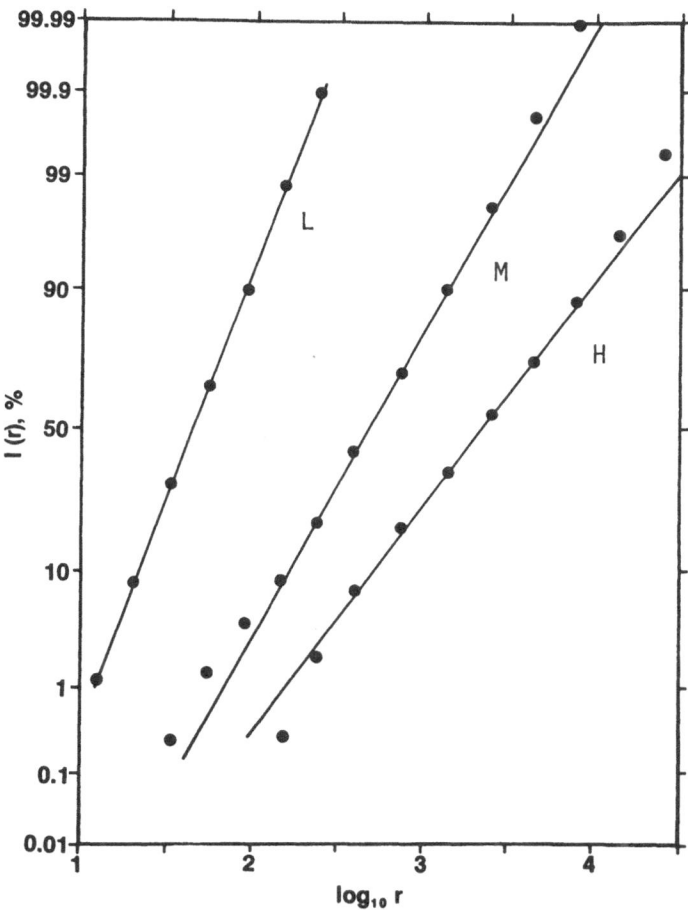

Fig. 8.  Fit of the three PMMA feedstocks to the log-normal distribution function. From the attendant lines, the following parameters were computed (cf GPC, Fig. 1): H, $\overline{M}_n$ = 1.13X10$^5$, $\overline{M}_w/\overline{M}_n$ = 3.75; M, $\overline{M}_n$ = 3.48X10$^4$, $\overline{M}_w/\overline{M}_n$ = 1.98; and L, $\overline{M}_n$ = 3.84X10$^3$, $\overline{M}_w/\overline{M}_n$ = 1.35.

boundary composition. Since even slight variations in the high end of the molecular weight distribution curve would have a profound effect on the $\gamma$ values theorized as $\bar{x}_n \to \epsilon$, and since the H component differed most from the log-normal distribution, an exact fit was not anticipated for the H/L blend. Such a consideration had no physical meaning for the H/M blends, however, in which x = 368 and $\gamma_c$ = $3.3 \times 10^5$ erg/cm$^2$.

The schematic of Fig. 9 clarifies the observations noted in Fig. 7. As the top frame shows, each polymer component has a distribution curve which, in the most elementary form, may be treated as having only a mean degree of polymerization, $\bar{x}_{n_1}$. When the fracture toughness of a base component polymer is desired, however, knowledge of the distribution curve is necessary in order to determine the fraction of mers which can participate in plastic deformation,[1-3] i.e., those in the shaded region, or

$$1-w_1 = 1-\int_0^x W(r)dr \qquad (7)$$

When each base composition of a polymer is treated individually, that set has a common value for x (for PMMA, x = 900). However, when $\bar{x}_{n_1}$ is blended with another base polymer, for example $\bar{x}_{n_2}$, the $\gamma$ for the $\bar{x}_{n_1}/\bar{x}_{n_2}$ blend cannot be evaluated using x. Instead the blend shown in the second frame and reproduced in the third frame, must be considered as though it had an effective mean degree of polymerization, $x_n'$, and an apparent distribution--the latter of which is assumed to mimic its constituents. From these same base components several other alloy compositions could be formulated, one of which is depicted in the last frame, having effective mean molecular weights, $M_0\bar{x}_n''$, $M_0\bar{x}_n'''$, ..., $M_0\bar{x}_n^j$. Invariably the critical degree of polymerization necessary for plastic deformation shifts from the base polymer value, x, to a new value, $x'$, $x''$, ..., $x^j$ for each composition $\bar{x}_{n_1}/\bar{x}_{n_2}$. As x was found to remain constant for the homologous series of base components, likewise $x' = x'' = ... = x^j$ for each set of binary blends, $\bar{x}_{n_1}/\bar{x}_{n_2}$.

Returning to Fig. 7, the values of x are interpreted as an index of chain mobility. This so-called plasticization effect is noted to a different degree for each binary, being greatest for the bimodal blends, H/L. That the effect should be the most pronounced for these blends is sensible, since small amounts of a low molecular weight polymer are expected to at least maintain $\gamma$ (although it may be at the expense of other properties, e.g., ultimate tensile strength, elongation, and wear[14,34,35]). Although based upon this single criterion, the unimodal (H/M) and not the mixed (M/L) binary blends provided the second greatest plasticization effect, the downward translation of $\gamma_c$ for H/M cannot be ignored. With that consideration, the actual H/M binary blends fall below the upper hypothetical trace of M/L, too. When both property sensitive parameters are

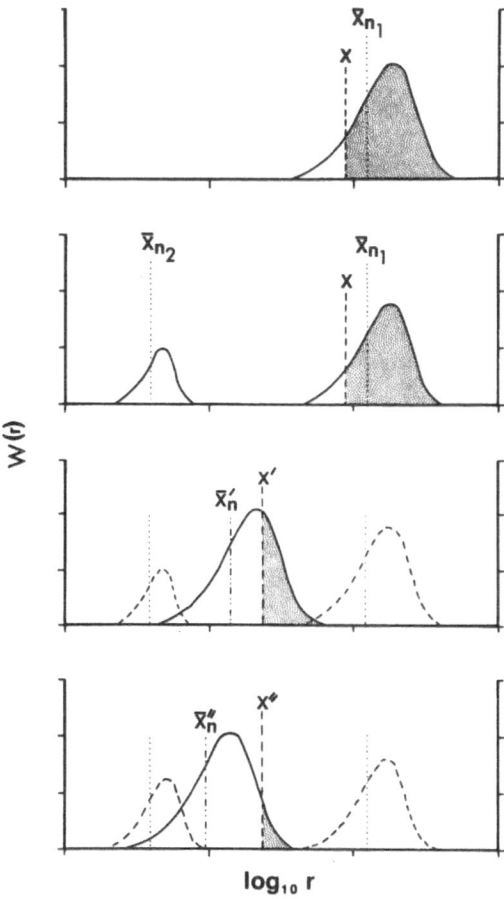

Fig. 9.  Schematic drawings illustrating the approach taken to
compute the fracture toughness for a fraction (uppermost
frame) versus a blend (remaining frames).  In the latter
interpretation, assignment of an effective mean degree of
polymerization (e.g., $\bar{x}_n'$), an apparent molecular weight
distribution, and a critical degree of polymerization
necessary for plastic deformation (e.g., $x'$) is critical.

viewed, the theory substantiates the general viewpoint that small
additions of low molecular weight fractions should enhance energy
absorption capabilities--even when those fractions behave in a
brittle manner in the bulk.

Substantiation of the Fracture Toughness Model:  Fatigue Crack
Propagation (FCP) Tests

    Although the data base was not nearly so extensive, the fatigue
crack propagation work of Manson and co-workers provided further
support.[36,37]  In that report seven PMMA control feedstocks were
polymerized, five of which were evaluated for their stress intensity
factors, $K_{I_c}$.  From three of these controls ($\emptyset$-0, $\overline{M}_n$ = 4.2x10[5];
$\emptyset$-2, $\overline{M}_n$ = 2.9X10[5]; and $\emptyset$-6, $\overline{M}_n$ = 2.9X10[4]), seven binary blends were
polymerized from monosols, machined into single-edge-notched compact
tensile specimens, and tested in cyclic loading.  For the $\emptyset$-0 base
alloys this included a 5, 10, 98, and 99% $\emptyset$-6 group; whereas for the
$\emptyset$-2 base alloys this included a 98, 99, and 99.5% $\emptyset$-6 group.  Instead
of using the GPC data presented for the blends, each $\overline{M}_n$ was calculated using constituent $\overline{M}_n$ values and equation 2.  $K_{I_c}$ was estimated
using equation 8 (plane strain),[9]

$$K_{I_c}^2 = 2\gamma E \cdot (1-\nu^2)^{-1} \tag{8}$$

in which E was assigned a constant value of 4.42X10[5] psi (cf Results
Section) and $\nu$ = 0.32.

    Using the same format as Fig. 7, Fig. 10 plots the controls (□)
along with $\emptyset$-0/$\emptyset$-6 (+) and $\emptyset$-2/$\emptyset$-6 (x) blends.  Since both the test
geometry and methodology differ significantly from the data reported
in Fig. 7, $\gamma_c$ decreases to 2.0X10[5] erg/cm[2].  This solution is
compatible with low velocity, stable crack growth[2,24,38,39] as is the
previous measurement of 5.0X10[5] erg/cm[2] reasonable for high velocity,
unstable fracture.[1,31,38,39]  Having made this substitution into
equation 5 the dashed line resulted (x = 900, again!  See also Fig.
2, ref. 3).  And even though the information is quite limited,
analysis of the blends indicates a significant displacement of the $\gamma$
function to lower $\overline{x}_n$.  Treating $\emptyset$-0 and $\emptyset$-2 data as being similar in
molecular weight ($\emptyset$-0/$\emptyset$-6 and $\emptyset$-2/$\emptyset$-6 equal 14.5 and 10.0,
respectively), an envelope could be defined having as its upper and
lower boundaries, x = 235 and x = 500 ($\gamma_c$ = 2.0X10[5] erg/cm[2]).  From
the results of Fig. 7 and 10 then, comments voiced by Manson's
group[36,37] that for a given molecular weight the blend machined
better than the component polymer, that the binary could be fabri-
cated to lower molecular weights than the base constituents, and that
the resistance to fatigue crack propagation could be enhanced by
small additions of a high molecular weight species (cf Fig. 9 lower,
ref. 18) are all a consequence of the increased $\gamma$ via the controlled
release of entanglements.  This manipulation of energy barriers to
influence chain mobility, or equivalently "... the relative ability
to stabilize or destabilize the entanglement network formed during
crazing prior to the advance of the main crack"[36] should be apparent
in the fracture morphology.

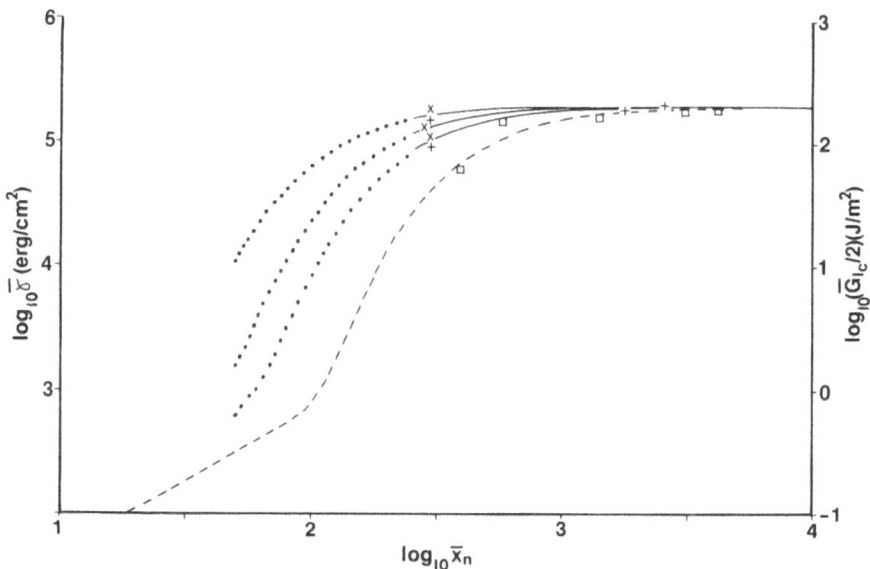

Fig. 10. Logarithmic dependence of $\bar{\gamma}$ on $\bar{x}_n$: $\emptyset\text{-}0/\emptyset\text{-}6$ (+) and $\emptyset\text{-}2/\emptyset\text{-}6$ (x) blends. Control or base constituents are identified by open symbols (□). As before the dashed curve denotes the $\gamma$ function for PMMA fractions (x = 900); while from top to bottom, the three solid curves represent the $\gamma$ function for x = 235, 368, and 500. In all cases, $\gamma_c = 2.0 \times 10^5$ erg/cm$^2$.

## Further Substantiation of the Model:  Fracture Morphology

Previous investigations have attributed the bulk of the energy dissipated in the failure of PMMA to the crazing process.[3,40,41] Two fracture features of importance in this process are the formation of parabolic markings[40,42,43] and interference colors.[40,44-46] The former occur as the triaxial stress state causes excessive cavitation in the crazed layer, thereby opening up secondary fracture sites whose subsequent growth eventually intersects the primary fracture front--the loci of which form a conic section. The latter appear because of the extension of the stressed polymer into an oriented foam ("craze") which even after failure has a different refractive index than the bulk polymer--the color of which indicates the relaxed thickness. In conventional PMMA which has a polydispersity index of about 2, both these features become sporadic at about $M_v = 1.0 \times 10^5$ and disappear by about $M_v = 8.5 \times 10^4$.[47] This last observation coincides with the precipitous drop in $\gamma$ from ~ $10^5$ erg/cm$^2$ (cf Fig. 14, ref. 47). If a search is limited to just these two discriminating features, then the fracture morphology may be correlated with $\gamma$

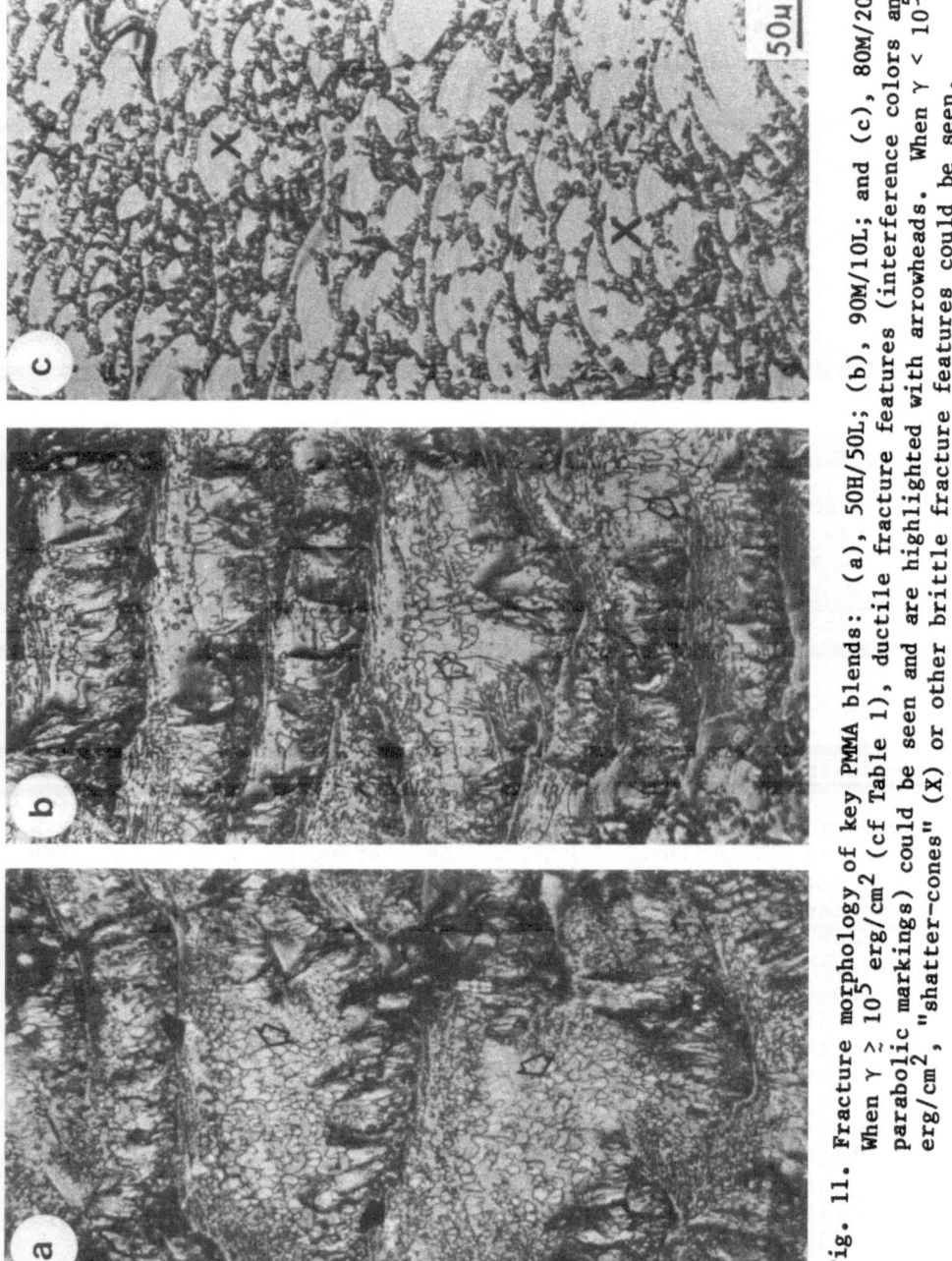

Fig. 11. Fracture morphology of key PMMA blends: (a), 50H/50L; (b), 90M/10L; and (c), 80M/20L. When $\gamma \gtrsim 10^5$ erg/cm$^2$ (cf Table 1), ductile fracture features (interference colors and parabolic markings) could be seen and are highlighted with arrowheads. When $\gamma < 10^5$ erg/cm$^2$, "shatter-cones" (X) or other brittle fracture features could be seen.

and not $\bar{x}_n$. For every H/L blend, interference colors and parabolic markings were observed that were characteristic of high molecular weight acrylic. Even the 50H/50L composition, whose $\bar{M}_n$ equalled only $7.56 \times 10^3$, exhibited significant plastic deformation (cf Table 1 and Fig. 11a). For the M/L series the break is actually bracketed by the compositions, 90M/10L and 80M/20L. In the first, interference colors and the last vestiges of conical markings can be seen, although the $\bar{M}_w$ is less than $8.5 \times 10^4$ (Fig. 11b). With only a small decrease in $\bar{M}_w$, however, the colored patches and parabolic features vanish in 80M/20L, the microstructure looking very similar to a fractionated PMMA of $\bar{M}_v = 6 \times 10^4$ (cf Fig. 11c with Fig. 9 of ref. 47). Further analysis of these blend morphologies with particular emphasis on "rib spacings" and their relationships to $\gamma$ is beyond the scope of this paper (for single component work, cf refs. 48 and 49). What these selected fracture details have confirmed, however, is that the appearance of the fractures coincide with the physical measurements themselves. On this basis alone, acrylics with $\gamma < 10^5$ erg/cm$^2$ will be progressively more susceptible to failure, irrespective of their molecular size distribution.

## ACKNOWLEDGMENTS

This investigation was supported by RCDA No. 00052 (R.P.K.) and research grant No. DE-02668 from the National Institute of Health.

## REFERENCES

1. R. P. Kusy and D. T. Turner, Polymer 17:161 (1976).
2. R. P. Kusy and M. J. Katz, J. Mater. Sci. 11:1475 (1976).
3. R. P. Kusy and M. J. Katz, Polymer 19:1345 (1978).
4. R. E. Robertson, "Toughness and Brittleness of Plastics," R. D. Deanin and A. M. Crugnola, eds., American Chemical Society, Washington, D.C., p. 89 (1976).
5. Use of Acrylic Monomers in the Preparation of Low Number Average Molecular Weight Polymers, Rohm and Haas Technical Report TMM-23, March 1965.
6. S. R. Sandler and W. Karo, "Polymer Synthesis--Vol. 1," Academic Press, New York, p. 297 (1974).
7. T. G. Fox, J. B. Kinsinger, H. F. Mason, and E. M. Schuele, Polymer 3:71 (1962).
8. R. P. Kusy, W. F. Simmons, and A. R. Greenberg, Polymer 22:268 (1981).
9. J. E. Srawley and W. F. Brown, Jr., NASATND-2599 (1965).
10. M. Dole, "The Radiation Chemistry of Macromolecules," Academic Press, New York, Vol. 2, Ch. 6. (1973).
11. H. J. Cantow and G. V. Schulz, Z. Phys. Chem. 2:117 (1954).

12.  F. A. Bovey, "The Effects of Ionizing Radiation on Natural and Synthetic High Polymers," Interscience, New York (1958).
13.  P. J. Flory, Ind. Eng. Chem. 38:417 (1946).
14.  R. P. Kusy and W. F. Simmons, J. Biomed. Mater. Res., submitted.
15.  T. G. Fox, Jr. and P. J. Flory, Am. Chem. Soc. 70:2384 (1948).
16.  T. G. Fox, Jr. and P. J. Flory, J. Appl. Phys. 21:581 (1950).
17.  A. M. Sookne and M. Harris, J. Res. Nat. Bur. Stand. 30:1 (1943).
18.  A. M. Sookne and M. Harris, Ind. Eng. Chem. 37:478 (1945).
19.  P. J. Flory, J. Am. Chem. Soc. 67:2048 (1945).
20.  J. A. Yanko, J. Polym. Sci. 3:1576 (1948).
21.  E. H. Merz, L. E. Nielsen and R. Buchdahl, Ind. Eng. Chem. 43:1396 (1951).
22.  H. R. Jacobi, Kunststoff 43:9 (1953).
23.  P. I. Vincent, Polymer 1:425 (1960).
24.  J. P. Berry, J. Polym. Sci. (A) 2:4069 (1964).
25.  R. P. Kusy and D. T. Turner, Polymer 15:394 (1974).
26.  A. A. Griffith, Philos. Trans. R. Soc. (A) 221:163 (1921).
27.  F. Bueche, J. Appl. Phys. 26:738 (1955).
28.  F. Bueche, "Physical Properties of Polymers," Interscience, New York, Ch. 3 (1962).
29.  R. S. Porter and J. F. Johnson, Chem. Rev. 66:1 (1966).
30.  D. T. Turner, Polymer 19:789 (1978).
31.  J. P. Berry, J. Polym. Sci. 50:107 (1961).
32.  J. J. Gilman, J. Appl. Phys. 31:2208 (1960).
33.  L. H. Peebles, Jr., "Molecular Weight Distributions in Polymers," Interscience, New York, Ch. 1 (1971).
34.  B. L. Johnson, Ind. Eng. Chem. 40:351 (1948).
35.  H. Wakeham, "Cellulose and Cellulose Derivatives--Part III," E. Ott, H. M. Spurlin and M. W. Grafflin, eds., Interscience, New York, p. 1334 (1955).
36.  S. L. Kim, J. Janiszewski, M. D. Skibo, J. A. Manson, and R. W. Hertzberg, Polym. Eng. Sci. 19:145 (1979).
37.  S. L. Kim, summary of progress report, private communication.
38.  G. P. Marshall, L. E. Culver, and J. G. Williams, Plast. Polym. 37:75 (1969).
39.  A. K. Green and P. L. Pratt, Eng. Fract. Mech. 6:71 (1974).
40.  J. P. Berry, J. Appl. Phys. 33:1741 (1962).
41.  R. P. Kambour and R. E. Barker, Jr., J. Polym. Sci. 4:359 (1966).
42.  J. A. Kies, A. M. Sullivan, and G. R. Irwin, J. Appl. Phys. 21:716 (1950).
43.  F. Schwarzl and A. J. Staverman, "Die Physik der Hochpolymeren," H. A. Stuart, ed., Springer-Verlag, Berlin, Ch. 3 (1956).
44.  M. Higuchi, Rep. Res. Inst. Appl. Mech., Kyushu Univ. 6:173 (1958).

45.  I. Wolock, J. A. Kies, and S. B. Newman, "Fracture," John
     Wiley, New York, p. 250 (1959).
46.  J. P. Berry, Nature 185:91 (1960).
47.  R. P. Kusy and D. T. Turner, Polymer 18:391 (1977).
48.  R. P. Kusy and D. T. Turner, J. Dent. Res. 54:1233 (1975).
49.  R. P. Kusy, H. B. Lee, and D. T. Turner, J. Mater. Sci.
     12:1694 (1977).

MELT RHEOLOGY OF BLENDS OF SEMI-CRYSTALLINE POLYMERS;
PART II, DYNAMIC PROPERTIES OF POLY(ETHYLENE TEREPHTALATE) -
POLY(AMIDE - 6,6) MOLTEN BLENDS

L. A. Utracki and G. L. Bata

National Research Council of Canada
Industrial Materials Research Institute
750, Bel-Air, Montreal, Quebec, H4C 2K3

ABSTRACT

Melt rheology of poly(ethylene terephthalate), PET, poly(amide - 6,6), PA, and their blends were investigated at 240, 260, 280 and 300°C on Rheometrics Mechanical Spectrometer in a dynamic mode within the frequency range from $10^{-1}$ to $10^2$ (rads/sec). Dynamic viscosity, storage and loss shear moduli were recorded.

First, the isothermal rate of the overall thermal degradation process for each sample was determined, from which the activation energy of the process, $E_D$, as a function of the composition was computed. The determined values of $E_D(\pm 2)$ = 30 and 47 (kcal/mole) for PET and PA respectively are in agreement with the literature values. The rates of degradation as well as $E_D$'s of the blends were found to be larger than those calculated from the properties of PET and PA by using an additivity rule.

Next, substracting the degradation effect from the recorded rheological signals, the true flow curves result. The dynamic viscosity, $\eta^*$, and the dynamic measure of the primary normal stress difference coefficient, $\psi^*$, as functions of the frequency, $\omega$, were computed for each temperature, T, and composition, c. It was found that at T $\geqslant$ 260°C the addition of small amounts (c $\leqslant$ 10 wt%) of PA lowers the viscosity in full range of $\omega$. This effect was not observed at 240°C, i.e. for supercooled blends. Similar behaviour is observed for $\psi^*$; here, however, the minimum occurs at a lower concentration, c $\approx$ 5%. As in the case of $\eta^*$, the depression of $\psi^*$ due to PA presence is strongest at the highest temperature, 300°C, where $\psi^*$ is about ten-fold lower than that for PET. Regarding the structure of the blends, it has been observed that molten sample containing 5% PA was transparent, and that containing c > 10%, milky.  The ATR-FTIR data shows that at least up to 30% PA, the continuous phase is made up of PET.


## INTRODUCTION

Blending and alloying polymers represents an increasingly important segment of the plastics industry.  The commercial production of blends in 1983 is projected[1] to reach 600,000 tons.  The advantages of blending and the resulting blend properties have been summarized in a series of recent reviews, which suggest [2-8] that most of the work done in this field is related either to the materials of commercial importance ( e.g. rubber modified polystyrene, poly(phenylene oxide) /polystyrene, ABS, PVC blends...), or to miscible ("compatible") polymer mixtures.

In most cases miscibility of polymers, as defined by thermodynamics, leads to additivity of properties.[4,8] Synergistic miscible blends tend to be rare; i.e. polyblends of this type do not offer "unexpected" technological advantages to the manufacturer.  On the other hand properties of the miscible systems can be easily calculated from those of the pure polymers, and theoretically treated in an exact manner.[9]

Much greater challenge and potentially greater rewards are to be found in the field of incompatible polymer blends.[2,3,8]  In this case, final properties will be determined by those of the pure polymers, geometrical arrangements of the two phases, their morphology, and the nature (and thickness) of the interface (interphase?).  The fundamental treatment of these systems is further complicated by their non-equilibrium nature.[10-11]

In the first paper of this series[12] we reported on the kinetics and activation energy of thermal degradation of poly(ethylene terephtalate), PET, poly(amide-6,6), PA, and their blends, as well as on the melt viscosity of these materials.  The degradation was studied at temperatures ranging from 260 to 300°C, whereas the viscosities were measured between 240 and 300°C at the rate of shear, $\dot{\gamma}$, from about $10^{-2}$ to $10^5$ $(s^{-1})$.

The degradation studies lead to the following conclusions: (i) the degradation process does not depend on this parameter within the studied range of frequencies, $\omega$ ;  (ii) the overall thermal degradation process follows first order kinetics; (iii) the isothermal degradation rates for the blends are higher than can be calculated from the results obtained for the two neat polymers using an additivity rule; (iv) the activation energy of the overall degradation process, $E_D(\pm 2) = 30$ and $47$ (kcal/mole), calculated respectively for PET and PA, is in good agreement with the literature data; (v) for the blends the dependence $E_D$ vs. c closely follows the second order relation with the value of the interaction parameter, $\Delta E = 20$ (kcal/mole), indicating a strong binary effect.

A good agreement between the rotational steady state and dynamic viscosities was observed for all samples. In addition, for neat polymers and blends containing $c \geqslant 30$ wt% of PA a good agreement between the rotational and capillary rheometer data was obtained. The results from these two types of instruments did not agree for samples containing $10 \leqslant c \leqslant 30\%$ PA, thus indicating that the melt structure in the two instruments was different and resulted in a non-viscometric flow. It has also been noted that the flow curve, $\log \eta^*$ vs. $\log \omega$, for these samples, has one extra region of flow inserted between the customary two neighbouring ones: Newtonian and transitory. In this new region $\eta^*$ slowly decreases with the increase of frequency over one-to-two decades of $\omega$, at all temperatures. The proposed interpretation of this phemomenon is a deformation of dispersed PA droplets in PET matrix. The plots of zero-shear and non-zero-shear viscosities vs. c were found to be qualitatively the same, having a local minimum at a specific PA concentration, $c_m$. For $\dot{\gamma} \rightarrow 0$, $c_m \simeq 10\%$ was found. With an increase in $\dot{\gamma}$, $c_m$ was observed to increase.

The results of this study also indicate a considerable degree of compatibility between PET and PA. A gel-like association between their molecules in the melt, with a "gel melting" temperature $280 < T_{gm}$ (°C) $< 285$, was observed especially for the higher PA - content blends. It should be noted that $T_{gm}$ is about 20°C higher than the melting point of PA ($T_m = 264$°C), and about 15°C higher than any endothermic process detectable by differential scanning calorimetry (DSC).

The melting behaviour of PET/PA blends was found to be quite different than that reported[13,14] for PET/poly(epsilon caprolactam). $T_m$'s were virtually independent of composition in the later case, while for PET/PA a systematic variation of $T_m$ was observed; one set of the DSC-maxima followed a straight line dependence in $T_m$ vs. c graph, while the second closely followed a second-order curve. The X-ray diffraction studies[14] did not reveal any significant variation of the crystalline cells parameters with change of composition.

EXPERIMENTAL

Two commercial resins, listed in Table 1 were used.

TABLE 1

Properties of the Resins
(As Quoted in Commercial Literature)

| Commercial Name<br>Symbol | Kodapak 7352<br>PET | Zytel 42<br>PA |
|---|---|---|
| $\overline{M}_w$ | 55,000 | $(36,000)^a$ |
| $\overline{M}_w/\overline{M}_n$ | 2.2 | - |
| Melting point (°C) | 257 | 258 |
| "softening" point (°C) | 245 | - |
| Melt viscosity at<br>275°C (poise) | - | 25,000 |

Note:  a – estimated from the melt viscosity[16] at 275°C.

Prior to blending, PET was dried for 24 hours at 135°C, and PA for the same period but at 85°C. Weighed amounts of the resins were tumble-mixed in a double-cone blender for 5 minutes and immediately loaded into a hopper of a 1-inch diameter screw, Metalmeccanica-6SRE injection molding machine. The temperatures of the 3 zones of the barrel were kept at 280, 260 and 200°C. The moldings were ground in a granulator Alsteel (with 3/16" diameter hole) screen.

The Rheometrics Mechanical Spectrometer, Model RMS-605, was used in a dynamic (parallel-discs) mode with both 25 amd 50 mm diameter testing fixtures. The measurements at 300 and 290°C were done isothermally, i.e. the test fixtures were preheated to these temperatures; for 280, 260, 240°C the fixtures were preheated to 290°C, the sample was then loaded, trimmed, and subsequently the temperature was adjusted to the desired level. The total waiting time (from the time of loading until the start of measurement), or the soak time, was kept at a minimum, normally from 40 to 180 seconds. The testing was carried out in a hot air medium. The temperature was controlled by an environmental chamber surrounding the test fixtures. Initial tests indicated that the thermal degradation effect is not significantly different if nitrogen was used instead of air. The details of the procedures were previously[6,12,17] published.

The storage shear modulus G', the loss shear modulus G", and the dynamic viscosity,

$$\eta^* = (G'^2 + G''^2)^{\frac{1}{2}}/\omega \tag{1}$$

(where $\omega$ is the frequency in radians/sec) were printed and plotted by RMS. The data were corrected for the degradation effects. The units of $\eta^*$ and G', G" are respectively: (poise) and (dynes/cm$^2$).

In addition, Fourier Transform Infrared spectra were taken[18], using Nicolet 7000 FTIR spectrometer in a reflective mode (ATR).

The list of samples used in this study is given in Table II.

TABLE II

Composition and Newtonian viscosities, $\eta_o^*$, of the PET-PA blends as a function of temperature, $T^{\circ}C$.

| NO. | SAMPLE CODE | PA CONTENT c (wt %) | $\eta_o^* \times 10^{-3}$ (poise) | | | |
|-----|-------------|---------------------|-----------|-----|-----|-----|
|     |             |                     | T = 300 | 280 | 260 | 240 |
| 1 | PET | 0 | 1.6 | 2.2 | 2.9 | 3.8 |
| 2 | B 2/1 | 5 | 0.83 | 1.5 | 2.4 | 4.8 |
| 3 | B 2/2 | 10 | 0.71 | 1.3 | 2.6 | 6.0 |
| 4 | B 2/3 | 25 | 1.2 | 2.0 | 3.5 | 7.3 |
| 5 | B 2/4 | 30 | 1.6 | 2.4 | 4.1 | 8.7 |
| 6 | B 2/5 | 35 | 2.0 | 2.8 | 5.1 | (13)* |
| 7 | PA 42 | 100 | 6.1 | 16 | 37 | (130)* |

Note:  * calculated by extrapolation using Arrhenius plot.

RESULTS AND DISCUSSION

In Figure 1 the isothermal (uncorrected for the degradation) data of the steady shear viscosity $\eta$ (poise) as a function of the rate of shear $\dot{\gamma}$ (sec$^{-1}$) and the dynamic viscosity $\eta^*$ (P) as a function of frequency $\omega$ (rad/sec) were plotted for PET at 280°C. The dynamic measurements were carried out at $\epsilon$ = 60% strain, selected on the basis of strain scan (at $\omega$ = 1 and 10) as the optimum $\epsilon$ - value for the stability of the three dynamic functions: $\eta^*$, storage modulus G', and loss modulus G". A good superposition of the $\eta(\gamma)$ and $\eta^*(\omega)$ is observed. The

slight decrease of the viscosity with ($\omega, \dot{\gamma}$) is primarily due to
the thermal degradation (the scans were always done from the
low $\omega$-value up). In effect, for all the samples tested (see
Table II) the Newtonian behaviour was observed up to about 1
(rad/sec). It can also be seen that the values of G' in Figure 1
are one-to-two decades lower than those of G". This means that
the magnitude of $\eta^*$ is determined primarily by that of G", i.e.
the rheological behaviour of PET/PA blends show similarities to
those of the dilute polymer solutions or polymer melts in the
terminal zone.[19]

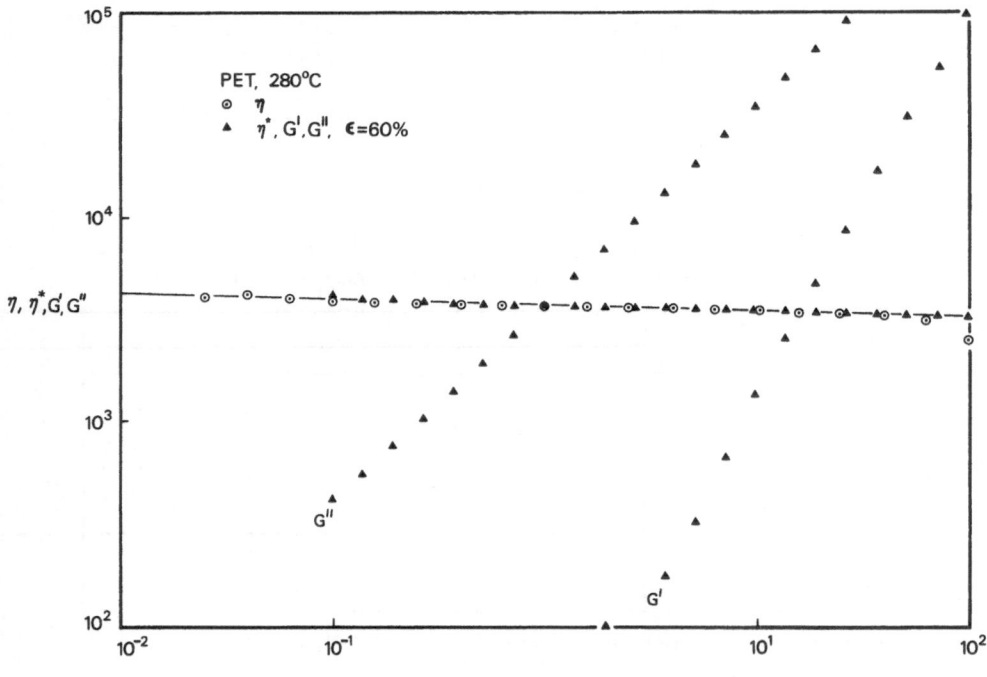

Fig. 1.   Rheological properties of PET at 280°C.   Circles:   RMS,
         cone-and-plate in steady;   Triangles:   RMS,   parallel
         disc in oscillation.   Uncorrected data.

Under these circumstances one can write:

$$(\sigma_{11} - \sigma_{22})/\dot{\gamma}^2 = 2G'/\omega^2, \tag{2}$$

where $N_1 = (\sigma_{11} - \sigma_{22})$ is the first normal stress
difference, as determined in a steady state rheological
experiment. However, instead of $N_1$, it is advantageous[20] to

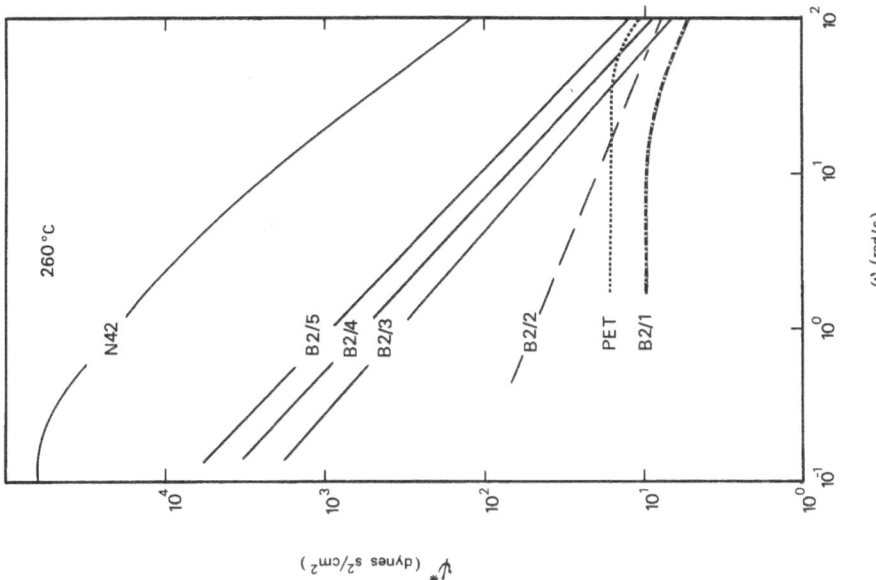

Fig. 3. $\psi^*$ vs. $\omega$ at 260°C. For simplicity the data points are omitted.

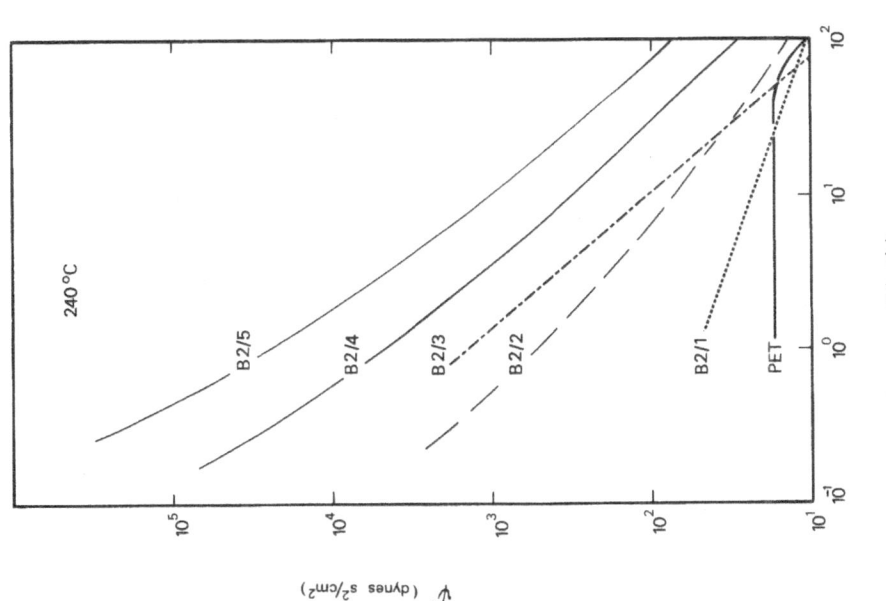

Fig. 2. $\psi^*$ vs. $\omega$ at 240°C. For simplicity the data points are omitted.

use the first normal stress difference coefficient:

$$\psi = N_1/\gamma^2 = (\sigma_{11} - \sigma_{22})/\gamma^2 \quad . \tag{3}$$

The reason for this is, that for simple viscoelastic liquids the coefficient behaves in a similar manner to that of $\eta$, giving a constant value,

$$\psi_0 = \lim_{\gamma \to 0} \psi \tag{4}$$

characteristic of the original, undisturbed structure of the melt. From eqs. (2) and (3) one can write:

$$\psi^* = 2G'/\omega^2 \tag{5}$$

where * indicates that the values of the first normal stress difference coefficient were calculated from the dynamic test data.

Figures 2-5 show the $\psi^*$ dependencies on $\omega$, at 240, 260, 280 and 300°C respectively. In all four figures for the neat resins, a plateau of $\psi_0^*$ is visible. At 260°C, $\psi^*$ for B 2/1, the blend containing the lowest (5%) admixture of PA, also shows the leveling effect for low values of $\omega$. As temperature of the sample decreases (crystallisation of PA) or increases (phase separation) the effect disappears. For all the other blends the values of $\psi^*$ systematically decrease with increase of $\omega$. Also the initial decrease of $\psi^*$ at low values of $\omega$ is steeper here, thus resembling the behaviour of lightly cross-linked polymer melts.[20]

A cross-plot of data is shown in form of $\psi^*$ ($\omega$ = 10 rad/sec) vs. c dependence in Figure 6. The observed relation is qualitatively similar to that reported[12] for $\eta_0$, with two notable exceptions: the minimum at $c_m$ is much more accentuated, and it occurs at lower concentration of PA: $c_m \simeq$ 5 wt%.

The activation energy of the normal stress difference coefficient:

$$E_\psi = R d\ln\psi^*/d(1/T) \tag{6}$$

was calculated from the data of Figure 6. The results are shown in Figure 7. It can be seen that by contrast with the results published for polystyrene/ high density polyethylene blends[21], here $\psi^*$ shows a strong temperature dependence. In most cases the values of $E_\psi$ for a given sample, are larger than those of the activation energy of the viscous flow, $E_\eta$. The values of $E_\eta$ in Figure 7 are replotted from the reference.[12] Since, according

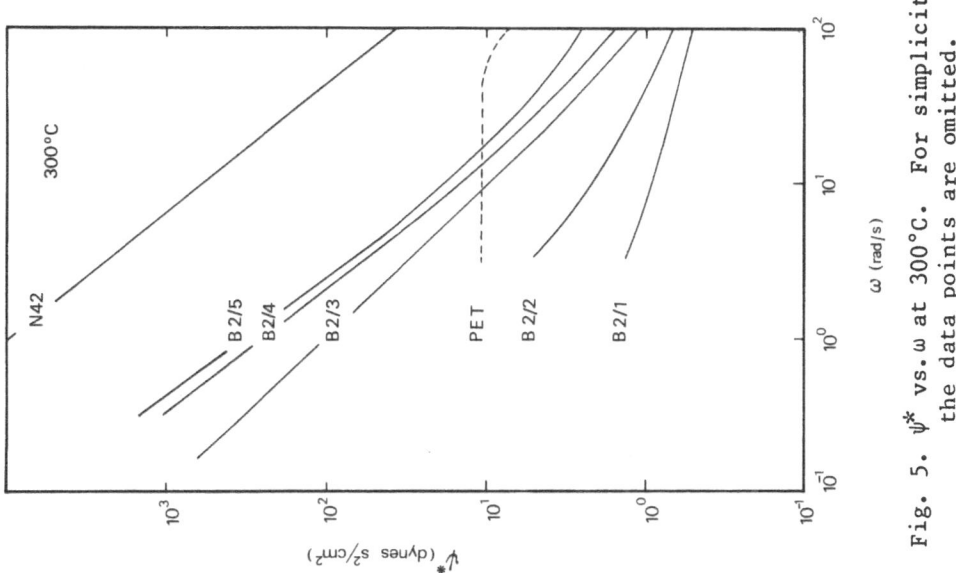

Fig. 5. $\psi^*$ vs. $\omega$ at 300°C. For simplicity the data points are omitted.

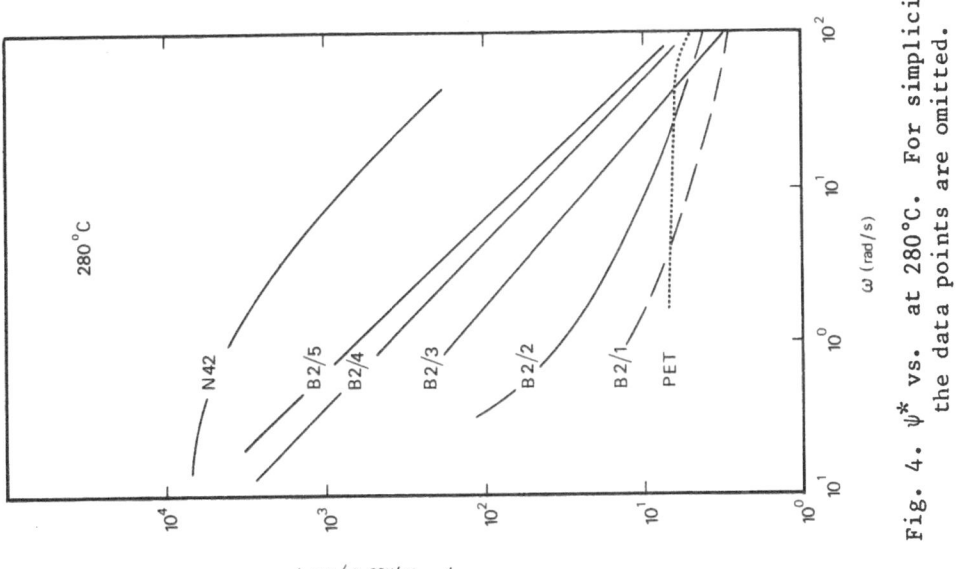

Fig. 4. $\psi^*$ vs. at 280°C. For simplicity the data points are omitted.

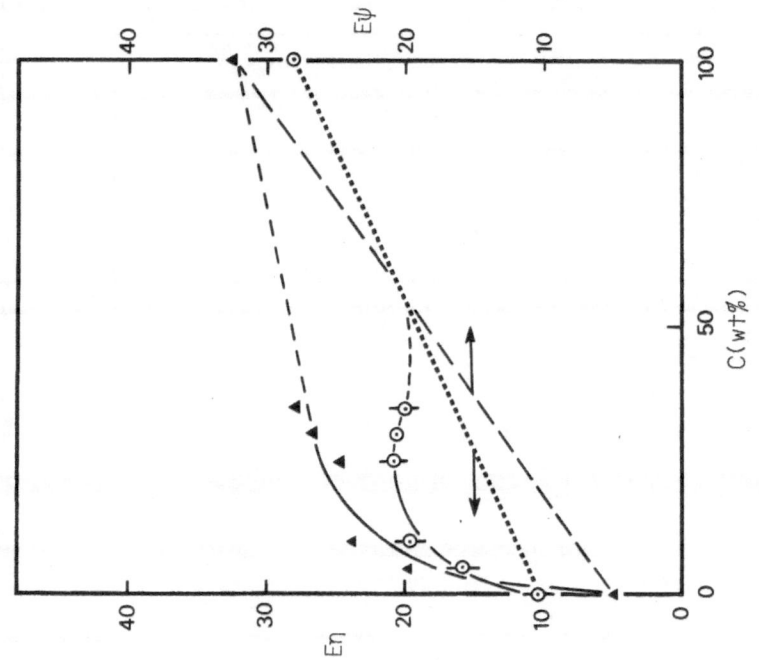

Fig. 7.  The concentration dependence of
        the activation energies of viscous
        flow, $E_\eta$, and of the normal stress
        difference, coefficient $E_\psi$.

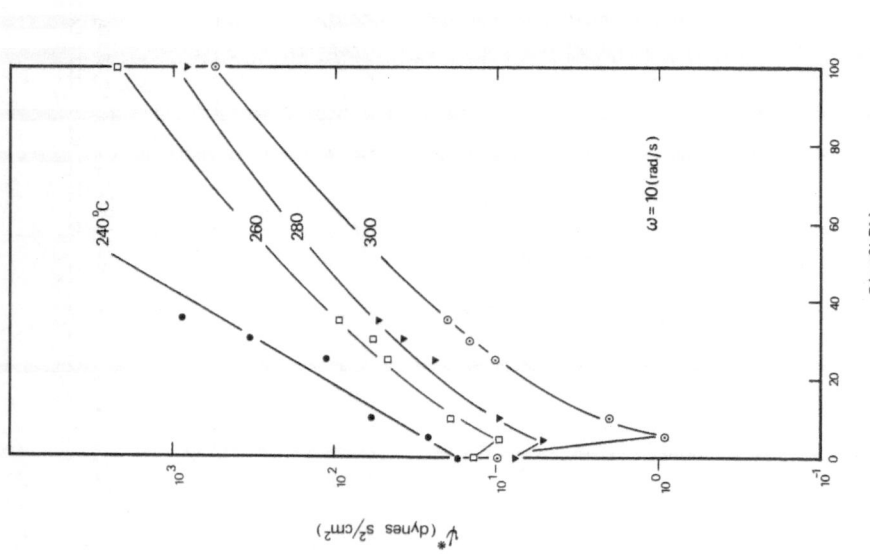

Fig. 6.  The concentration dependence of $\psi^*$
        at $\omega = 10$ (rad/sec) and at the four
        temperatures indicated.

to the second order fluid theory, the steady state compliance is expressed as:

$$J_e^o = \psi/2\eta^2,\qquad\qquad\qquad (7)$$

and since according to Rouse $J_e^o$ is a slowly varying function of T, the theory predicts that $E\psi \simeq 2E_\eta$. It can be seen however that for the investigated systems this prediction is not fulfilled. Even for PET, where eq. (7) should be strictly valid, $E_\eta$ and $E_\psi$ are equal within the range of experimental errors: $\pm 2$ (kcal/mole). This implies a relative constancy not of $J_e^o$ but of the recoverable shear strain:

$$S = J_e^o\,\eta\dot{\gamma}\qquad\qquad\qquad (8)$$

The constancy of S was previously[20] observed for rigid and plasticized samples of poly(vinyl chloride).

Finally a few words on the structure of the blends in the melt. The molten samples of PET, B 2/1 and PA are transparent, indicating an "optical homogeneity" of the liquids, i.e. if the phase separation in B 2/1 does occur, the domain size must be smaller than 0.2 μm. The first sign of "milkiness" was observed for B 2/2, containing 10% PA. All the other blends were opaque. The reflective mode ATR-FTIR spectra at ambient temperature for PET, B 2/2 and B 2/4 are virtually identical, indicating that at least up to 30% PA the composition of the continuous phase is practically the same – it is PET. In other words, the changes in rheological behaviour, observed within the range $0 \leqslant c \leqslant 30$, seem to coincide with the phase separation (detected by visual method), and not with a phase inversion.

As reported[6] the viscosity vs. composition for most of the immiscible polymer blends demonstrate a negative deviation from the log-additivity rule. The information on $N_1$ is much more scarce, and while in some cases (e.g. polystyrene / polycarbonate [22]) parallels the $\eta$ behaviour, in some others (e.g. polystyrene/polypropylene containing 10% $CaCO_3$ [23]) it indicates an opposite effect.

CONCLUSION

The PET/PA blends show a similar behaviour to most of the immiscible blends – their $\eta^*$ and $\psi^*$ vs. c curves both pass through a local minimum at $c_m \simeq 10$ and 5% respectively. The concentration seems to coincide with the onset of opacity; i.e. the depression in the rheological functions is probably due to a decrease of the intermolecular interaction in PET melt by addition of PA macromolecules. It very well may be that the effect is due to a simple disruption of the regularity of interactions in PET melts. Studies on the system are being continued to answer this and other questions.

REFERENCES

1.  Anonym., Polymer News, 7 (2): 86 (1981).
2.  M. Shen and H. Kawai, AIChE J., 24 (1): 1 (1978).
3.  T. Alfrey, Jr. and W. J. Schrenk, Science, 208: 813 (1980)
4.  D. R. Paul and J. W. Barlow, J. Macromol. Sci., Rev. Macromol. Chem., C18 (1): 109 (1980).
5.  J. Roovers, Meth. Experim. Phys., 16C: 275 (1980).
6.  L. A. Utracki and M. R. Kamal, "Melt Rheology of Polymer Blends", Paper presented at IMRI mini-symposium series "Polymer Blends - 1981" Montréal, Québec. To be published in Polymer Eng. Sci.
7.  L. A. Utracki and G. L. Bata, "Polymer Alloys", CAN-PLAST - 1981, Montréal, November 1981.
8.  O. Olabisi, "Polyblends" prepared for Kirk-Othmer Encyclopedia of Chemical Technology, April 1982. See also: O. Olabisi, L. M. Robeson and M. T. Shaw, "Polymer-Polymer Miscibility", Academic Press, New York, 1979.
9.  R. Simha, Paper presented at the 1st IMRI Mini-symposium on Phase Equilibria in Polymeric Systems, Montréal, 1980, NRCC spec. ed. 1981. To be published in Polymer Eng. Sci.
10. B. A. Thornton, R. G. Villasenor and B. Maxwell, J. Appl. Polymer Sci., 25: 653 (1980).
11. M. Kapuścinski and H. P. Schreiber, Polymer Eng. Sci., 19: 900 (1979).
12. L. A. Utracki, M. R. Kamal, V. Tan, A. M. Catani and G. L. Bata, J. Appl. Polymer Sci., to be published.
13. V. Garg, C. L. Nagendra, A. Misra, V. Chandhary and D. S. Varma, Ang. Makrom. Chem., 90: 57 (1980).
14. K. Dimov, M. Savov and J. Georgiev, Ang. Makromol. Chem., 84: 119 (1980).
15. A. Garton, private communication.
16. P. Parrini, D. Romanini and G. P. Righi, Polymer, 17: 377 (1976).
17. C. Macosko and J. M. Starita, SPE J., 27: 38 (1971); F. G. Mussatti and C. W. Macosko, Polymer Eng. Sci., 13: 236 (1973).
18. J. -J. Hechler, private communication.
19. J. D. Ferry, "Vicoelastic Properties of Polymers", J. Wiley & sons, New York, 2nd edition, 1970, Chs. 9 and 10.
20. L. A. Utracki, Polymer Sci., Polym. Phys. Ed., 12: 563 (1974).
21. C. D. Han and Y. W. Kim, Trans. Soc. Rheology, 19: 245 (1975).
22. Y. S. Lipatov, V. F. Shumsky, A. N. Gorbatenko, Y. N. Panov

and L. S. Bolotnikova, *J. Appl. Polymer Sci.*, 26: 499(1981).

23. Y. W. Kim and C. D. Han, *J. Appl. Polymer Sci.*, 20: 2905 (1976), ibid. 21: 515 (1977).

# DOMAIN STABILITY DURING CAPILLARY FLOW OF WELL DISPERSED TWO PHASE POLYMER BLENDS. POLYSTYRENE/POLYMETHYLMETHACRYLATE BLENDS

J. Lyngaae-Jørgensen, F.E. Andersen, and N. Alle*

Instituttet for Kemiindustri, Building 227
The Technical University of Denmark
2800 Lyngby - Denmark

SYNOPSIS

Hypotheses for domain stability, i.e. the conditions where each domain of a dispersed phase remains as a coherent domain during capillary flow is tested with data for polystyrene - polymethylmethacrylate melts. It is found that stable domains exist when the ratio between the (Trouton) zero shear viscosity of the discrete phase to the continuous phase is larger than approximately one. Blends of two thermodynamically incompatible polymers, polystyrene (PS) and poly(methylmethacrylate) (PMMA), were melt blended in a Brabender Plasticorder over the compositions: 100% PS, 75%, 50% 25% and 0% PS. Viscous and die swell behaviour of pure polymers and the blends at different temperature were obtained using an Instron capillary rheometer and the microstructure of the melt blends was studied by transmission electronmicroscopy.

The question of domain stability is "tested" by a) comparing the average volume of coherent discrete domains before and after extrusion through a capillary and b) by an indirect method where a model prediction for the viscosity for stable domain flow is compared with experimental data.

It has been found that viscosity-shear rate data

---

* Present address: Dow Chemical (Netherland) B.V., 4530AA, Terneuzen, the Netherlands.

of PS-PMMA blends at different temperatures constitute
master curves at constant blend composition when plotted
as $(\eta/\eta_o)$ versus $(\dot{\gamma}\eta_o/\rho T)$.  Furthermore all the master
curves of the blends and homopolymers with different
molecular weight distributions superposed into a single
master curve when plotted as $(\eta/\eta_o)$ versus $(\eta_o M_C H/\rho RT)$
where $M_C$ is twice the molecular weight between entanglements and H is the heterogeniety $(\bar{M}_w/\bar{M}_n)$.

From the observed melt rheology results and the
agreement of experimental results with model predictions
supported by direct evidence of electron photomicrographs,
it is concluded that the morphology of incompatible
polymer blends depends on the composition ratio and
Newtonian (or Trouton) viscosity ratio of the components.
When the blend has the component with low Newtonian
viscosity as dispersed phase, the dispersed droplets
seem to be unstable and to break up into smaller droplets
in capillary flow.  On the other hand if the blend has
the higher Newtonian viscosity component as a dispersed
phase the dispersed droplets appear to form a stable
morphology with continuous threads or a fibrillar pattern or elongated droplets in capillary flow.

In order to explain the experimental observations a
hypothesis is formulated for the stability of coherent
domains in the socalled relaxation region near the inlet
region in the capillary for viscoelastic domains in
viscoelastic media at high shear rates.

According to this hypothesis a necessary criterion
for domain stability for a domain near the capillary
axis (critically condition) is that the ratio between
the relaxation time $\lambda_D$ of the domains to that of the
continuous phase $\lambda_K$ is larger or equal to 1.

The relaxation time is given by $\lambda = \dfrac{a\ \eta_o \cdot M_C\ H \cdot \rho}{c^2\ RT}$

where $\eta_o$ is the zero shear viscosity, $M_C$ is twice the
molecular weight between entanglements, $H = \bar{M}_w/\bar{M}_n$, $\rho$ is
the polymer density, c is the polymer concentration, R
is the gas constant, T is the absolute temperature and
a is a constant.

## INTRODUCTION

In recent years there has been a great deal of
interest in the studies of the structure and properties
of polymer blends.  The practical motivation for this

interest is the achievement of desirable mechanical, rheological or other properties. Because of the generally immiscible nature of polymers, polymer blends usually exhibit microphase separations with two-phase structure. The morphologies of these two-phase polymer blends are determined not only by the composition ratio of the system, but also by processing conditions[1]. A better understanding of rheological behaviour and rheology-morphology relationships of two-phase polymeric material in the molten state would be important to exploit the unique properties for various engineering applications and for determining the optimal process conditions.

Most of the recent work on two-phase polymer systems has been made in establishing the relationship between physical properties and morphology in the solid state. But so far relatively little has been reported to describe the morphology from the melt rheology of the components and principles which govern the morphology developments[2-13] in flow fields for dispersed two-phase polymer systems. Stability criteria for dispersed droplets were reviewed by Han and Funatsu[14] in a recent paper. Based on Taylors analysis[15] Han and Funatsu proposed an empirical criteria function drawn in a delineation of

$\dfrac{\eta_B \dot{\gamma}_c a}{\delta}$ against log $\dfrac{\eta_A}{\eta_B}$ where $\eta_A$ and $\eta_B$ are viscosities

of domain phase and suspending medium, respectively, $\dot{\gamma}_c$ is the critical value of apparent shear rate, a is the domain radius and $\delta$ the interfacial tension. Stable

domains were found for low values of $\dfrac{\eta_B \dot{\gamma}_a a}{\delta}$ and for

$\dfrac{\eta_A}{\eta_B} \geq 3$.

Chin and Han[16] have recently analyzed the flow behaviour of a two phase viscoelastic system at the inlet region to a capillary and performed a stability analysis of an elongated droplet in the fully developed shear flow region in a capillary.

Recently Tsebrenko, Rezanova and Vinogrado[17] investigated blends of 20% polyoxymethylene in ethylenevinylacetate copolymers.

Their main conclusion was that ultrathin fibers of "unlimited" length of POM could only be formed at a viscosity ratio close to unity.
for the cooled and uncooled extrudates respec-

The purpose of the present work has been to try to develop criteria for the stability of domains of well-dispersed (domains with characteristic dimensions much smaller than characteristic process equipment dimensions), two phase blends and to investigate the morphology during capillary flow.

Another goal is to study the flow properties of such melts.

Table 1.   Zero Shear Viscosities ($\eta_o \cdot 10^{-4}$ poise) of PS-PMMA Blends Determined by using Ferry's Equation

| Temperature | Composition (wt% of PMMA) | | | | | PS $\eta_o/\eta_o$ |
|---|---|---|---|---|---|---|
|  | 0 | 25 | 50 | 75 | 100 |  |
| 180°C | 88.12 | 61.36 | 57.05 | 54.66 | 52.12 | 1.7 |
| 200°C | 25.12 | 14.02 | 12.40 | 12.15 | 10.44 | 2.4 |
| 220°C | 9.87 | 3.93 | 3.79 | 3.34 | 2.34 | 4.2 |
| 240°C | 5.13 | 1.40 | 1.35 | 0.97 | 0.62 | 8.3 |

A system of polystyrene-polymethylmethacrylate blends were chosen for this study because the ratio of zero shear viscosities of polystyrene to polymethylmethacrylate measured at different temperatures varied between 1,8 and 8,4 as seen in table 1.  Furthermore the system show crossover behavior, meaning that a ratio between viscosities larger than one at low shear rate change to a ratio lower than one at higher shear rate. Finally microtoming these samples gave very satisfactory contrasts in transmission electron microscopy pictures.

Table 2.   Starting Polymers for PS-PMMA Blends

|  | Polystyrene (PS) | Polymethyl-methacrylate PMMA) |
|---|---|---|
| Commercial name | Hoechst Hostyren N 7000 | ICI Diakon LO 951 |
| Density $(g/cm^3)$ | 1.05 | 1.17 |
| Melt index (g/10 min) | 2 | 16 |
| $\bar{M}_W$ | $3.67 \cdot 10^5$ | $8 \cdot 10^4$ |
| $\bar{M}_n$ | $1.17 \cdot 10^5$ | $3.76 \cdot 10^4$ |
| $\bar{M}_w/\bar{M}_n$ | 3.14 | 2.13 |

Table 3.   Sample Codes and Composition of Poly-styrene-Polymethylmethacrylate Blends

| Sample Code | Composition (wt %) |
|---|---|
| PS | 100% polystyrene |
| PMMA | 100% polymethylmethacrylate |
| PS/PMMA = 25/75 | 25% polystyrene and 75% polymethylmethacrylate |
| PS/PMMA = 50/50 | 50% polystyrene and 50% polymethylmethacrylate |
| PS/PMMA = 75/25 | 75% polystyrene and 25% polymethylmethacrylate |

EXPERIMENTAL

Materials

    The materials used in the present study for making
of the polymer blends were commercial grades of polystyrene (PS) and poly(methylmethacrylate) (PMMA) which
were supplied in pellet form.  The sources and molecular

characteristics of these polymers are summarized in
Table 2.  Compositional characteristics and sample codes
are given in Table 3.

<u>Blend Preparation</u>

     All the blends described in Table 2 were prepared
by melt mixing at 200°C in a Brabender Plasticorder
(with a Walzerkneter Type 30 mixing head having mixing
chamber volume of approximately 33 cm³) using the following procedure.

     The two pelletized PS and PMMA polymers in the
desired composition were manually mixed.  This mixture
was added gradually to the mixing bowl of Brabender
which had been preheated to 200°C.  The speed of the
mixing blades was set at 20 rpm and mixing continued
till a constant torque was reached, which took about 12-
15 min.  The well mixed blends from the Brabender were
compression moulded into 1-mm thick sheets at 200°C in
a hydraulic press, followed by air cooling to room
temperature.  All the samples cut from these sheets were
used for rheological and morphological characterization.
More details on blend preparation are described in Ref.
18.

Table 4.   Molecular Weight Characteristics of
           PS-PMMA Blends

| SAMPLES | Before mixing in the Brabender | | | | After mixing in the Brabender[*] | | |
|---|---|---|---|---|---|---|---|
| | $\bar{M}_w \cdot 10^{-5}$ | $\bar{M}_n \cdot 10^{-4}$ | $\bar{M}_c \cdot 10^{-4}$ | $\bar{M}_w/\bar{M}_n$ | $\bar{M}_w \cdot 10^{-5}$ | $\bar{M} \cdot 10^{-4}$ | $\bar{M}_w/\bar{M}_n$ |
| PS [*] | 3.67 | 11.69 | 3.12 | 3.14 | 3.12 | 11.06 | 2.82 |
| PMMA [*] | 0.80 | 3.76 | 1.05 | 2.13 | 0.82 | 4.02 | 2.04 |
| PS/PMMA=75/25 [**] | 2.96 | 7.65 | 2.09 | 3.86 | 2.64 | 7.92 | 3.33 |
| PS/PMMA=50/50 [**] | 2.24 | 5.69 | 1.57 | 3.93 | 2.39 | 7.65 | 3.12 |
| PS/PMMA=25/75 [**] | 1.52 | 4.52 | 1.26 | 3.35 | 1.99 | 5.99 | 3.32 |

[*]   Determined by GPC technique.

[**]  Determined by Equations (5)-(6) only before mixing in the Brabender.

    Pure PS and PMMA were also subjected to the same
mixing condition in order to study their possible degradation during mixing.  The molecular characteristics of
pure polymers determined by Gel Permeation Chromatography
(GPC) technique before and after melt mixing in the Brabender are shown in Table 4.  From Table 4 it is confirmed that no appreciable degradation took place at $200^\circ$C
during melt blending of PS and PMMA.

Gel Permeation Chromatography (GPC)

    The molecular weight averages ($\bar{M}_w$, $\bar{M}_n$ and $\bar{M}_z$) of PS
and PMMA and PS-PMMA blends after processing in the
Brabender Plasticorder were determined, using the GPC
technique.  All GPC measurements were made with a Waters
Associates Model 200 instrument which is equipped with
four polystyrene gel columns of $10^6$, $2 \cdot 10^4$ and $10^3$ and
$10^3$ Å nominal porosities.  The GPC instrument was run
under the following conditions:

    Flow rate: 1 ml/min
    Injection volume: 2 ml
    sample concentration: $2 \cdot 5 \cdot 10^{-3}$ gm/ml

    Tetrahydrofuran (THF) was used as a solvent at
room temperature in all the measurements.

    The GPC columns were calibrated with narrow distri-
bution polystyrene standards from Pressure Chemical Co
and from National Bureau of Standards under the above
mentioned conditions except the injection volume which
was 0.5 ml.  All the molecular weight averages summarized in Table 4 were calculated using the calibration
found for PS standards and approximate "Q factor correction".

Rheometry

    Melt flow behaviour and die swell behaviour of PS,
PMMA and their blends in the range of 0.3 to 1000 sec
were determined by using an Instron Constant Speed Rheo-
meter at four temperatures, $180^\circ$C, $200^\circ$C, $220^\circ$C and $240^\circ$C
respectively.  The capillary used has length/diameter
(L/D) ratio of 81.79 with a diameter of 1.245 mm and
$90^\circ$ entrance angle.  The Rabinowitsch correction was
applied to all data in calculating the wall shear rate
($\dot{\gamma}$).  {However, the end correction used by Bagley has
been neglected in calculating the wall shear stress
because the capillary with a large L/D ratio was used
in this study.}

For the die swell measurements, the polymer melt
from the Instron rheometer was extruded downward and
directly into air at room temperature. When steady
state conditions have been achieved, the extrudate of
about 3 to 4 cm in length was cut and used as a specimen
for die swell measurements. All diameter measurements
were made approximately at about 5mm to 10mm from the
lower end, where elongation was negligible, with a micro-
meter to the nearest 0.01 mm. Unannealed die swell
ratios reported in this study are corrected for the
density difference between room temperature and extrusion
temperature.

Morphology

A JEOL transmission electron microscope (Japan
Electronics and Optics Laboratories, CO. LTD, Tokyo,
Model JEM-100B-TR) was used for studying the morphology
of all PS-PMMA blends having different blend compositions. The original melt blends obtained from Brabender
Plasticorder and the extrudates of each blend obtained
from the Instron Capillary Rheometer at $200^{\circ}C$ and at a
crosshead speed of 0.05 cm/min ($\dot{\gamma} = 3.14$ sec$^{-1}$) and 0.5
cm/min ($\dot{\gamma}_{app} = 31 \cdot 38$ sec$^{-1}$) were used for morphological
characterization. Efforts were made to preserve the
morphology by rapid cooling of the extrudate with a
stream of cold $N_2$ near the exit of the capillary while
the polymer melt is being extruded through the capillary.

Samples were sectioned by using an Ultramicrotome
(Reicht OM U3, Austria) at room temperature. No stai-
ning was found necessary as the contrast between PS
and PMMA was sufficient to distinguish the two-phases
with dark and light regions corresponding to the PS and
PMMA respectively.

FLOW BEHAVIOUR ANALYSIS

The rheological behaviour of polymer blends differs
from that of parent polymers because of the generally
incompatible nature of polymers. In the present work
we compare the flow properties of the blends with the
behaviour expected for homopolymers in order to gain
information on specific flow mechanisms.

Shear Rate-Temperature Superposition

For many polymer systems which are amorphous one-phase polymers, the shear rate dependence of viscosity can be expressed in reduced variable form[19-22].

$$\frac{\eta}{\eta_o} = V \ (\lambda_o \dot{\gamma}) \tag{1}$$

where $\eta$ is the non-Newtonian viscosity
$\eta_o$ is the Newtonian viscosity
$\dot{\gamma}$ is the shear rate
$\lambda_o$ is the relaxation time constant
$V$ is the viscosity master function which depends
on the molecular theory considered.

The two theories that mostly agree with the experimental data are those af Bueche and of Graessley[23-25]. Bueche's and Graessley's thesis assume that the polymer is monodisperse and the question arises as to whether these theories are applicable to a polydisperse material. Many workers[25-29] have suggested different mothods for constructing master viscosity curves with polydispersity as parameter. In general the theoretical forms of relaxation time constant ($\lambda_o$) are expected to be proportional to $\eta_o$ and inversely proportional to $\rho T$ for a given material. Thus ($\lambda_o$) for the same homopolymer melt at different temperatures can be expressed as[22]

$$\lambda = \frac{\eta_o F}{\rho T} \tag{2}$$

and from equation (1) it follows that

$$\frac{\eta}{\eta_o} = V \left( \frac{\eta_o \dot{\gamma}}{\rho T} \cdot F \right) \tag{3}$$

where F is a property which depends on molecular structure but is independent of temperature. Based on equation (3), one would expect for a given homopolymer that the reduced curves of $\eta/\eta_o$ versus $\eta_o \dot{\gamma}/T$ should superimpose.

Experimental evidence on many polydisperse polymers has indicated that the polydisperse material departs earlier from Newtonian flow and shows more reduced rate of viscosity reduction with shear rate than the narrow distribution polymer[25]. This may be attributed to the greater degree of chain entanglements that occur with a broad distribution molecular weights of polymers.

Recently, Lyngaae-Jørgensen[30] has divided linear polymers into roughly two groups. He predicted and observed that all polymers of a particular group fall into a single master curve when plotted as $\eta/\eta_o$ versus $\dot{\gamma}\eta_o M_c H/\rho RT$ in which relaxation time constant is defined as

$$\lambda_o = \frac{\eta_o M_c H}{\rho RT} \qquad (4)$$

$M_c$ is twice the molecular weight between entanglements and H is the heterogeneity index ($H = \bar{M}_w/\bar{M}_n$).

## Extension to Two-Phase Polymeric Materials

The above molecular theories and master curve principles are established for one-phase thermoplastic polymers. In order to investigate whether it is possible to distinguish the flow behavior of a two phase system from the flow behavior of a homogeneous one phase system of the same components by use of these master curve principles, the data for the blends was treated as if they were homogeneous single phase systems.

Based on Equation (3), if the reduced plots of $\eta/\eta_o$ versus $\dot{\gamma}\eta_o/\rho T$ for a given two-phase polymeric material at different temperatures superimpose onto a master curve, the two-phase polymeric material in the molten state behaves as a single-phase homopolymer in this respect or exhibits a morphology in the capillary which seems to be independent of temperature. Furthermore, it is known that F in Equation (3) changes with the heterogeneity (H) of the sample for homopolymer melts. The weight and number average molecular weights of a two-phase binary blend ($\bar{M}_{wB}$ and $\bar{M}_{nB}$) of two polydisperse polymers can be estimated by the following relations[31, 32].

$$\bar{M}_{wB} = w_1 \bar{M}_{w1} + \bar{w}_2 M_{w2} \qquad (5)$$

$$\frac{1}{\bar{M}_{nB}} = \frac{w_1}{\bar{M}_{n1}} + \frac{w_2}{\bar{M}_{n2}} \qquad (6)$$

where $w_1$, $w_2$, $\bar{M}_{w1}$, $\bar{M}_{w2}$, $\bar{M}_{n1}$ and $\bar{M}_{n2}$ are the weight fractions, the weight and number average molecular weights of component 1 and 2 respectively.

The critical molecular weight ($M_c$) of polymer may be interpreted as a material constant signifying the lower limit of molecular weight for which non-Newtonian flow can be observed and which is strongly dependent on the molecular structure [31,32]. It is possible to determine $M_c$ experimentally for homopolymers but it is very difficult to determine critical molecular weight of polymer blends ($M_{cB}$) experimentally due to two-phase structure.

A simple relation between the critical molecular weight of a binary polyblend ($M_{cB}$), critical molecular weights of two components ($M_{c1}$ and $M_{c2}$) and volume fractions may be obtained on the assumption that there are no variations in volume during the mixing of the components

$$\frac{1}{\rho_B} = \frac{w_1}{\rho_1} + \frac{w_2}{\rho_2} \qquad\qquad (7a)$$

and within the limits of this assumption

$$\frac{\rho_B}{M_{cB}} = \frac{v_1 \rho_1}{M_{c1}} + \frac{v_2 \rho_2}{M_{c2}} \qquad\qquad (7b)$$

where $\rho_1$, $\rho_2$, $v_1$ and $v_2$ are the densities and volume fractions of components 1 and 2 respectively.

According to Equations (5) and (6), it is expected that two-phase polymer blends have higher heterogeneity ($H=M_w/M_n$) than their respective homopolymers if the polyblend exhibits miscibility on a molecular scale in the molten state under shear flow conditions.

Thus for a two-phase structure in the molten state one should at least expect a deviation from the master curve $\eta/\eta_o$ against $\dot{\gamma}\eta_o M_c H/\rho RT$ caused by a heterogeneity effect.

## Stability of Domains During Capillary Flow

The question of whether a dispersed droplet remains as a coherent entity during capillary flow is investigated by application of electron microscopy before and after extrusion and the criterion for domain stability used is that the average volume of the dispersed phase is the same before and after extrusion: $\bar{v}_B = \bar{v}_A$. (8).

Furthermore a rather uncertain indirect approach has been applied in order to enlighten the stability and to gain information concerning the morphology of the melt during flow in the capillary. This last approach consists of comparing the measured viscosity function with a predicted ideal flow behavior applying a theoretical mixing rule.

## Mixing Rule

In a previous paper[12], a mixing rule based on model morphologies proposed earlier by Vinogradow[33,34] was derived in order to establish a rheology-morphology relationship for dispersed two-phase polymer blends flowing through a capillary.

The following mixing rule was derived for a two-phase polymer blend having stable domains and which forms a morphology with a large number of continuous long thin layers or fibrils in the capillary:

$$\frac{1}{\eta_B(\tau_w)} = \frac{v_1}{\eta_1(\tau_w)} + \frac{v_2}{\eta_2(\tau_w)} \qquad (9)$$

where $\eta_B$ is the viscosity of the blend, $\eta_1$, $\eta_2$ and $v_1$, $v_2$ are the viscosities and volume fractions of the components 1 and 2 respectively.

If a two-phase polymer blend is to form a homogeneous mixture with break-up phenomena of dispersed droplets in the capillary entrance regions or unstable morphology in capillary, it is expected that the experimental results of this blend deviate from the calculated results by Equation (9). On the contrary if the experimental results are in good agreement with the calculated results by Equation (9), it is possible that the blend system forms a morphology with continuous layers or fibrils or elongated droplets in capillary flow.

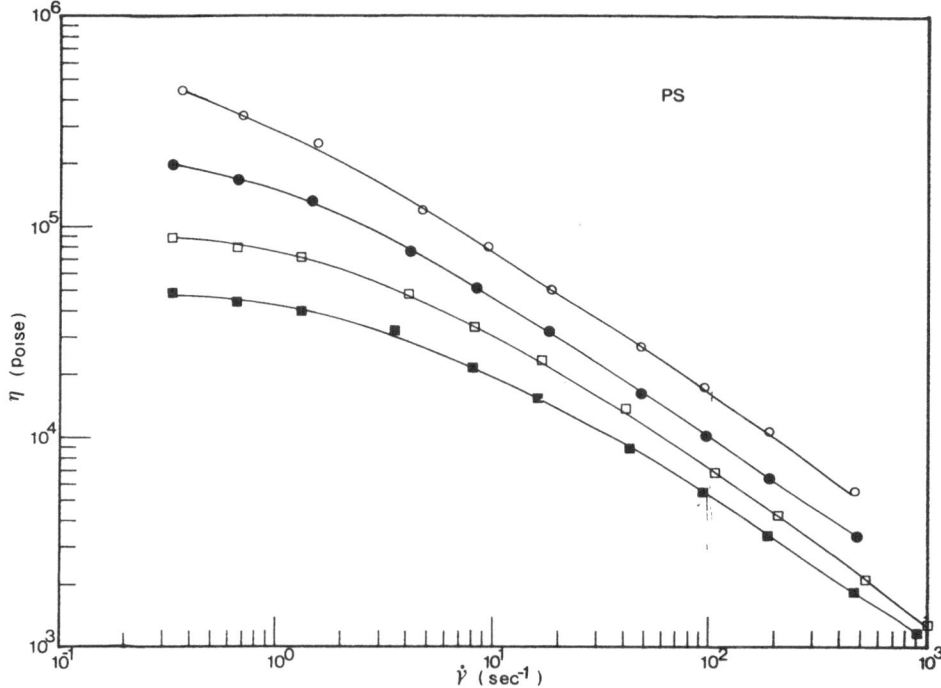

Fig. 1.   Viscosity versus shear rate for polystyrene:
          (○) 180°C; (●) 200°C; (□) 220°C; (■) 240°C


RESULTS

Viscous Behaviour

        Figures 1-5 show the plots of steady state viscosity
as a function of shear rate at 180°C, 200°C, 220°C and
240°C for the two pure polymers and the three blends of
PS and PMMA respectively.  The Rabinowitch[35,36] correction was applied to all data in calculating the true
wall shear rate ($\dot{\gamma}$).  However, the end correction used
by Bagley[37,38] has been neglected.  Values of zero shear
viscosity ($\eta_o$) for PS, PMMA and PS-PMMA blends at all
temperatures were obtained through extrapolation from
the Newtonian range by using Ferry's Equation[39].  Such
plots gave good straight lines.  It is seen from Figures
1-5 that the three PS-PMMA blends show the same kind of
flow behaviour as pure homopolymers at all temperatures.

It has been suggested that the use of shear stress instead of shear rate is more appropriate for correlating the viscoelastic properties of dispersed two-phase polymer melts because of discontinuity of shear rate at the

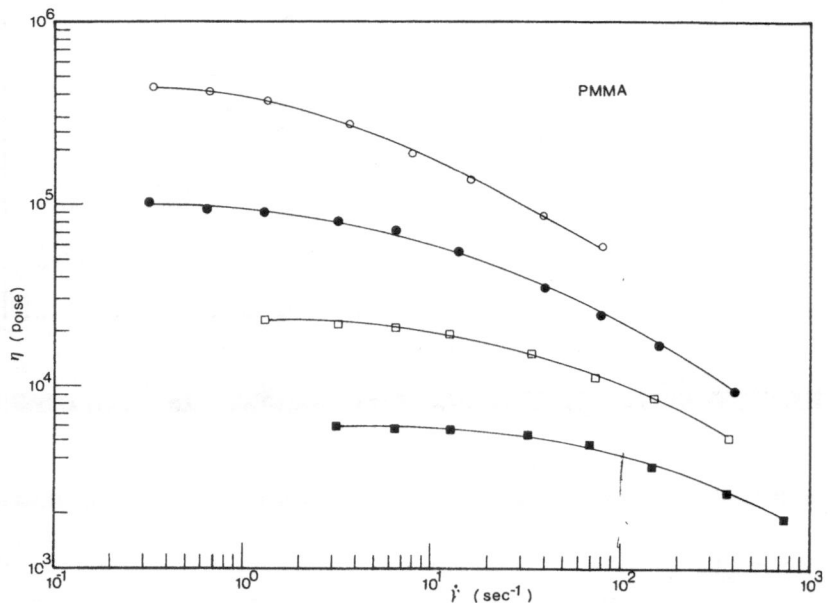

Fig. 2.   Viscosity versus shear rate for polymethyl-
          methacrylate:
          (O) 180°C; (●) 200°C; (□) 220°C; (■) 240°C

phase interfaces in two-phase flow through capillaries.[4,31] Figure 6 shows typical plots of viscosity versus shear stress at 200°C for PS, PMMA and their blends.  An interesting observation may be made from Figure 6.  For

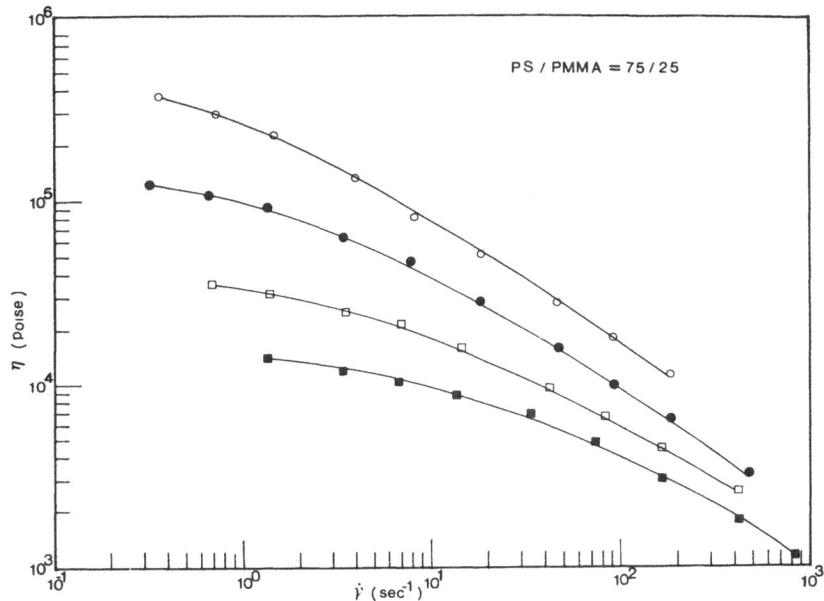

Fig. 3.   Viscosity versus shear rate for the blend
          PS/PMMA = 75/25:
          (O) 180°C; (●) 200°C; (□) 220°C; (■) 240°C

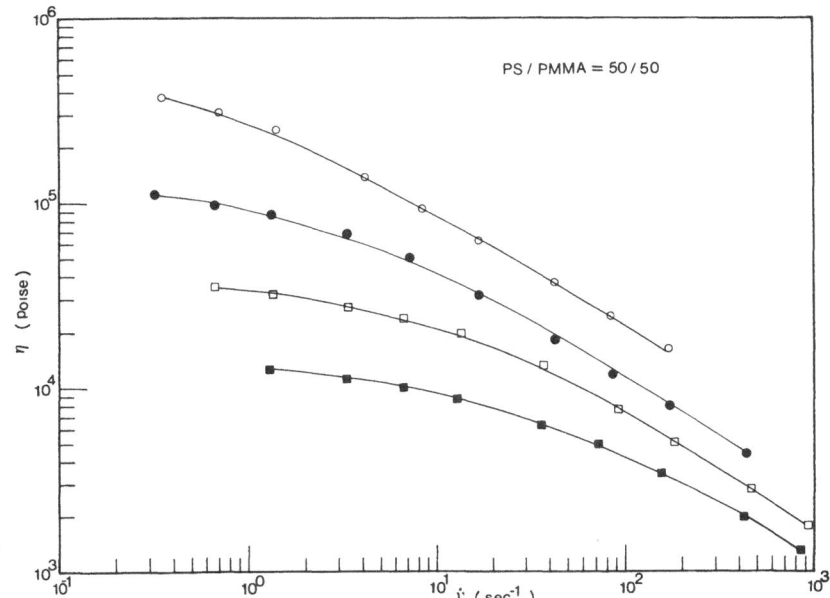

Fig. 4.   Viscosity versus shear rate for the blend
          PS/PMMA = 50/50:
          (O) 180°C; (●) 200°C; (□) 220°C; (■) 240°C

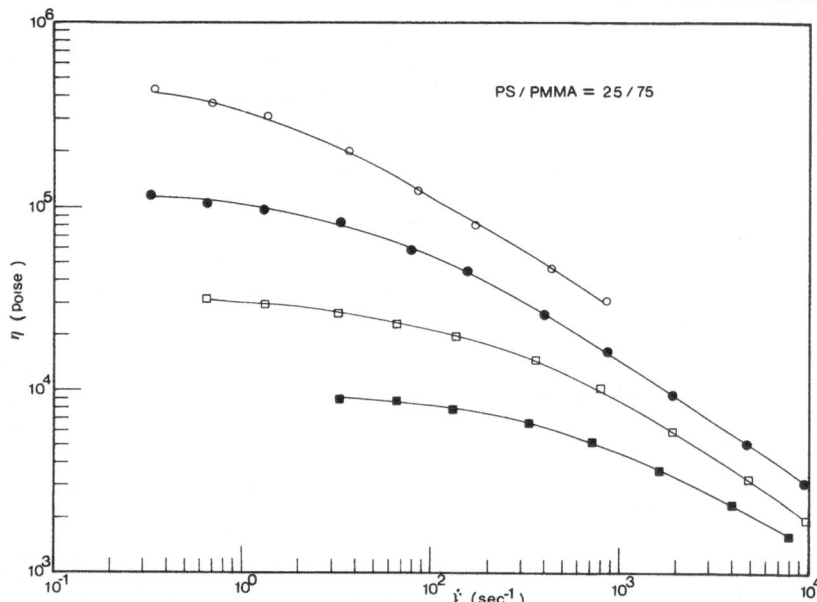

Fig. 5.  Viscosity versus shear rate for the blend
         PS/PMMA = 25/75:
         (O) 180°C;  (●) 200°C;  (□) 220°C;  (■) 240°C

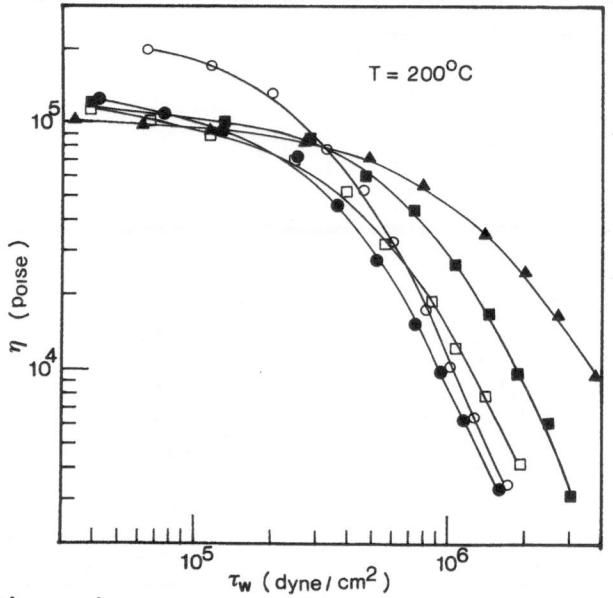

Fig. 6.  Viscosity versus shear stress for PS-PMMA
         blends at 200°C:
         (O) PS; (●) PS/PMMA = 75/25;  (□) PS/PMMA = 50/50;
         (■) PS/PMMA = 25/75;  (▲) PMMA

PS/PMMA = 75/25 blend the melt viscosities are lower than that of the molten PS and PMMA over the entire range of high shear stresses. For the PS/PMMA = 50/50 blend, the viscosities are lower than molten PS and PMMA over a certain range of shear stresses. However, the viscosties of PS/PMMA = 25/75 blend lie in between those of homopolymers over the entire range of shear stresses. Furthermore it is seen in Figure 6 that the viscosity of PS is higher than PMMA at low shear stresses but a viscosity reversal takes place at high shear stresses.

Figure 7 shows the composition dependence of melt viscosity of PS-PMMA blends at 200°C at $\tau = 5 \cdot 10^5 \frac{dyn}{cm^2}$ and $\tau = 1 \cdot 10^6 \frac{dyn}{cm^2}$ respectively. It is seen from Figure 7 that the viscosity goes through a minimum at a blending ratio of about 25 wt % of PMMA. Similar observations of viscosity minimum were reported by some authors[31,34,40] for other types of two-phase polymer blends.

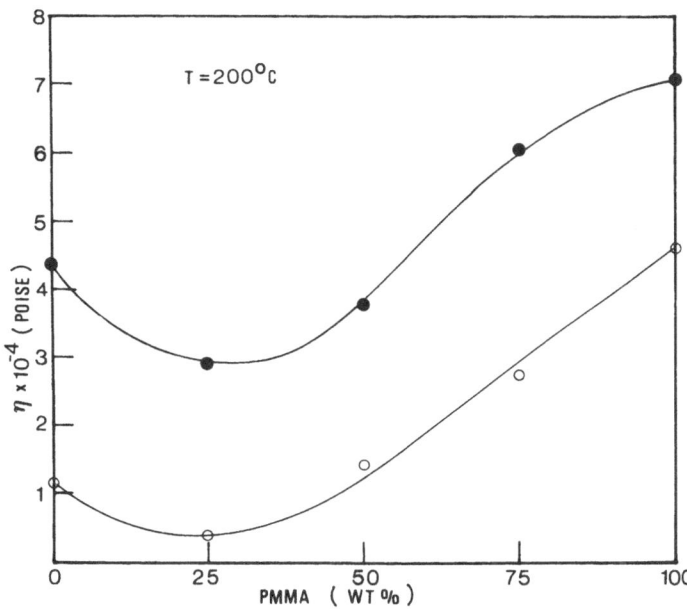

Fig. 7. Composition dependence of the viscosity of PS-PMMA blends at 200°C:
(●) $\tau_w = 5 \cdot 10^5$ dyne/cm²; (O) $\tau_w = 1 \cdot 10^6$ dyne/cm²

Shear Rate-Temperature Superposition

Figure 8 shows the reduced viscosity ($\eta/\eta_o$) versus reduced shear rate ($\dot{\gamma}\eta_o/\rho T$) of the viscosity curves (Figs. 1-4) for PS, PMMA and PS-PMMA blends respectively. It is obvious from Figure 8 that the superposition is achieved for all the three blends within the experimental uncertainty. Furthermore it is seen from Figure 8 that the master curves of pure polymers and the blends of PS and PMMA do not fall onto a single master curve but do fall in between those of parent polymers. Figure 9 shows the reduced plots of ($\eta/\eta_o$) versus ($\eta_o\dot{\gamma} M_c H/\rho RT$) for PS, PMMA and their blends.

It appears from Figure 9 that the master curves of PS, PMMA and PS-PMMA blends fall onto a single master curve. The shifts in the master curves of the blends to the right relative to the single master curve defined by PS and PMMA are not observed in Figure 9. (However, the PS/PMMA = 75/25 blend appears to show such tendency in comparison to other blends). The average molecular weights and critical molecular weights of PS-PMMA blends calculated by using Equations 5-8 are shown in Table 3.

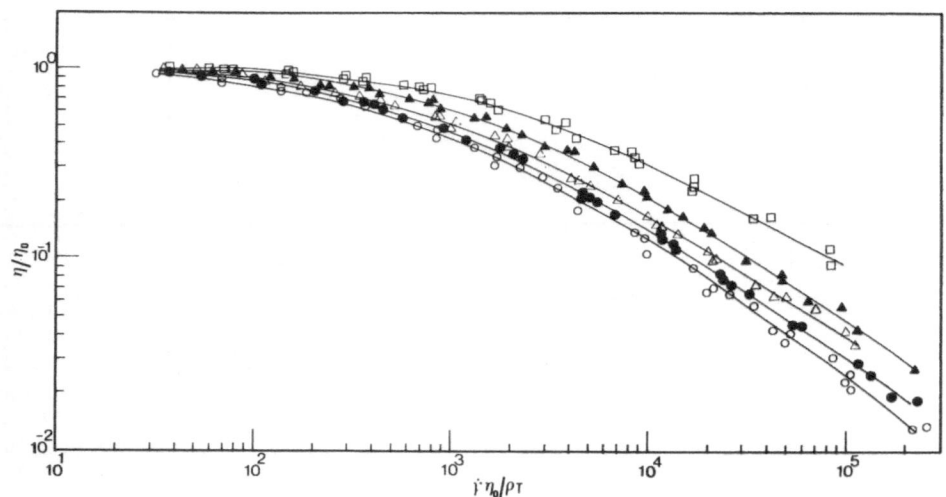

Fig. 8.    Master curves for PS-PMMA blends obtained by plotting the reduced viscosity ($\eta/\eta_o$) versus reduced shear rate ($\dot{\gamma}\eta_o/\rho T$) at temperatures of 180°C, 200°C, 220°C and 240°C respectively: (O) PS; (●) PS/PMMA = 75/25; (△) PS/PMMA = 50/50; (▲) PS/PMMA = 25/75; (□) PMMA

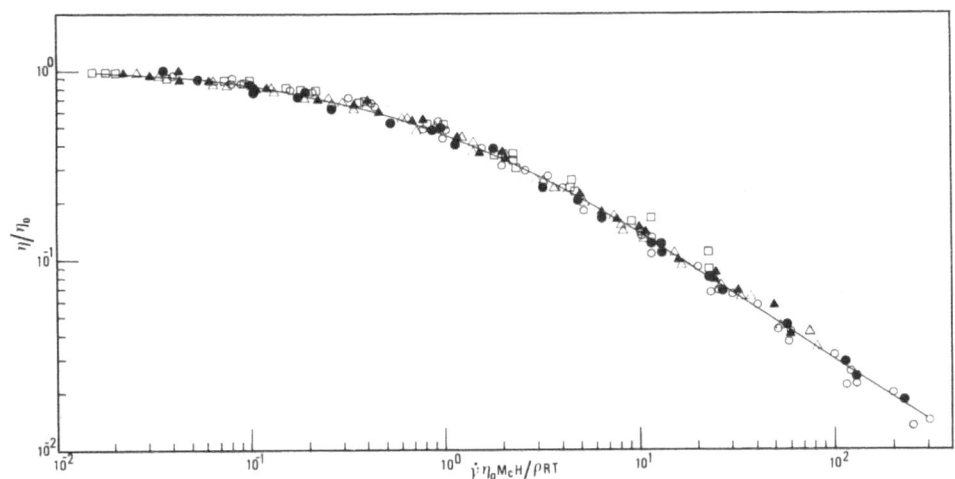

Fig. 9.    Master curves for PS-PMMA blends obtained by plotting the reduced viscosity ($\eta/\eta_o$) versus reduced
shear rate ($\dot{\gamma}\eta_o M_c H/\rho RT$) at temperatures of 180°C,
200°C, 220°C and 240°C respectively: (O) PS;
(●) PS/PMMA = 75/25; (△) PS/PMMA = 50/50; (▲)
PS/PMMA = 25/75; (□) PMMA

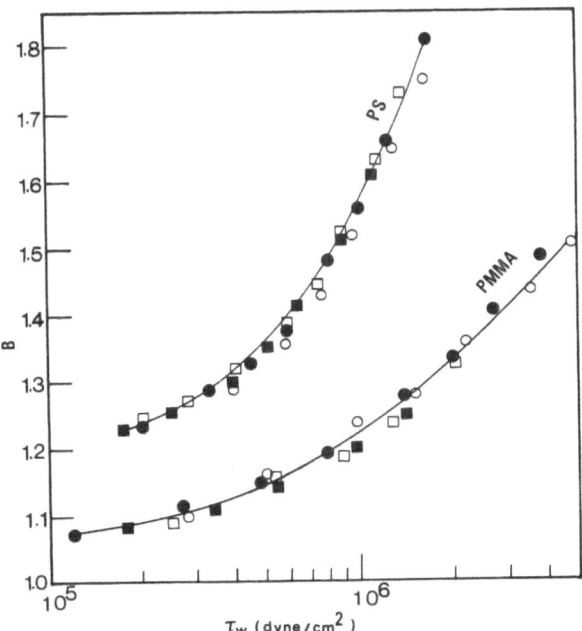

Fig. 10.   Die swell ratio as a function of shear stress
for PS and PMMA: (O) 180°C; (●) 200°C; (□) 220°
C; (■) 240°C

Die Swell Behavior

    In capillary flow, die swell or extrudate swell
ratio (B) is defined as the ratio of (completely relaxed)
extrudate diameter (De) at the extrusion temperature to
the capillary diameter (D)

$$B = \frac{De}{D} \tag{10}$$

    Figure 10 represents the plots of die swell ratio
(B) versus capillary wall shear stress ($\tau_w$) for PS and
PMMA at four different extrusion temperatures.  It is
apparent from Figure 10 that PS exhibits higher die
swell in comparison with PMMA.  Furthermore, it is seen
in Figure 10 that all the die swell data of PS and PMMA
at different temperatures collapse to single curves.

    Figures 11-13 show the plots of die swell ratio (B)
against shear stress at different extrusion temperatures
for PS-PMMA blends.  It is apparent from Figures 10-13

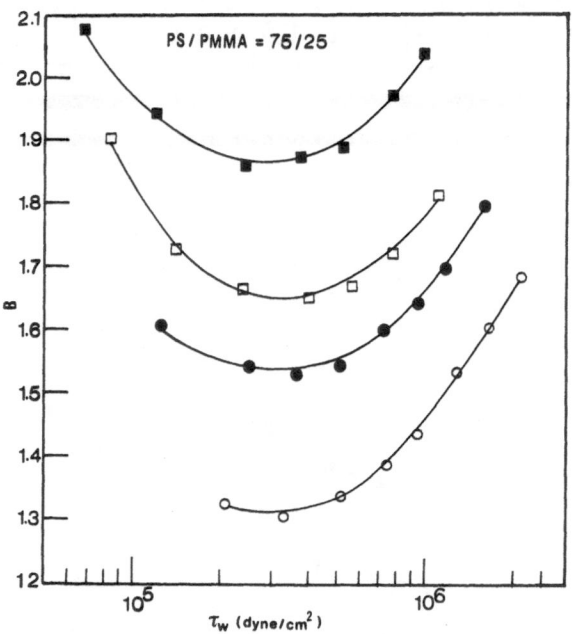

Fig. 11.  Die swell ratio as a function of shear stress
          for the blend PS/PMMA = 75/25: (O) 180°C;
          (●) 200°C; (□) 220°C; (■) 240°C

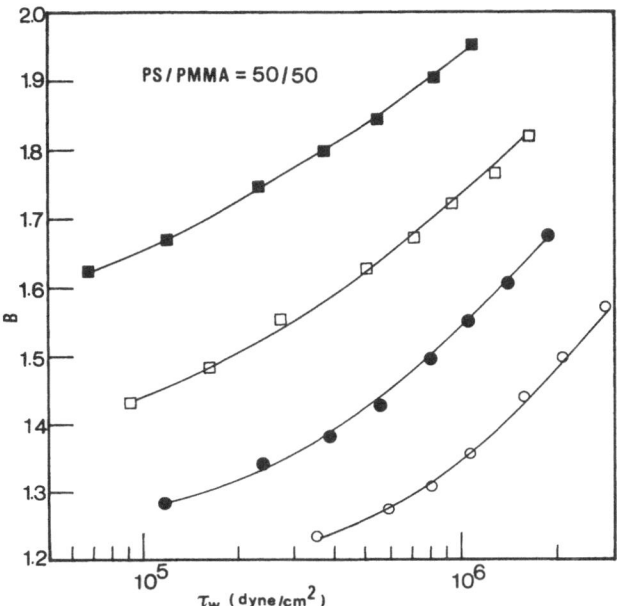

Fig. 12.   Die swell ratio as a function of shear stress
           for the blend PS/PMMA = 50/50: (O) 180°C;
           (●) 200°C; (□) 220°C; (■) 240°C

that the die swell behaviour of the blends is not the
same as for homopolymers.  Die swell data at different
temperatures do not superpose to single curves but increase with temperature at fixed shear stress.   An
interesting observation may be made in Figure 11 for the
PS/PMMA = 75/25 blend system.  A minimum in die swell
ratio is observed around a shear stress of $2 \cdot 5 \cdot 10^5$
dyn/cm$^2$ only for the PS/PMMA = 75/25 blend.

     The composition dependence of melt elasticity for
PS-PMMA blends is shown in Figure 14 at 200°C in terms
of die swell ratio versus blending ratio using a fixed
value of shear stress as a parameter.  It is interesting
to observe from Figure 14 that the die swell ratio goes
through a maximum at a blending ratio of about 25 wt %
of PMMA.  Recently, Han and Coworkers[31,41] have reported
a very similar observation to that shown in Figure 14
for two-phase polymer melts.

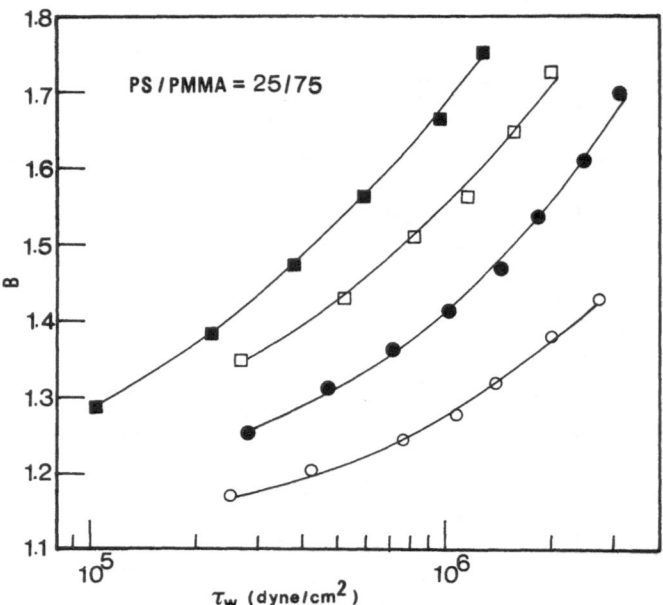

Fig. 13  Die swell ratio as a function of shear stress for
the blend PS/Pmma = 25/75; (O) 180°C; (●) 200°C;
(□) 220°C; (■) 240°C

## Mixing Rule

Figure 15 shows a typical plot of the theoretical
and experimental relationship between the viscosity and
shear stress for PS-PMMA blends at 200°C.  The solid
curves in Figure 15 were predicted according to the mixing rule given by Equation (9), whereas filled symbols
represent the experimentally observed values of viscosity
at different chosen shear stresses.  It is seen from
Figure 15 that the viscosity curves theoretically predicted by Equation (9) coincides closely with the
experimental results for the PS/PMMA = 25/75 blend.  In
the case of PS/PMMA = 75/25 and PS/PMMA = 50/50 blends,
the experimental values deviate from the theoretical
values and are lower than those of theoretical values.
The results suggest the existence of a transition in the
type of morphology depending upon the composition ratio
and relative zero shear viscosity (or Trouton viscosity)
of the components.

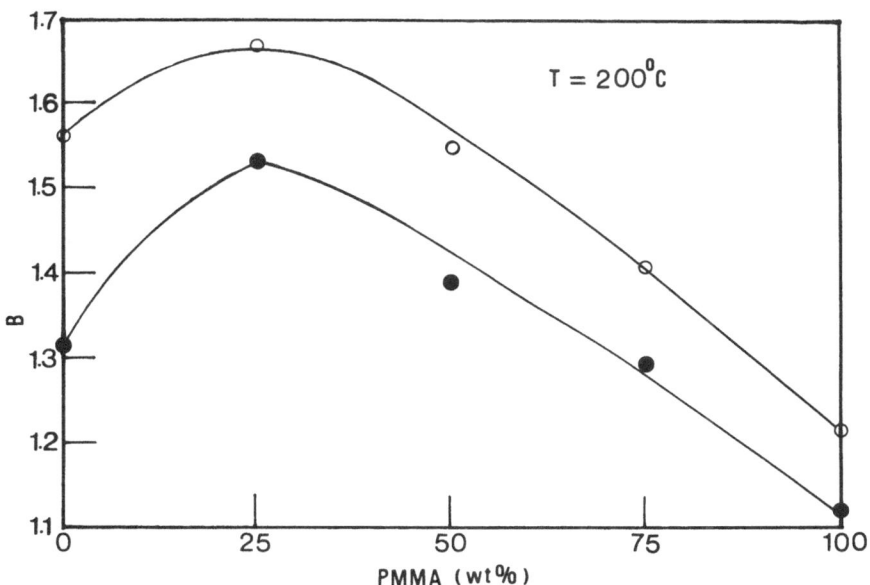

Fig. 14   Die swell ratio versus blending ratio for PS-PMMA
          blends at 200°C: (O) $\tau_w$ = 4 • 10$^5$ dyne/cm$^2$;
          ● $\tau_w$ = 1 • 10$^6$ dyne/cm$^2$

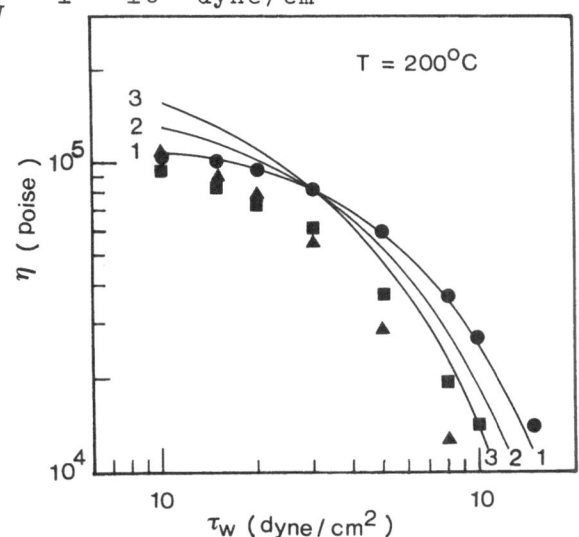

Fig. 15   Comparison of experimental results with the theore-
          tical predictions for the melt viscosity of PS-
          PMMA blends at 200°C: ( ) Predicted from the de-
          rived blending Equation 9 (1-) PS/PMMA = 25/75;
          (2-2) PS/PMMA = 50/50; (3-3) PS/PMMA = 75/25.
          Experimental data: (▲) PS/PMMA = 25/75; (■) PS/
          PMMA = 50/50; (●) PS/PMMA = 75/25

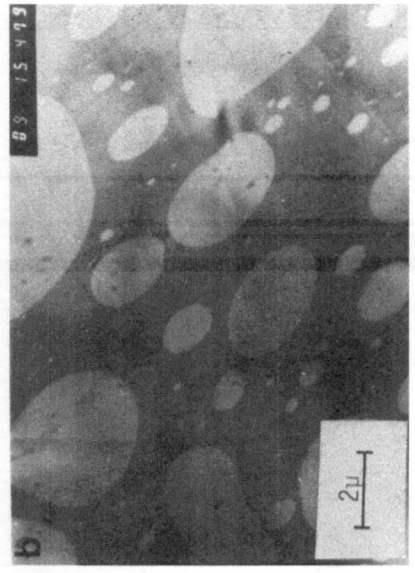

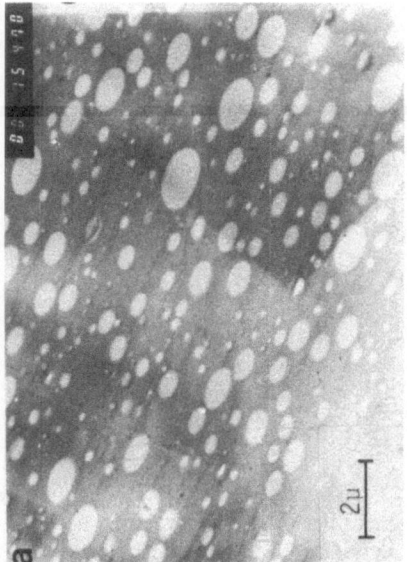

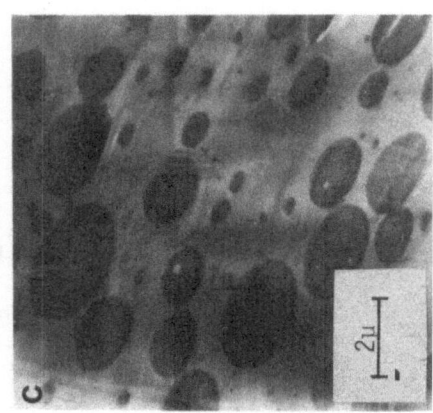

Fig. 16. Electron micrographs of the PS/PMMA blends obtained after melt mixing in Brabender Plasticorder at 200°C: (a) PS/PMMA = 75/25, (b) PS/PMMA = 50/50 (c) PS/PMMA = 25/75

Morphological Characterization

Figure 16 shows the electron micrographs of PS-PMMA blends obtained after melt mixing in the Brabender Plasticorder at $200^{\circ}$C at different blending ratios. In Figure 16 the dark and light regions correspond to PS and PMMA respectively. It is seen from Figure 16 that PS (black) forms the continuous phase and PMMA (white) forms the dispersed phase for PS/PMMA = 75/25 and PS/PMMA = 50/50 blends. Phase inversion has taken place in PS/PMMA = 25/75 blend where PMMA (white) forms the continuous phase with PS (black) as the discrete phase. The scale of phase separation estimated from Figure 16 shows that the size of domains of the PS/PMMA = 75/25, PS/PMMA = 50/50 and PS/PMMA = 25/75 blends are in the range of 0.2-0.6 µm, 0.4-3.5 µm and 0.5-2 µm respectively. Furthermore it is noted from these micrographs that the dispersed domains are nearly spherical and the phase boundaries between PS phase and PMMA phase is relatively sharp.

The morphological structures of extrudate cross sections of three PS-PMMA blend systems without cooling and with cooling of extrudates by a stream of $N_2$ at $6^{\circ}$C are shown in Figures 17 and 18 respectively. These extrudates were obtained at $200^{\circ}$C and at a cross head speed of 0.5 cm/min. ($\dot{\gamma}_{app}$ = 31·38 sec$^{-1}$).

From Figures 17(a) and 18(a) it is apparent for PS/PMMA = 75/25 blend that the crosssections of the dispersed domains after extrusion are smaller than the original structure (Fig. 16a). The domain sizes after the extrusion are in the range of 0.01-0.2 µm and 0.1 µm-0.3 µm for the cooled and uncooled extrudates respectively.

In the case of the PS/PMMA = 50/50 blend system both the state of dispersion as well as the mode of dispersion changes due to the extrusion process as can be seen in Figures 17(b) and 18(b). When compared with the original structure (Fig. 16b), the dispersed PMMA (white) domains form an interconnected structure after the extrusion process and tend to become the continuous phase. For the PS/PMMA = 25/75 blend system it is seen in Figures 17(c) and 18(c) that the sizes of the PS domains (dark) are in the range of 0.3-1.5 µm.

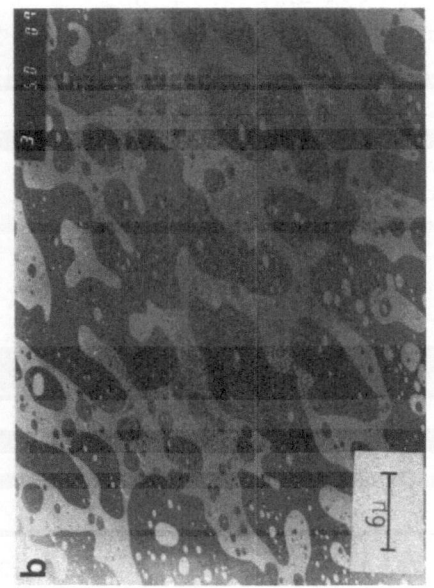

Figure 17. Electron micrographs of extrudate samples cross section of the
PS-PMMA blends (extrusion temperature
$200^{\circ}$C, crosshead speed 0.5 cm/min;
$\dot{\gamma}_{app} = 31.38$ sec$^{-1}$)

(a) PS/PMMA=75/25,
(b) PS/PMMA=50/50,
(c) PS/PMMA=25/75.

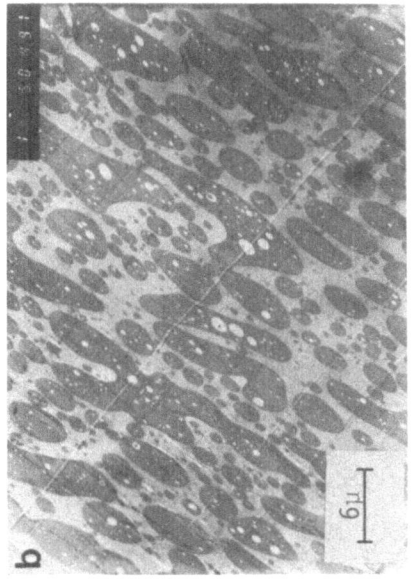

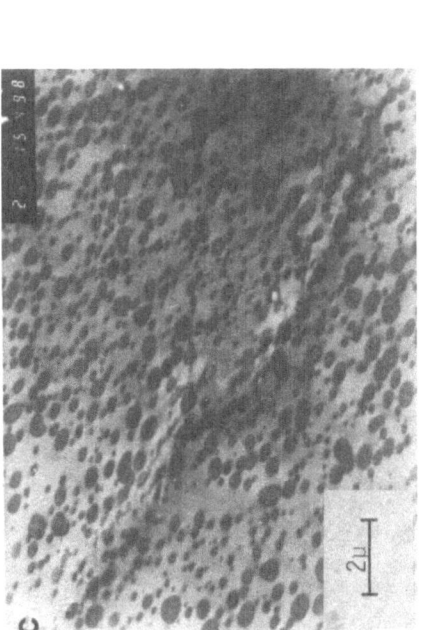

Figure 18. Electron gicrographs of extrudate samples cross section of the PS/PMMA blends obtained by cooling the extrudates with stream of $N_2$ at 6°C (temp. 200°C; cross-head speed = 0.5 cm/min, $\dot{\gamma}_{app}$=31·38 sec).

(a) PS/PMMA=25/25,
(b) PS/PMMA=50/50,
(c) PS/PMMA=25/75.

Since evaluation of the number of elongated domains was experimentally very difficult to obtain, complete relaxation of the fibrillar domains to spherical domains were obtained by annealing in silicone oil at $200^{\circ}C$. Five minutes were found to be sufficient for the PS/PMMA 25/75 blend and 2 min. for the PS/PMMA 75/25 blend. At these temperatures the elongated droplets relax to nearly spherical form by virtue of the interfacial tension.

Fig. 19 shows electron micrographs on the nearly spherical domains(diameter independent of the cutting direction) after complete annealing (PS/PMMA = 25/75).

The average domain volume measured from all pictures available of the blend PS/PMMA = 25/75 is within experimental uncertainty the same before and after extrusion. D before $\tilde{=}$ $1.4_6$ μm, D after = $1.5_3$ μm. The morphology of the 50/50 blend is clearly changed and the average volume of the PS/PMMA 75/25 blend has decreased by extrusion $\frac{\underline{v} \text{ before}}{v \text{ after}} \sim 4$.

PS/PMMA = 25/75 blend and the PS/PMMA = 75/25 have been elongated to ellipsoidal or fibrillar structures with length to diameter ratios more than 3-4 in the capillary. These findings correspond reasonably well to the swelling of the extrudates caused by annealing in the sence that the ratio between the cross section area of the unannealed sample and the annealed sample is roughly proportional to the ratio between the average domain diameter of the unannealed sample to the domain diameter of the almost spherical domains in the fully relaxed sample. Annealing of the homopolymers extruded at the same conditions gave surface tension induced swelling less than 1.05.

Since the average volume ratio $\frac{\overline{v}_{before}}{v_{after}}$ for the blend PS/PMMA = 75/25 is ~4 the number of domains have increased by a factor 4 by extrusion at $200^{\circ}C$ with shear rate: 31.38 $sec^{-1}$.

Unfortunately, annealing and long residence time in the rheometer (reservoir) barrel give rise to the so-called Ostwald ripening, meaning that the smallest domains are unstable and disappear by diffusion through the continuous phase to the larger domains[42]:

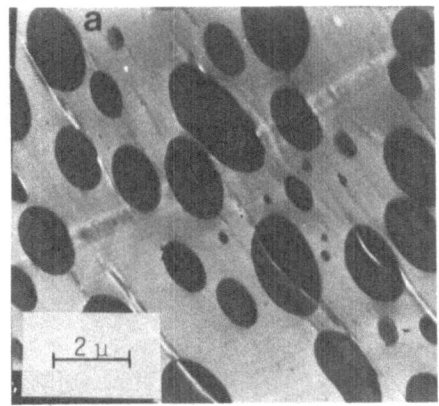 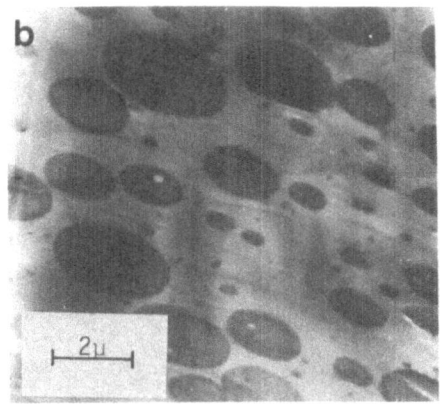

Fig. 19.            Comparison of the original domain struc-
               ture for the PS/PMMA = (25/75) blend with the
               structure (b) obtained after annealing for 5
               minutes at 190°C in silicone oil after extrusion
               ($\dot{\gamma}$ = 31 · 38 sec$^{-1}$).


        This effect was experimentally verified for the
investigated blends.  For the PS/PMMA = 25/75 blend, this
effect was so slow that the conclusion reached above
still holds, whereas the conclusion for the PS/PMMA =
75/25 blend may be questionable.  In fact an evaluation
of many ultramicrotomic cuts both perpendicular and parallel to the flow direction for extrudates of PS/PMMA =
75/25 indicate that samples extruded at 200°C at $\dot{\gamma}_{app}$ =
31.4 sec$^{-1}$ are indeed unstable.  But even though
the number of domains increase approximately 4 times
during flow, the larger domains seem  to have grown (may
be due to diffusion processes in the rheometer reservoir).
The domains in both the PS/PMMA = 25/75 and the PS/PMMA
= 75/25 blend have been elongated to ellipsoidal forms
as seen on fig. 20 showing ultramicrotomic cuts parallel
to the flow direction on cooled samples.

        The conclusion reached on annealed samples is nearly
verified from these investigations except from the observations that near the center axis of the extrudate
many tiny droplets are some times formed and the surface
layers are heavly distorted for the PS/PMMA = 25/75 blend
at 200°C and $\dot{\gamma}$ = 31.4 sec$^{-1}$.

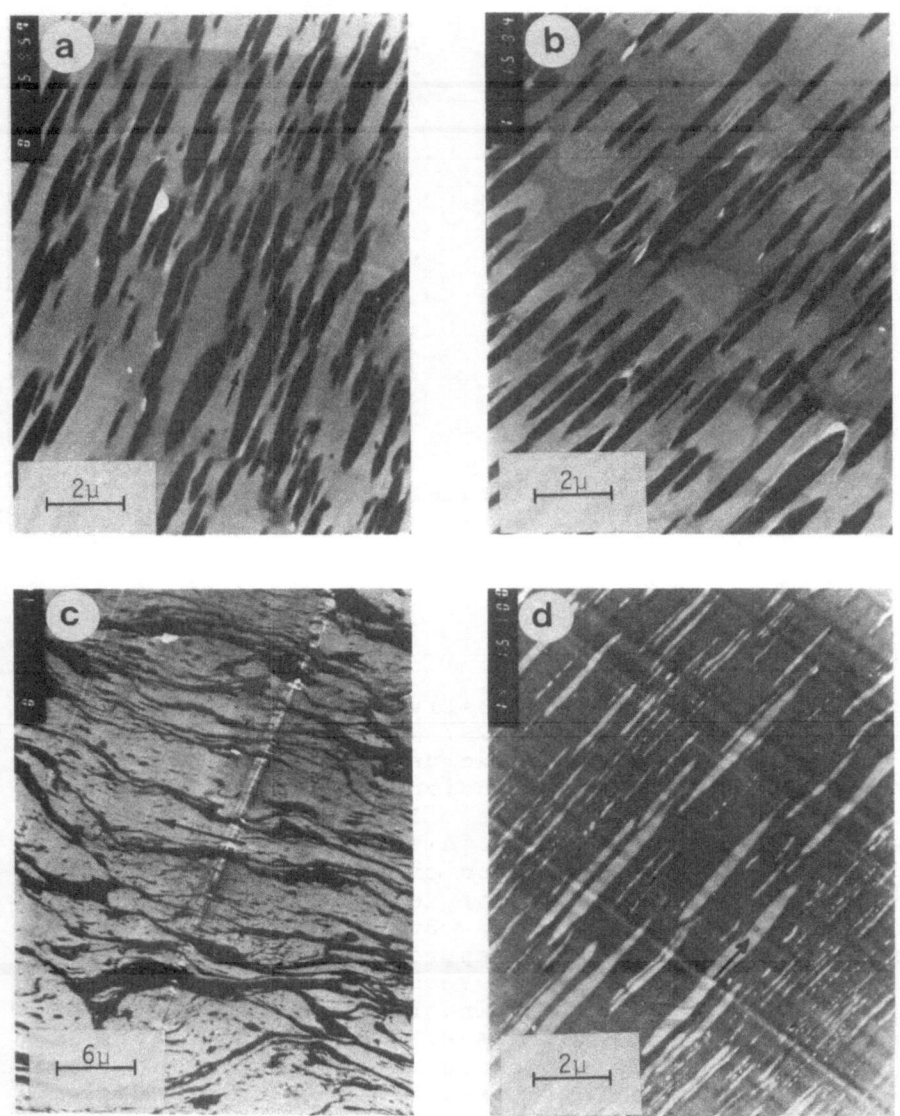

Fig. 20.          Microtomic cut parallel to the flow
direction (arrow) at the center (a), at half radius
distance from center (b), and at the periftri (c) for
PS/PMMA: (25/75) cooled extrudates ( $\dot{\gamma}$ = 31.38 sec$^{-1}$,
200$^{\circ}$C).   Fig. 20c show a cut parallel to the flow direction (arrow) at the center for an extruded sample of
PS/PMMA: (75/25) ( $\dot{\gamma}$ = 31.38 sec$^{-1}$, 200$^{\circ}$C).

## DISCUSSION

In addition to the hazy appearance of presently investigated melt mixed PS-PMMA blends, our electron micrographs confirm that PS-PMMA blends form a two-phase structure with rather sharp phase demarcation (Fig. 16). The nature and fineness of the microstructure of two-phase PS-PMMA blends in terms of mixing history, composition ratios and melt rheology of the components is discussed below.

### Melt Rheology of Blends

The experimentally observed results of melt rheology measurements indicate that the three blends of PS and PMMA follow the same type of flow behaviour as pure polymers at all temperatures (Figs. 1-5). As shown in Figure 6, the melt viscosities of PS/PMMA = 75/25 and PS/PMMA = 50/50 blends are lower than that of the molten PS and PMMA and the viscosity goes through minimum at 25 wt% of PMMA (Fig. 7). The occurrence of viscosity minima in polymer-polymer blends is a feature that is not understood so far.

The shear rate-temperature superposition principle is commonly found for homopolymers which are homogeneous and single phase in nature. The superposition of steady state viscosity data for blends onto master curves (Fig. 8) shows that one cannot distinguish between two phase flow and flow of homogeneous materials by application of the master curve treatment for the investigated PS/PMMA blends.

### Melt Elasticity of Blends

The die swell or extrudate swell encountered in the extrusion of polymeric material may be taken as a characteristic elastic property. As shown in Figure 10 there is no effect of temperature on the extrudate expansion when die swell data are plotted as a function of shear stress. However, there is an effect of temperature on the extrudate expansion of the PS-PMMA blends, where the die swell increases with increasing temperature. A minimum in die swell ratio was observed in a delineation of die swell against shear stress for the PS/PMMA = 75/25 blend (Figs. 11-13). One might assume that the melt elasticity minimum indicated a change in the relaxation process within molten blends due to structural arrangements in capillary flow. The die swell at $200^\circ C$ and

constant shear stress as a function of composition exhibit a maximum as shown on fig. 14.

The observed die swell behaviour is in reasonable
accordance with the expected behaviour of a two phase
extrudate with thin fibrillar domains.

The total die swell of such a system would consist
of a nearly momentary elastic recovery and a superimpo-
sed slower swelling process induced by the interphase
tension which seek to minimize the interphase area.

The isothermal surface induced die swell, $B_s$, of a
homopolymer extrudate is given by

$$B_s = \frac{\alpha}{2\eta_o R_o} \cdot t + 1 \text{ for } \frac{R_o}{L_o} << 1$$

where $\alpha$ is the surface tension, $R_o$ and $L_o$ are the initial
radius and length of an entity, t the    time    of recovery and $\eta_o$ is the sample viscosity.

For a rough analysis of the blend swell behaviour,
this expression is sufficient. For $\alpha$ and $R_o$ constant,
an increasing temperature means simply that the viscosity, $\eta_o$, falls and that the time increases since all effects unchanged the time to cool the sample below the
glass transition temperature increase with increasing
extrudate temperature.

Thus at constant shear stress, normal elastic response, which is independent of temperature, is super-
imposed by an increasing interface induced swell when
the temperature is increased.

The maximum in die swell found at a composition
with 25% PMMA at $200°C$ and constant shear stress may be
caused by the smaller radii of these domains and the
lower domain viscosity.

Domain Stability

At high shear rates, Han and Funatsu[14] concluded
that most authors find that domains are stable if the
ratio of the shear viscosity of the discrete phase to

the viscosity of the continuous phase $\frac{\eta_1}{\eta_2}$ is higher than
approx. 3.

If this prediction were correct, one should expect that polymer blends which show a behaviour where this viscosity ratio at low shear rates are higher than approx. 3 but which show ratios less than one at high shear rates should give stable domains at low shear rates but unstable domains at high shear rates (or shear stresses).

This behaviour was not observed. The mixing rule, Equation 9, was shown to be obeyed at all temperatures for the composition PS/PMMA = 25/75. Fig. 15 shows the results at T = 200$^{\circ}$C. The zero shear viscosities for all blends are shown in table 4.

For the blend PS/PMMA = 25/75 at T = 200$^{\circ}$C, it was shown that the domains remained constant under conditions ($\dot{\gamma}$ = 31.38 sec$^{-1}$) where the shear viscosity ratio was approx. 0.3. The zero shear viscosity ratio for this material was 2.5.

We interpret our results as follows: if the zero shear viscosity ratio $\eta_1/\eta_2$ is larger than approx. 1 the discrete domains are constant during capillary flow; if the zero shear viscosity ratio is smaller than 1, droplet breakup is expected.

This observation seems to be in accordance with the observation of Tsebrenko et al. that long microfibrils can only be obtained for a viscosity ratio $\eta_1/\eta_2$ close to one as discussed in the following text. We rationalize these findings as follows.

Vinogradov and Han have described the deformation of domains at the inlet region of a capillary rheometer. Fig. 21 give a principal sketch of two possible situations.

According to these investigations the break up seemingly takes place in the so-called recoil area near the inlet.

A hypothesis for droplet break up

In order to rationalize the experimental observations, the following oversimplified analysis is performed.

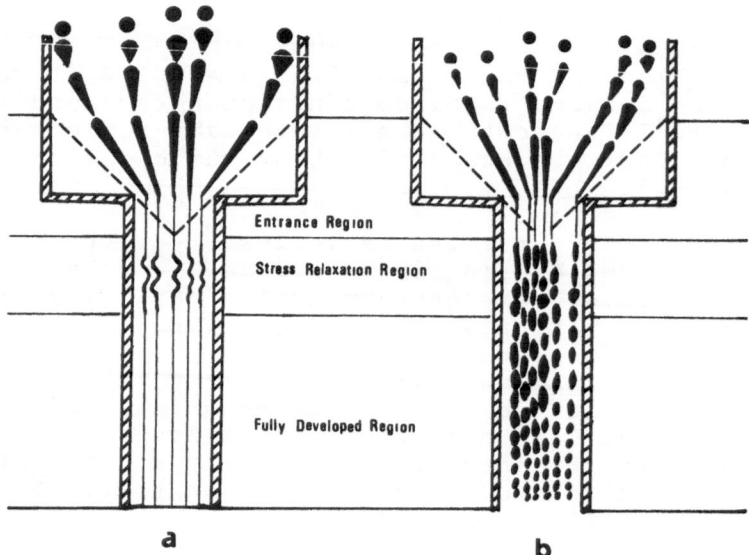

Fig. 21.    Simplified representation of molten two-phase
dispersed polymer blend flowing through a
capillary.   The black portion represents the
deformation of the dispersed droplets of one
polymer in the continuous matrix of other
polymer.

THEORETICAL

A stability criterion for the stability (meaning in this context that a discrete domain remain as a coherent but of course deformed entity) through the relaxation zone is sought.

Taylor's criterion[15] for droplet stability can be stated:

$$\sigma(\frac{1}{r_1} + \frac{1}{r_2}) \geqq \text{constant} + p_i - P_o$$

where $r_1$ and $r_2$ are the principal radii of curvature of a deformed drop, $p_i$ and $p_o$ are the pressures inside and outside the drop.

For a viscoelastic droplet suspended in a viscoelastic medium an equivalent criterion may be stated:

$$\sigma(\frac{1}{r_1} + \frac{1}{r_2}) \geq (N_K - N_D)_{max} \tag{11}$$

where $N_K$ and $N_D$ are the maximum normal stress differences acting outside and inside the domain, respectively.

For a long cylindrical domain the left side is: $\frac{\sigma}{r_1}$ where $r_1$ is the cylinder radius.

The following analysis applies at the capillary inlet and the beginning of the capillary. Furthermore to simplify the analysis, a domain placed at the <u>capillary axis</u> is considered. The shear stress in fully developed capillary flow is therefore zero.

Equation (11) can be written for $\lambda\dot\gamma > 1$ where $\lambda$ is a characteristic relaxation time and $\dot\gamma$ is the shear rate[30].

$$\frac{\sigma}{r_1} \geq K_K(\alpha_{x,K}^2 - \alpha_{z,K}^2) - K_D(\alpha_{x,D}^2 - \alpha_{z,D}^2) \tag{12}$$

for a cylindrical domain where K is a constant and $\alpha_x$ describes the deformation of the polymer molecules in the flow direction while $\alpha_z = \alpha_y$ describes deformations in mutually perpendicular directions, both perpendicular to the flow direction. Subscripts D and K refer to the domain and the continous phase respectively. In the following the case $\lambda\dot\gamma > 1$ for both phases is considered.

The boundary conditions used by Cox[43], continuous velocity and stress components over the interfaces used for the essentially elongational-deformation dominated flow at the entrance into the capillary, requires that the average normal stress difference on both phases at the capillary inlet are balanced by the surface tension and that eq. 12 is followed.

Entering the capillary, the deformed polymer molecules start to relax with a characteristic relaxation time

$$\lambda = \frac{a\eta_o \, M_c \, H\rho}{c^2 RT}$$

where $\eta_o$ is the zero shear viscosity, $M_c$ is twice the molecular weight between entanglements, $H$ is $\bar{M}_w/\bar{M}_n$, $\rho$ is the density, c the polymer concentration, R is the gas constant and T the absolute temperature and a is a dimensionless constant[30].

As a necessary (and sufficient?) criterion for domain stability we chose the criterion:

$$\frac{\sigma}{r_1} + N_D \geq N_K \tag{13}$$

implying that a necessary but not sufficient condition for domain instability is that we have a finite rate of deformation of the domain in the flow direction given by the equation:

$$-\eta_D \dot{\epsilon}_D = \frac{\sigma}{r_1} + N_D - N_K \tag{14}$$

and that $\dot{\epsilon}$ is positive (elongation).

At the capillary inlet we analyze the case

$$\frac{\sigma}{r_{1,o}} + N_D - N_K \cong 0 \tag{15}$$

giving for t the residence time in the capillary:

$$\dot{\epsilon} = o, \quad r_1 = r_{1,0}, \quad N_D = N_{D,0}, \quad N_K = N_{K,0} \quad \text{and}$$

$$\frac{N_{D,0}}{N_{K,0}} = 1 - \frac{\sigma}{r_{1,0} N_{K,0}} \,, \quad 0 \leq \frac{N_{D,0}}{N_{K,0}} \leq 1 \quad \text{for } t = 0$$

In order to investigate whether (13) is satisfied
it is sufficient to investigate whether or not $N_D - N_K$ is
an increasing or decreasing function of t in the neigh-
bourhood of t = 0. An exact simulation of the normal
stress relaxation is not easy since there are too many
unknown variables in the problem. However, the sign
of the variation near t = 0 should be relatively insen-
sitive to the exact form. Assume that

$$N_D = N_{D,0} \cdot \exp\left(-\frac{t}{\lambda_D}\right) \text{ and}$$

$$N_K = N_{K,0} \; \exp\left(-\frac{t}{\lambda_K}\right) \tag{16}$$

The function $F = N_D - N_K$ is an increasing function of
time near t = 0 for

$$\frac{\lambda_D}{\lambda_K} \geq 1 \quad \text{and for} \tag{17a}$$

$$1 - \frac{\sigma}{r_{1,0} N_{K,0}} \leq \frac{\lambda_D}{\lambda_K} \leq \sqrt{1 - \frac{\sigma}{r_{1,0} N_{K,0}}} \tag{17b}$$

which are the conditions where stable domains are predicted. At time t = 0 we may write[30] :

$$N_{K,0} \cong \frac{c^2 RT}{b M_c H \rho}\left(\alpha^2_{X,K,0} - \frac{1}{\alpha_{X,K,0}}\right)$$

where b is a constant a little larger or equal to 1,
which for

$$\alpha_{X,K,0} \gg 1 \text{ gives } N_{K,0} \gtrsim \frac{c^2 RT}{M_c H \rho} \tag{18}$$

For an undiluted system the right side of (18) is
often $\sim 10^6 \frac{dyn}{cm^2}$, $\sigma$ is often around 5 $\frac{dyn}{cm}$ meaning that
only for very thin domains will (17b) deviate from (17a).

If (15) is not followed exactly, an eq. like (14)
must be followed. A material balance over the inlet
zone shows that the average velocity of the domains and

the continuous media must be the same, since accumulation
of one phase is impossible. If the domain viscosity is
higher than the viscosity of the media, the rate of deformation will be less than in the continuous zone. Consequently any relative movement of an interphase in the
flow direction will always be towards the domain phase
(compression) thus stabilizing that phase and vice-versa.

Therefore the stability criterion (17a)
$\frac{\lambda_D}{\lambda_K} \geq 1$ is still valid whereas the stability in the interval

$$1 - \frac{\sigma}{r_{1,0} N_{K,0}} \leq \frac{\lambda_D}{\lambda_K} \leq \sqrt{1 - \frac{\sigma}{r_{1,0} N_{K,0}}}$$

may depend on whether eq. 15 is reasonably followed (or
not) for t = 0.

However, for $\lambda \dot{\gamma} > 1$ and $\frac{\lambda_D}{\lambda_K} > 1$ droplet formation may
eventually lead to a negative change in free energy.

If droplet formation from a nearly cylindrical
domain with radius $r_1$ reduces the elastic free energy
$\Delta F_{el}$ of the domain phase to that of the continuous phase,
one has, in analogy with van Oene's analysis, that the
total change in free energy $\Delta F_T$ by formation of a droplet
with diameter $\sim r_1$ is approximately:

$$\Delta F_T = \Delta F_{el} + \Delta F_{surface\ tension} \simeq -\frac{1}{2}(N_D - N_K) + \sigma \frac{K}{r_1}$$

where K is a constant close to 1. Thus if $N_D - K_K \gtrsim \frac{2}{r_1}$

droplet formation is accompanied by a negative change in
free energy. If $N_D$ and $N_K$ follows eq. 16 a maximum in
$N_D - N_K$ is obtained at

$$t_{max} = \frac{\lambda_D \lambda_K}{\lambda_D - \lambda_K} \ln \left( \frac{\lambda_D \cdot N_{K,0}}{\lambda_K \cdot N_{D,0}} \right) \tag{20}$$

For $N_{K,0} \cong N_{D,0}$,

$$(N_D - N_K)^{max} = N_{D,0} \left\{ \left( \frac{\lambda_K}{\lambda_D} \right)^{\frac{1}{\frac{\lambda_D}{\lambda_K} - 1}} - \left( \frac{\lambda_K}{\lambda_D} \right)^{\frac{1}{1 - \frac{\lambda_K}{\lambda_D}}} \right\} \quad \text{implying that}$$

only for $\dfrac{\lambda_D}{\lambda_K} \approx 1$ can domain stability be expected at high shear rates if "spontaneous" droplet formation plays a part.

An estimation af $r_1$ can easily be obtained in the limiting case that the Trouton viscosities of both phases are constant and $N_D \gg \dfrac{\sigma}{r_1}$. $\eta_D$ and $\eta_K$ are constants. At the inlet to the capillary we have $N_D = N_K \approx \tau = \eta_D \dot{\epsilon}_D = \eta_K \dot{\epsilon}_K$ where $\dot{\epsilon}_D$ and $\dot{\epsilon}_K$ are the rates of elongation of the domain and continuous phase respectively. A material balance over the inlet region to the capillary shows that the ratio between the volumetric flow rates of the two phases are equal to the mixing ratio, implying that the average residence time of the two phases is equal. The total deformation of each phase is

$$\epsilon_D = \int_o^t \dot{\epsilon}_D(t)\,dt = \frac{1}{\eta_D} \int_o^t \tau(t)\ dt \ \text{ and}$$

$$\epsilon_K = \int_o^t \dot{\epsilon}_K(t)\,dt = \frac{1}{\eta_K} \int_o^t \tau(t)\,dt$$

and $\dfrac{\epsilon_D}{\epsilon_K} = \dfrac{\eta_K}{\eta_D}$

For an undisturbed domain with original radius R,

$\dfrac{R}{r_1} \approx \dfrac{D_r}{D_c}$, if the domains have the same rheological properties as the suspending medium. $D_r$ is the reservoir diameter and $D_c$ is the capillary diameter.

Thus

$$\frac{\ln\left(\dfrac{R}{r_1}\right)}{\ln\left(\dfrac{D_r}{D_c}\right)} \approx \frac{\eta_K}{\eta_D}$$

or

$$r_1 \cong \left(\frac{D_c}{D_r}\right)^{\frac{\eta_K}{\eta_D}} \cdot R \qquad\qquad (21)$$

The hypothesis to be tested is that: $\frac{\lambda_D}{\lambda_K} \gtrsim 1$ (17a) will give stable domains.

Tsebrenko, Rezanova and Vinogradov's[17] observation that the optimum conditions for production of very long microfibrils is $\eta_D = \eta_K$ is in reasonable accordance with eq. 17 a and eq. 21.

In reference 12 it was shown that eq. 12 is follo-wed for $\frac{\eta_{0,D}}{\eta_{0,K}} \approx \frac{\lambda_D}{\lambda} > 1$ for HDPE/PP blends.

It is shown above that for PS/PMMA blends, criterion (8) and (9) are followed when $\frac{\lambda_D}{\lambda_K} \gtrsim 1$ also for systems where the ratio between the shear viscosities $\frac{\eta_D}{\eta_K}$ is higher than one, at low shear rates but less than one at high shear rates.


CONCLUSIONS

From the results of this study, the following con-clusions may be drawn for initially well mixed two-phase PS-PMMA blends obtained at the same processing history:

1.  Master curves with reduced viscosity ($\eta/\eta_0$) against reduced shear rate ($\eta\dot{\gamma}/\rho T$) are obeyed at fixed composition.

2.  All data fall on a single master curve when reduced viscosity ($\eta/\eta_0$) is plotted against ($\eta_0 \dot{\gamma} M_c H/\rho RT$). It was thus not possible to distinguish the flow behaviour of this blend from the expected behaviour of a homogeneous melt with the same com-position.

3.  The mixing rule: $\dfrac{1}{n_B(\tau_w)} = \dfrac{v_1}{n_B(\tau_w)} + \dfrac{v_2}{n_1(\tau_w)}$ eq. 9 is

    followed for blends where the ratio between the
    zero shear viscosity of the domain phase to the

    continuous phase $\dfrac{n_{1,o}}{n_{2,0}}$ is larger than one. This is

    independent of the actual viscosity ratio at
    higher shear stresses.

4.  Apart from the material in the surface layer and at
    the center of the extrudates, the number of domains
    are constant, meaning that the domains are stable

    when the viscosity ratio $\dfrac{n_{1,0}}{n_{2,0}}$ is larger

    than one - in our interpretation when the ratio
    between the relaxation times of the domain material
    to the relaxation time of the continuum is larger
    than one. All domains seems to be stable at low
    flow rates.

5.  Discrete spherical domains are deformed in the flow
    direction to fibrillar or elongated droplets.

6.  The die swell behaviour is determined by the elastic
    recovery of the deformed polymer molecules and the
    interfacial tension induced retraction of the fi-
    brillar domains.

7.  In the molten state, during the capillary flow, the
    dispersed droplets of the lower Newtonian or Trouton
    viscosity minor component break up into smaller
    elongated droplets. On the other hand the disper-
    sed droplets of the higher Newtonion or Trouton
    viscosity minor component appear to form a morpho-
    logy with continuous threads or elongated droplets.

ACKNOWLEDGEMENTS

    The authors wish to express their gratitude to
Danish Council for Scientific and Industrial Research
for financial support of the project.

    This work is partly based on the Ph.D. (lic. techn.)
thesis of Narasaiah Alle, submitted to the Faculty of
Chemical Engineering, Technical University of Denmark,
March 1980.

REFERENCES

1.  M. Shen and H. Kawai, AIChE J., 24, 1(1978).
2.  H. Van Oene, J. Colloid. Interfac Sci., 40, 448
    (1972).
3.  J.M. Starita, Trans. Soc. Rheol., 16, 339(1972).
4.  C.D.Han, Trans. Soc. Rheol., 19, 245(1975).
5.  G.N. Argeropoulos, F.C. Weissert, P.H. Biddison and
    G.G.A. Böhm, Rubber Chem. Technol., 49, 93(1976).
6.  A.P. Plochocki, Trans. Soc. Rheol., 20, 287(1976).
7.  C.J. Nelson, G.N. Argeropoulos, F.C. Weissert and
    G.G.A. Böhm, Angew. Makromol. Chem., 60/61, 49
    (1977).
8.  S. Danesi and R.S. Porter, Polymer, 19, 448(1978).
9.  J. Lyngaae-Jørgensen and N. Alle, Polym. Prep., 19,
    103(1978).
10. J. Lyngaae-Jørgensen, N. Alle and F.L. Marten, Adv.
    Chem. Series, 176, 541(1979).
11. N. Alle and J. Lyngaae-Jørgensen, Rheol.Acta., 19,
    94(1980).
12. N. Alle and J. Lyngaae-Jørgensen, Rheol. Acta., 19,
    104(1980).
13. N. Alle and J. Lyngaae-Jørgensen, Proc. VIII Int.
    Congr. Rheol., Naples, II, 521(1980).
14. C.D. Han and K. Funatsu, J. Rheol., 22, 113(1978).
15. G.F. Taylor, Proc. Roy. Soc., A 138, 41, (1932).
16. H.B. Chin and C.D. Han, J. Rheol., 23, 557, (1979)
    and J. Rheol., 24, 1, (1980).
17. M.V. Tsebrenko, N.M. Rezanova and G.V. Vinogradov,
    Polym. Eng. & Sci., 20, 1023(1980).
18. N. Alle, Ph.D. (lic.techn) Thesis, The Technical
    University of Denmark, Lyngby, Denmark 1980.
19. G.V. Vinogradov and A.Y. Mallin, J.Polym.Sci.,
    A-2, 2357(1964).
20. G.V. Vinogradov and N.V. Prozorovskaya, Rheol. Acta.,
    3, 156(1964).
21. G.V. Vinogradov and A.Y. Malkin, J. Polym. Sci.,
    A-2, 4, 135(1966).
22. R.C. Penwell, W.W. Graessley and A. Kovacs, J. Polym.
    Sci., Polym. Phys. Ed. 12, 1771(1974).
23. F. Bueche, J. Chem. Phys., 22, 603(1954).
24. W.W. Graessley, J. Chem. Phys. 43, 2696(1965).
25. S. Middleman, "The Flow of High Polymers", Inter-
    science, New York (1968).
26. J.F. Dunleavy and S. Middleman, Trans. Soc. Rheol.
    10, 157(1966).
27. D. Robinson, M.S. Thesis, University of Rochester,
    Rochester, New York (1967).
28. S. Middleman, J. Appl. Polym. Sci., 11, 412(1967).

29.  W.W. Graessley and L. Segal, AIChE J., 16, 261(1970).
30.  J. Lyngaae-Jørgensen, "Relations Between Structure
        and Rheological Properties of Concentrated Poly-
        mer Melts", Instituttet for Kemiindustri, The
        Technical University of Denmark, Lyngby, Denmark
        (1980).
31.  C.D. Han, "Polymer Processing", Academic Press,
        New York (1976).
32.  L.E. Nielsen, "Polymer Rheology", Marcel Dekker,
        New York (1977).
33.  G.V. Vinogradov, M. Yokob, M.V. Tsebrenko and A.V.
        Yadim, Int. J. Polym. Metals 3, 99(1974).
34.  A.P. Plochocki, "Polymer Blends", Vol. 2, p. 319,
        (D.R. Pauland S. Newman ed.) Academic Press, New
        York (1978).
35.  B. Rabinowitsch, Z. Phys. Chem., A145, 1(1929).
36.  I.M. Krieger and S.H. Maron, J. Appl. Phys., 23,
        147(1952).
37.  E.B. Bagley, J. Appl. Phys, 28, 624(1957).
38.  E.B. Bagley, Trans. Soc. Rheol., 5, 355(1961).
39.  J.D. Ferry, J. Am. Chem. Soc., 6, 1330(1942).
40.  H. Van Oene, "Polymer Blends", Vol. 1, p. 295,
        (D.R. Paul and S. Newman ed.) Academic Press,
        New York (1978).
41.  C.D. Han And T.C. Yu, AIChE J., 17, 1512(1971).
42.  L.P. McMaster, Adv. Chem. Series, 142, 143, (1975).
43.  R.G. Cox, J. Fluid. Mech., 37, 601, (1969).

# MELT COMPOUNDING OF PVC WITH ETHYLENE COPOLYMER RESINS

George H. Hofmann

E. I. Du Pont De Nemours & Company
Polymer Products Department
Wilmington, Delaware 19898

## INTRODUCTION

Ethylene Copolymer Resin (ECR) modifiers*, designed to be soluble in all proportions in PVC, form a wide variety of plasticized PVC blends. These solid, high molecular weight (Mw >250,000) resin modifiers, unlike conventional liquid plasticizers, do not migrate in PVC. Homogeneous blends of ECR and PVC are true polymer alloys that exhibit:

- Low extraction in soapy water, hexane and mineral oil
- Low volatility, migration and spew
- Outstanding low temperature impact resistance
- Excellent resistance to microbiological attack.

In order to take full advantage of these beneficial properties,[1,2] it is essential that the PVC/ECR blends are made homogeneous through effective melt compounding.

Conventional liquid plasticizer is completely absorbed by the PVC powder, in the dry-blending step, prior to melt compounding.[3-5] With the solid ECR pellets this cannot occur. A salt and pepper blend of pellets in PVC powder, therefore, is prepared prior to compounding.

During the melt compounding step, the low melting ECR (m.p. 66°C) melts first forming a relatively low viscosity phase in which the higher melting (m.p. >170°C) and more viscous PVC

---
*"Elvaloy" resin modifiers.

a                                   b

Fig. 1.   Unfluxed PVC gel particle in heterogeneous blend.
(a) scanning electron micrograph showing 0.12 mm gel
particle (magnified 200 times). (b) energy dispersive
X-ray scan showing higher Cl content, in the same gel
particle, relative to the surrounding matrix.

powder grains are suspended.  Under these conditions, it is
difficult to get enough shear energy into the system to
completely break down and disperse the PVC grains.  As a result,
many PVC grains can pass through the melt compounding step
intact.  These unfluxed PVC grains, referred to as gel, have been
positively identified as the heterogeneities in PVC/ECR blends
that can produce rough, pimpled extrudates with reduced physical
properties (Fig. 1).

This paper discusses means of reducing or eliminating gel in
PVC/ECR blends.  The morphology and rheology of PVC/ECR
compounding are considered which lead to conditions for
maximizing homogeneity (minimizing PVC gel).  Laboratory and
commercial scale, both batch and continuous melt compounding
equipment, were investigated.


EXPERIMENTAL

The Brabender Plasti-Corder (Model Pl-2, type 6 roller
mixer) was used as a simple, convenient method for scouting a
large number of formulation and processing variables.  Scale-up
was conducted on a Farrel Continuous Mixer (Size 4).

Compounding effectiveness (degree of blend homogeneity) was
quantified by counting the PVC gel in a 7 mil (0.18 mm)
compression molded film.  The gel count was the number of gels
per square cm obtained by magnifying 18 times with a microfiche
reader.  This proved to be a very sensitive test for gel in that
it was capable of detecting gel particles as small as 0.05 mm
diameter.  A typical sample had gel of all sizes ranging up to
0.2 mm diameter.

Our goal was to develop techniques for melt compounding
PVC/ECR into blends equivalent to PVC/DOP blends in homogeneity.
A gel count of 50 or less indicated a blend homogeneity suitable
for most applications (acceptable surface quality and physical
properties).  Blends plasticized with DOP (only) fell into this
category under most melt-compounding conditions.  The
formulations evaluated are found in Table 1.


DISCUSSION

Morphology of Melt Compounding

Unplasticized PVC.  Previous workers ,[6,7] characterized
unplasticized PVC melt compounding morphology by interpretation
of characteristic Brabender plastograms (Fig. 2).  Curves of

Table 1.  Formulations Used in Melt Compounding
          Study (PHR)

|                    | I      | II  | III | IV  |
|--------------------|--------|-----|-----|-----|
| PVC[1]             | 100    | 100 | 100 | 100 |
| ECR[2]             | VARIED | 100 | 70  | 70  |
| LIQUID PLASTICIZER | –      | 30  | –   | 30  |
| $CaCO_3$           | 25     | 25  | 25  | 25  |
| Ba/Cd SOAP         | 2      | 2   | 3   | 3   |
| PHOSPHITE CHELATOR | 1      | 1   | 1   | 1   |
| EPOXY SOYA OIL     | 10     | 10  | 10  | 10  |
| WAX                | –      | –   | 4   | 4   |

(1)  K VALUE OF 67 UNLESS NOTED OTHERWISE
(2)  ELVALOY® 741 RESIN MODIFIER

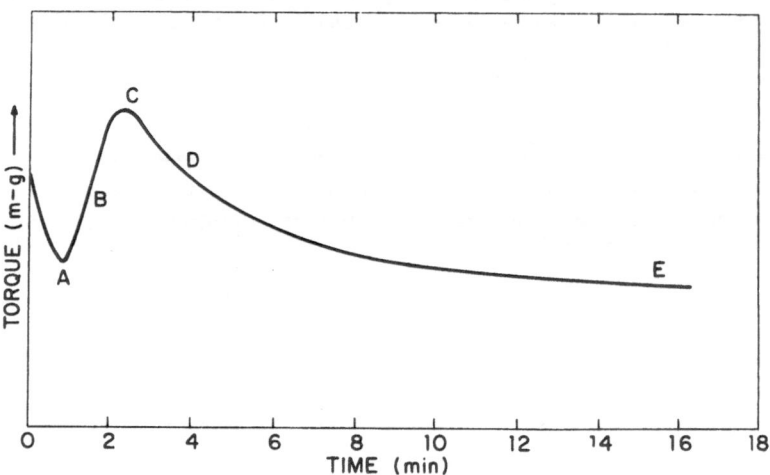

Figure 2.   Brabender Plastogram - PVC

essentially the same shape are produced by PVC/ECR blends and should, therefore, be instructive in the interpretation of PVC/ECR melt compounding morphology. Photomicrographs of the unplasticized samples, taken at points corresponding to the letters on the curve, showed the PVC morphology changes during the melting process. They showed that as the torque approaches minimum A, the 150 μm PVC grains (shown in Fig. 3a) are torn apart and most of them are broken into 1 μm primary particles. The temperature is low at this point in the process and little fusion interaction occurs between the primary particles as indicated by the low torque value. As the temperature increases, due to shear heating, the primary particles begin to melt and fuse together increasing the torque (B). At the maximum torque, C (fusion point), most of the primary particles are fused together, but are still distinguishable. After the fusion point (C) the melt temperature increases significantly (Fig. 4), causing a reduction in the melt viscosity and torque (D). At this point the particulate structure starts to disappear. The temperature continues to rise and the viscosity decreases until they level out to relatively constant values. At this equilibrium point (E) all primary particles have disappeared and a continuous melt is formed completing the process.

Liquid Plasticized PVC. Similar studies on DOP plasticized PVC [8] show that the same melt flow behavior pattern occurs. The only differences are a reduction in the temperature and torque at which these morphological transitions occur. The liquid plasticizer, which produces these improvements in processing characteristics, was shown to function by being absorbed into the 1 μm primary particles in the dry blending step, prior to melt compounding, causing a reduction in Tg and viscosity. [3,4]

ECR Plasticized PVC. Such an absorption mechanism is impossible with a solid high MW plasticizer such as ECR. When the PVC powder and ECR pellets are added to the melt compounding equipment, they are in fact mixed, but still two separate phases. Early in the melt compounding cycle before minimum torque A is reached, two processes are occurring. The 150 μm PVC grains are eroding into unmelted 1 μm primary particles and at the same time the ECR pellets are melting into a low viscosity melt phase. The low viscosity ECR melt encapsulates many of the 150 μm grains before they are eroded into primary particles. The primary particles that are formed, however, blend with the ECR and begin the fusion process (B) as a well dispersed phase. At the fusion point (C), however, many of the encapsulated 150 μm grains still remain intact. During further heating and melting (D-E) the dispersed 1 μm grains disappear forming a continuous PVC/ECR melt.

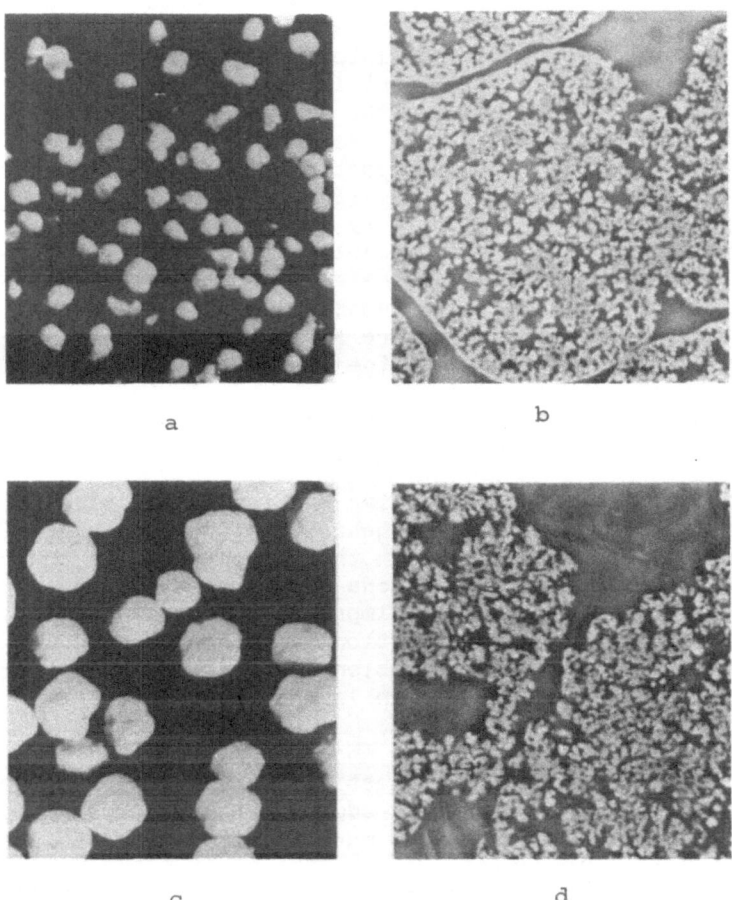

Fig. 3.  Photomicrographs of PVC resins[14].  (a) and (b) standard
         particle size (28X) and (415X), respectively; (c) and
         (d) large particle size (28X) and (415X), respectively.

A large number (up to 50 volume percent) of the 150 μm grains (or portions thereof) can remain, however, as discrete large particles in the blend.

The homogenization of this remaining high viscosity, but still thermoplastic, gel phase into the much lower viscosity PVC/ECR phase was treated as a problem analogous to difficult mixing problems described by Irving and Saxton.[9] This treatment allows the prediction of mixing behavior from rheological data.

## Rheology of Melt Compounding

High ECR Levels. A rheological description for rigid (unplasticized) PVC[10] was taken as an approximation for the rheological description of the thermoplastic PVC gel in heterogeneous PVC/ECR blends. These data are plotted in Figure 5 along with the rheology for the PVC/ECR continuous phase containing 100 phr ECR (ignoring effect of minor gel phase). The shear rate (37 $sec^{-1}$) on the Brabender (32 RPM) was calculated using the method reported by Goodrich and Porter.[11] At this shear rate a shear stress of approximately 40 kPa is imparted to the major component (PVC/ECR). Because shear stress is the same in both the major and minor (PVC gel) components, the shear stress isobar is followed until it intersects the line describing the minor component. As seen in Figure 5, the shear stress isobar intersects the PVC gel curve at a very low shear rate indicating that only a low rate of mixing will occur. This in fact was the case as seen in Table 2 and Fig. 6. A very high gel count remained even after 20 minutes of mixing at 185°C. As the temperature settings were lowered, however, the gel level decreased accordingly. This is the result of the increasing shear stresses generated by the increasingly viscous major component as its melt temperature decreases.

A further increase in shear stress was obtained by increasing the Brabender rotor speed to 64 RPM (shear rate of 73 $sec^{-1}$). As can be seen in Fig. 6, a further reduction in gel count was obtained but a goal of less than 50 was not achieved. Temperature settings had less influence on gel count at this higher RPM due to the higher level of shear heating. The shear heating at the 185°C setting led to degradation pointing out the desirability of achieving high shear stresses without generating excessively high melt temperature.

Shear rates in the 35–75 $sec^{-1}$ range of these experiments are representative of relatively low intensity mixing devices such as roll mills and single screw extruders. Even at the fusion point, relatively low (<75 kPa) shear stresses were developed (with high ECR levels) making high homogeneity difficult to achieve (Figure 7).

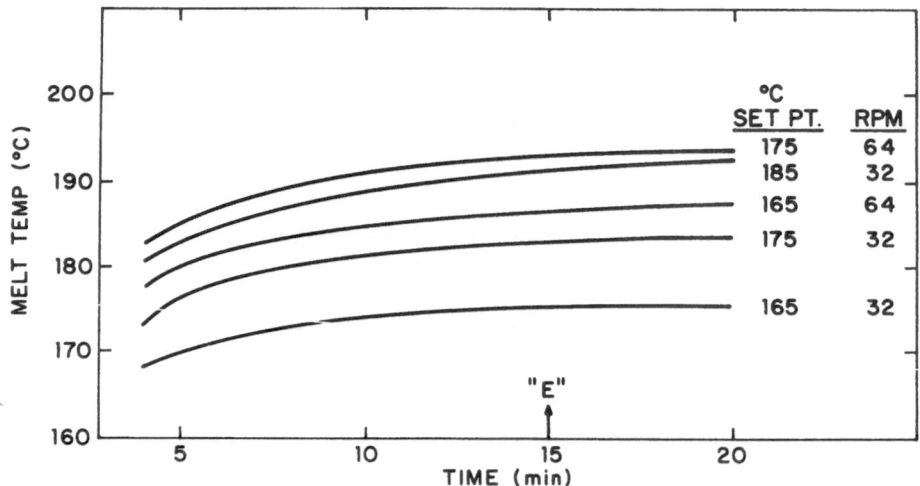

Figure 4.    Brabender melt temp. vs. time.

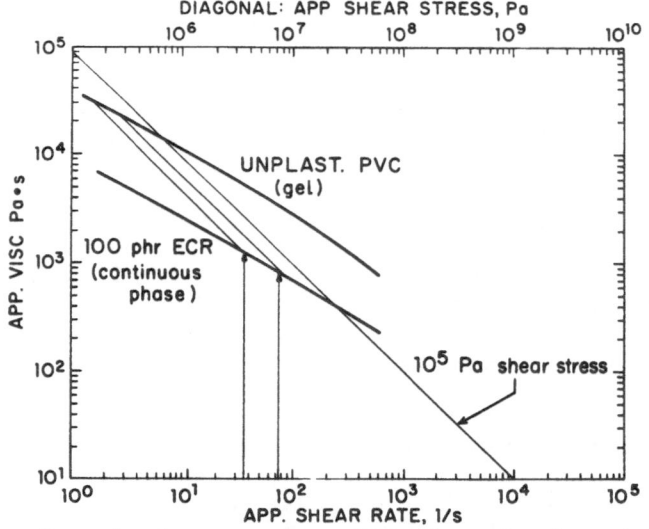

Figure 5.    Rheology PVC/ECR (190°C) Brabender mixing.

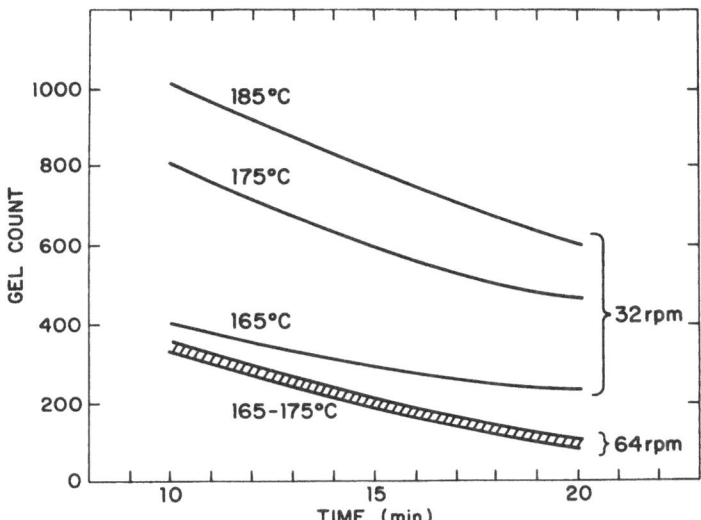

Figure 6.  Homogeneity vs. time, Brabender.

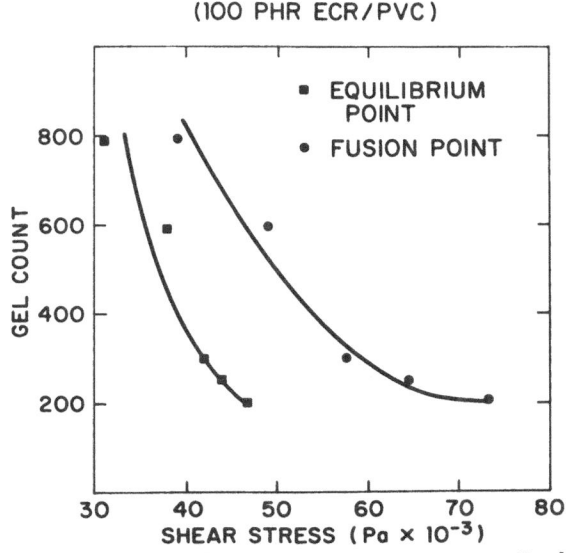

Figure 7.  Homogeneity vs. shear stress, Brabender.

Reduced ECR Levels. Homogeneous blends can be obtained on these devices at moderate to low ECR levels. Reduction in ECR level from 100 phr to 40 phr (Table 2) essentially eliminates gel due to the higher viscosity and shear stresses during the fusion and melting processes. The relative viscosities of the components and blends may be found in Fig. 8. As viscosity of the continuous phase approaches that of the unplasticized PVC gel phase, optimum homogeneity is approached.

## Sequential Addition

A technique found useful for minimizing the effect of encapsulating PVC grains with ECR at the beginning of the mixing cycle was to initially withhold a major portion of the ECR. As previously noted, the blend containing 40 phr ECR has a much higher initial viscosity and fusion torque than blends with 70 or 100 phr ECR (Table 2). This higher torque effectively breaks down the 150 $\mu$m grains. As seen in Table 3, after 10 minutes, the gel count was 10. At this time the additional ECR was added to produce a homogeneous blend, with slight additional mixing. This technique is very effective for batch compounding of PVC with high levels of ECR and is currently in use with Banbury mixers. [12]

## Optimum PVC

Table 4 summarizes data obtained with various types of PVC. Emulsion polymerized PVC, as expected, produces gel-free blends because the 150 $\mu$m grains, produced by suspension and bulk polymerization, are not present. The more economical suspension and bulk polymerized resins, however, show a wide variation in compounding performance with ECR. The standard suspension grades (C thru F) result in a variable but high gel level. Increasing MW correlates with increasing fusion times and torque but gel levels are unaffected. Bulk polymerized PVC, having essentially the same size grains but free of the peri-cellular membrane coating [5] of standard suspension grades, gave very short fusion times. This did not indicate, however, a more friable particle, since similarly high gel counts were obtained.

The best choice of resin, as indicated by the very low gel counts, is the large particle grade of suspension PVC. This material is polymerized under conditions that produce an average powder grain size 2-3 times larger than standard PVC (Fig. 9). Additionally, these grains are less dense, having a more open porous structure and do not have a surrounding membrane (Fig. 3b). This produces a more friable, shear sensitive powder that readily erodes into primary 1 $\mu$m particles early in the melt compounding process despite the presence of ECR. The fusion

Table 2. Compounding of ECR with Standard Suspension PVC[1]

| VARIABLES | ECR[2] phr | RPM | TEMP[3] (°C) | SHEAR[4] RATE (sec$^{-1}$) | FUSION POINT[5] | | | EQUILIBRIUM POINT[6] | | | |
|---|---|---|---|---|---|---|---|---|---|---|---|
| | | | | | (sec) | TORQUE (m-g) | SHEAR STRESS[4] (Pa x 10$^{-3}$) | TORQUE (m-g) | SHEAR STRESS[4] (Pa x 10$^{-3}$) | MELT TEMP(°C) | GEL COUNT |
| TEMP AND SHEAR RATE | 100 | 32 | 185 | 37 | 120 | 950 | 39 | 750 | 31 | 191 | 800 |
| | 100 | 32 | 175 | 37 | 150 | 1210 | 49 | 920 | 38 | 182 | 600 |
| | 100 | 32 | 165 | 37 | 180 | 1430 | 58 | 1040 | 42 | 175 | 300 |
| | 100 | 64 | 175 | 73 | 75 | 1590 | 65 | 1085 | 44 | 192 | 250 |
| | 100 | 64 | 165 | 73 | 90 | 1805 | 74 | 1155 | 47 | 186 | 200 |
| ECR LEVEL | 100 | 32 | 175 | 37 | 150 | 1210 | 49 | 920 | 38 | 182 | 600 |
| | 70 | 32 | 175 | 37 | 90 | 1660 | 68 | 1105 | 45 | 184 | 250 |
| | 40 | 32 | 175 | 37 | 75 | 2230 | 91 | 1340 | 55 | 186 | 0 |

(1) BRABENDER PLASTI-CORDER
(2) FORMULATION I, TABLE I
(3) MIXING CHAMBER TEMPERATURE AT START

(4) APPROXIMATION USING THE METHOD OF REFERENCE II
(5) POINT C IN FIGURE 1
(6) 15 MINUTES FROM START

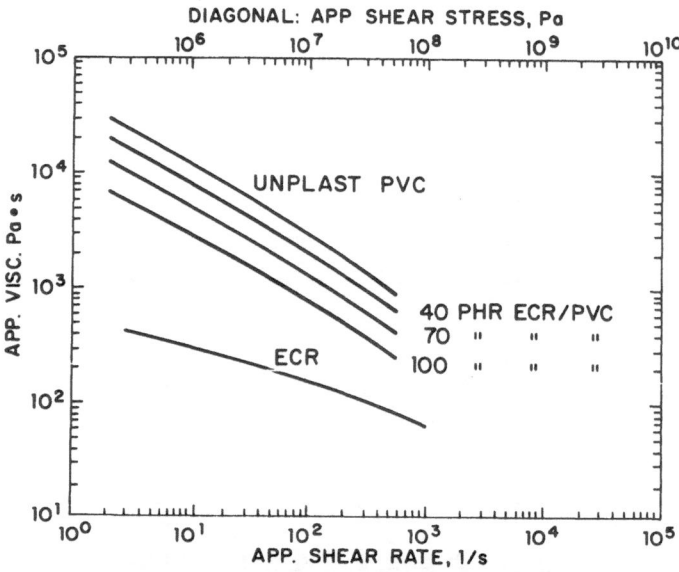

Figure 8.   Relative rheologies (190°C).

Table 3.   Sequential Addition, ECR to Brabender[1]

|  | FUSION POINT | | GEL COUNT | |
| --- | --- | --- | --- | --- |
|  | TORQUE | SHEAR STRESS | | |
|  | sec. | (m-g) | (Pa X 10^{-3}) | 10 mins. | 15 mins. |
| STANDARD RECIPE[2] WITH 40 phr ECR ADDED AT BEGINNING AND 60 phr ECR ADDED AT 10 mins | 75 | 2230 | 91 | 10[3] | 0 |
| STANDARD RECIPE[2] WITH 100 phr ECR ADDED AT BEGINNING | 150 | 1210 | 49 | 800 | 600 |

(1) 175° CHAMBER, 32 rpm

(2) USING STANDARD PARTICLE SIZE PVC

(3) PRIOR TO ADDITION OF SECOND CHARGE OF ECR.

Table 4.  Homogeneity vs. PVC Types [1]

| PVC TYPES | K VALUE | FUSION POINT | | | EQUILIBRIUM POINT | | |
|---|---|---|---|---|---|---|---|
| | | sec | TORQUE (m-g) | SHEAR STRESS (Pa X $10^{-3}$) | TORQUE (m-g) | SHEAR STRESS (Pa X $10^{-3}$) | GEL COUNT |
| EMULSION | | | | | | | |
| A | 68 | 105 | 1330 | 54 | 960 | 39 | 0 |
| B | 76 | 160 | 1310 | 53 | 1200 | 49 | 0 |
| STANDARD SUSPENSION [2] | | | | | | | |
| C | 54 | 45 | 1080 | 44 | 440 | 18 | >2000 |
| D | 67 | 150 | 1210 | 49 | 920 | 38 | 600 |
| E | 67 | 195 | 1010 | 41 | 900 | 37 | >2000 |
| F | 80 | 250 | 1575 | 64 | 1195 | 49 | 1300 |
| BULK [2] | | | | | | | |
| G | 55 | 15 | 1400 | 57 | 495 | 20 | >2000 |
| H | 67 | 15 | 1400 | 57 | 850 | 35 | >2000 |
| LARGE PARTICLE SUSPENSION [3] | | | | | | | |
| I | 69 | 95 | 1720 | 70 | 940 | 38 | 40 |
| J | 82 | 80 | 2400 | 100 | 1320 | 54 | 0 |

(1) 100 phr ECR, 175°C CHAMBER, 32 rpm, 15 min

(2) <5% RETAINED ON 60 mesh

(3) >90% RETAINED ON 60 mesh

times are considerably shorter and the fusion torque values
considerably higher, than comparable standard suspension grades,
supporting this mechanism.  The advantages of large grain PVC,
with conventional liquid plasticizers, also have been reported.[13]

## Coplasticizers

Liquid coplasticizers were assessed in the PVC/ECR
formulations (Table 5).  No significant reduction in gel count
could be obtained though a trend towards improvement with higher
MW coplasticizer was evident.  In general, long fusion times and
low broad fusion peaks were observed.  The absorbed liquid
plasticizer did not facilitate the erosion of the 150 μm grains,
into primary particles, in the presence of ECR.

## High Rate, High Shear Continuous Melt Compounding

Using Figure 10, it was predicted that the high shear rate
Farrel continuous mixer would produce low gel blends based on its
high shear (>300 kPa) capability.  The short residence time
(approximately 30 seconds), however, was a factor that could
negate the higher shear rate.  Formulations containing 70 phr ECR
were compared over a range of shear rates and temperatures (Table
6) on a No. 4, Farrel Continuous Mixer.  As can be seen, low gel
levels (5-20) were achieved on a composition containing a
standard grade of PVC providing that the melt temperature was
kept relatively low (160°C).  As on the Brabender, raising the
melt temperature (180°C) reduced the viscosity and shearing
action on the PVC particles causing the gel level to increase
(60-110).  The same composition, but modified by the inclusion of
DOP in the recipe, behaved in a similar manner and had comparable
gel levels.

Substitution of the standard PVC by the large grain PVC also
confirmed the Brabender results by producing an essentially
gel-free blend at low temperature with only very slight residual
gel remaining in the blends prepared at high temperatures.


SUMMARY AND CONCLUSIONS

Routine procedures used for liquid plasticized PVC are often
inadequate to produce homogeneous low gel alloys of PVC/ECR.
Melt compounding morphology and rheology were used in analyzing
this problem.  A Brabender Plasti-Corder technique, used to
evaluate the melt compounding of PVC/ECR, showed that standard
particle (grain) sized PVC suspension resin produces
heterogeneous blends, at shear stresses of 75 kPa or less.  The

Table 5. Homogeneity vs. Coplasticizers

| COPLASTICIZER[1] | FUSION POINT[2] | | | EQUILIBRIUM POINT | | GEL COUNT |
|---|---|---|---|---|---|---|
| | sec | TORQUE (m-g) | SHEAR STRESS (Pa x 10$^{-3}$) | TORQUE (m-g) | SHEAR STRESS (Pa x 10$^{-3}$) | |
| DOP | 180 | 575 | 23 | 475 | 19 | 1500 |
| BUTYL BENZYL PHTHALATE | 240 | 655 | 27 | 520 | 21 | 1000 |
| HEPTYL-NONYL-UNDECYL PHTHALATE | 240 | 550 | 22 | 480 | 20 | 1000 |
| DIMETHYL PHTHALATE | 240 | 550 | 22 | 435 | 18 | 1000 |
| DIBUTYL PHTHALATE | 240 | 715 | 29 | 485 | 20 | 700 |
| TRI(BUTOXYETHYL) PHOSPHATE | 150 | 820 | 33 | 520 | 21 | 300 |
| TRICRESYL PHOSPHATE | 210 | 810 | 33 | 580 | 24 | 650 |
| 2-ETHYLHEXYL DIPHENYLPHOSPHATE | 210 | 760 | 31 | 530 | 22 | 400 |
| DIISOBUTYL ADIPATE | 210 | 810 | 33 | 475 | 19 | 450 |
| EPOXIDIZED SOYBEAN OIL | 189 | 1020 | 42 | 590 | 24 | 150 |
| LOW MW POLYESTER | 150 | 890 | 36 | 560 | 23 | 800 |
| MED MW POLYESTER | 120 | 1020 | 42 | 580 | 24 | 300 |
| HIGH MW POLYESTER | 120 | 990 | 40 | 650 | 27 | 400 |

(1) 30phr, 100phr ECR

(2) 175°C, 32rpm, 15min

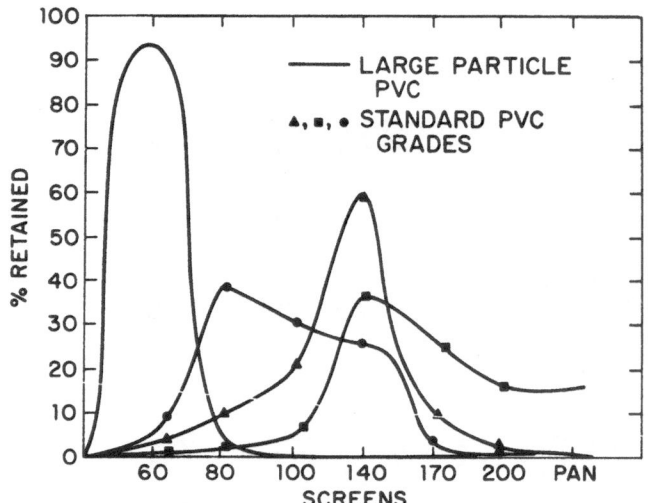

Figure 9.   Particle size distribution (Ref. 15).

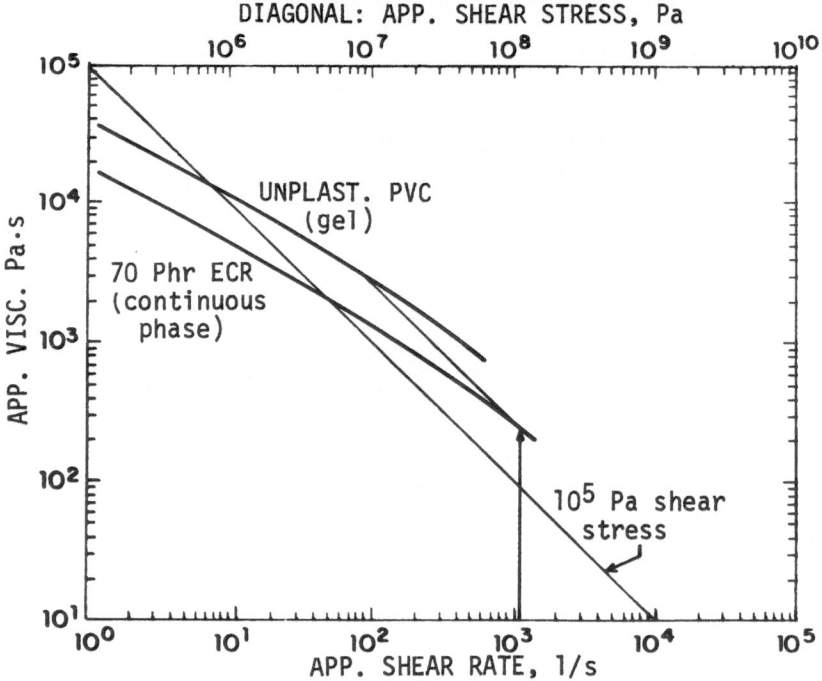

Figure 10.   Rheology PVC/ECR (190°C) FCM mixing.

Table 6.  High Intensity Continuous Mixer[1]

| FORMULATION[2] | MELT. TEMP. °C | SHEAR STRESS[3] (Pa × 10⁻³) | GEL COUNT[4] |
|---|---|---|---|
| III (STANDARD PVC) | 160 | 500 | 5-20 |
|  | 180 | 400 | 60-110 |
| IV (STANDARD PVC + DOP) | 160 | 300 | 20-40 |
|  | 180 | 200 | 40-130 |
| III (LARGE PARTICLE PVC) | 160 | 500 | 0 |
|  | 180 | 400 | <5 |

(1) FOUR INCH FARREL CONTINUOUS MIXER,
500 rpm (1125 sec⁻¹ SHEAR RATE), 544 kg/hr (1200 pph)

(2) EACH CONTAINS 70 phr ECR

(3) CONSTANT RATE CAPILLARY RHEOMETER

(4) CONTROL CONTAINING 60 phr DOP HAD A GEL COUNT RANGE OF 20-40

more shear-sensitive large particle PVC suspension resins, however, produce homogeneous blends under these conditions.

The high shear stresses (>300 kPa) of a Farrel continuous mixer produced homogeneous blends, at high rates, with both standard and large particle PVC. The large particle PVC, however, still showed the advantage of being homogeneous over a wider range of compounding conditions.

REFERENCES

1. C. F. Hammer, ACS, Coatings and Plastics Preprints, 37, 234 (1977).
2. J. P. Tordella, Modern Plastics, Jan. 1976, p. 64.
3. J. A. Wingrave, J. Vinyl Tech., 2, 204 (1980).
4. P. V. McKinney, J. Appl. Poly. Sci., 9, 3359 (1965)
5. J. R. Defiefe, J. Vinyl Tech. 2, 95 (1980).
6. E. B. Rabinovitch, J. W. Summers, J. Vinyl Tech., 2, 165 (1980).

7.   J. A. Wingrave, M. C. Peden, J. Vinyl Tech., 1, 107 (1979).

8.   T. F. Chapman, J. D. Isner, J. W. Summers, J. Vinyl Tech.,
     1, 131 (1979).

9.   V. W. Uhl, J. B. Gray, Mixing Theory and Practice, II, 193,
     Academic Press, New York, 1967.

10.  J. C. Chauffoureaux, C. Dehennau, J. Van Rijckevorsel , J.
     of Rheology, 23, 1 (1979).

11.  J. E. Goodrich, R. S. Porter, Polym. Eng. and Sci. 7, 45
     (1967).

12.  Technical Information Bulletin, "Using a Banbury to Com-
     pound PVC Blends Containing Elvaloy  Resin Modifiers,"
     Du Pont Company, Polymer Products Department, Wilmington,
     Delaware 19898.

13.  P. L. Shah and V. R. Allen, SPE Journal, 26, 56 (1970).

14.  Technical Information Bulletin, "Vygen 300 Series Resins,"
     General Tire & Rubber Co., Chemical, Plastics Div., Akron,
     Ohio.

15.  Technical Service Report, "Geon 90 Series Resins," TSR
     71-04, B. F. Goodrich Co., Chemical Group, Cleveland, Ohio.

POLYMER-PERFORMANCE ON THE DIMENSIONAL STABILITY AND THE MECHANICAL

PROPERTIES OF WOOD-POLYMER COMPOSITES PREPARED BY AN ELECTRON BEAM

ACCELERATOR

T. Handa, I. Seo, T. Ishii and Y. Hashizume

Dept. of Applied Chemistry, Faculty of Science
Science University of Tokyo
Kagurazaka, Shinjuku-ku, Tokyo, 162, Japan

ABSTRACT

The state of the art of polymers in wood and their performance
were studied for the wood-polymer composites produced by electron
beam irradiation.  The balance between the enhanced dynamic modulus
and the dimensional stability of the products were discussed to
prevent the formation of fatigue cracks during a long exposure to
repeated cycles of humid and dry weather.  The particular action of
the tight cross-linking bonds in enhancing the mechanical properties
as well as the dimensional stability of the products was recognized
with regard to the unsaturated polyester WPC.  In this regard, an
impregnation method in two steps was proposed for producing a polyester WPC fancy veneer with distinguished mechanical properties and
dimensional stability:  First, the styrene or acrylonitrile monomers
were impregnated in the moisture preswollen veneers and polymerized
under an accelerator.  Then, the styrene/unsaturated polyester system
was impregnated in the irradiated veneers successively and polymerized by the repeated irradiation.  The improvements of the dynamic
modulus and the dimensional stability of WPCs were coming from that
polymers of WPC in the first step were formed in cellular parts and
those in the second step were formed in the cellulosic chains.

INTRODUCTION

Wood has been used by human kind since ancient times not
simply as a fuel but also as a highly important construction material.  Wood is authentically a composite of crystalline cellulose

fibrils, amorphous hemicellulose and resinous lignin as a binder. The high mechanical strength of wood is coming from the highly order-ed and complex structure. However, it also has the devect that the elongation of wood by swelling induces a higher internal shear and results in the occurrence of surface crack by fatigue. Especial-ly, the surface crack by fatigue is related to the difference of elongation of wood between the longitudinal (L) and tangential (T) directions. A large number of investigations have been conduct-ed for reducing the dynamic anisotropy of wood with the wood-plastic composite. Polymers impregnated in wood can increase the dynamic modulus along the T-direction to wood with increasing polymer content. Thereby, the dynamical anisotropy of wood is reduced by the increase of dynamic modulus in T-direction with increasing polymer fraction. On the other hand, it has been shown that polymers of WPCs in cellu-lar parts do not contribute to the mechanical strength, but polymers of WPCs in cellulosic chain improve the dimensional stability of WPCs by suppressing the swelling. These results indicate that the material to be impregnated is controlled by the species of chemicals, the impregnating part of wood, and so on (1-7).

Although the literature of WPCs by γ-rays are abundant, there are a few by electron beam accelerators. The recent development of electron-beam curing has made it possible to have the surface-coat with a high dose-rate. It is very feasible for factory-producing concerning the impregnation of resin to wood and results in lowering the economic cost. For example, the surface-coat of wood can be finished for the monomer-prepolymer system within a few seconds by the high dose-rate (8-16).

The present paper pursues the impregnation conditions and the impregnation sites of WPCs concerning the veneer-monomer and veneer-monomer-prepolymer systems. Especially, the physical proper-ties of the impregnated WPCs were evaluated from the measurements of the dynamic modulus and the dimensional stability.


EXPERIMENTAL

Materials

Veneers used as specimens in the studies were rotary-cut ones of beech spawood (Tohoku original) and were carefully selected from factory products of similar qualities. All samples were oven-dried in vacuo at 80°C for 24 hr.

The chemicals used in the study were styrene (ST), methyl-methacrylate (MMA), acrylonitrile (AN), butyl acrylate (BA), acrylic acid (AA), polystyrene (PSt), polymethyl methacrylate (PMMA), paraffin, and unsaturated polyster resin especially provided by

NIHON Oil and Fats Co. Ltd. (Beam Coat 1000A-41) containing 75%
prepolymer of polyester (PE) and 25% styrene monomer (St).


## Experimental Procedures

Control of moisture content in veneers to various degrees
before impregnation of chemicals and dimensional measurements at
respective room humidity (R.H) were conducted in desiccators
standarized by inorganic solutions of 43, 65, 81, 93, and 98%
R.H. respectively.  The samples were placed in the chamber at 20°C
and were conditioned to be equilibrated to respective R.H. for the
impregnation and dimension measurement.

Impregnation of chemicals to veneers thus conditioned was pur-
sued by repeating the freeze-thaw cycles using ice-methanol temperatures.  After the impregnation process, the samples were wrapped
in bags of aluminum foil and replaced with nitrogen gas for the
irradiation process.

Polymerizations for monomer- and prepolymer/monomer-W.P.C. (irradiation) systems were carried out at a total dosage of 6 Mrad to the
respective surface of the sample by I.C.T-type electron accelerator
(300 kV, 50 mA).  After these processes, the residual monomer in the
system was expelled in vacuo at room temperature for 5 hr and then
the samples were dried in vacuo at 80°C for 24 hrs.

For W.P.C (injection) systems, benzene solution of PSt or PMMA
at various concentrations was introduced into wood by the repeated
cycles of freeze and thaw processes and then dried at 80°C for 24
hrs in vacuo to expel benzene solvent completely.

For AN (irradiation)/PSt (impregnation)-W.P.C systems, benzene
solution of PSt at various concentrations was introduced into AN-
W.P.C (irradiation) by impregnation method and then the solvent was
removed by successive drying.

Solid-paraffin W.P.C. was prepared by dipping the veneers in a
hot molten paraffin  bath.

Measurement of the bulking degree of the samples in oven dried
condition was conducted in the tangential direction of rotary-cut.
Polymer content (P.C) and bulking due to WPC were estimated by the
following relations.

$$\text{Polymer Content (P.C)} = \frac{W_p - W_o}{W_o} \times 100 \text{ \%}$$

where $W_o$ is the weight of the dried original wood and $W_p$ is the
weight of the dried WPC.

$$\text{Bulking} = \frac{L_p - L_o}{L_o} \times 100 \ \%$$

where $L_o$ is the tangential length of the dried original wood and $L_p$ is the tangential length of the WPC.

## Measurements

Test samples of 10 cm (tangential) x 2.5 cm (longitudinal) x 0.065 cm (radial) were placed in the said humidity chamber standarized to 93% R.H. and exposed to atmosphere in the vessel for 15 days at 20°C. The antishrinkage efficiency (A.S.E.) was calculated from the difference of dimensional change between the leached swelled condition and the final oven dry condition.

$$\text{A.S.E.} = \frac{L - T}{L} \times 100 \ \%$$

where L is the % swelling in the control sample and T is % swelling of the treated sample.

The dynamic modulus (E') and the loss modulus (E") of specimens have been measured by Rheovibron DDV-II and the vibration-reeds method.

The test samples of WPC were produced by the following scheme.

veneers of beech sapwood
↓
impregnation
↓
electron irradiation (total dose 6 Mrad)
↓
after treatment (105°C 12 hr oven dry)
↓
plywood (veneers of WPC were bonded on 9 mm thick plywood in the parallel directions.)
↓
test 1) wet and dry cycle test
     2) anti-abrasion test

RESULTS AND DISCUSSION

## Monomer, Polymer and Polymer-Monomer Systems

The reaction spaces for the polymerization and graft-copolymerization of synthetic monomers in wood can be conventionally classified into the authentic outer space (A) and the artificial inner-space (B). The former consists of vessels, tracheae and lumens

which are channeled by pits and voids, whereas the latter consists
of tiny capillaries which are created mainly in cell walls by swelling due to moisture and polar organic solvents.  When either monomers
or polymer solutions are impregnated into oven-dried woods, they
stay mostly in the space A except those monomers such as acrylonitrile (AN) and acrylic acid (AA), which have a high swelling capabilities can be distributed in the space B by dissolving them in
polar organic solvents which have the higher swelling power and the
lower boiling temperatures than those of monomers.  It should be
noted thereby, that those monomers were partitioned in the space B
with those solvent molecules.  The high molecular weights polymers
impregnated to the space A as a solution can never be distributed
in the space B after the removal of solvents, even if polar solvents
which can swell the wood system very strongly are used.

     In this context, there must be no alternative means to encapsu-
late polymers inside of the cell walls of wood or to the space B
other than to achieve it first by impregnating monomers to the
space B and then by polymerizing them there.

The major characteristics specific to the radiation-induced
polymerization of monomers in wood under electron beam accelerator
is that the polymerization proceeds preferentially for the monomers
in the space B than those in the space A under a moderate dose.
Thereby, it can be said that the growth of propagating chains must
occur preferentially in the space B involving those monomers al-
igned along the capillaries, whereas the combination type termination of irradiated monomers must occur priorily in the space A
under a high dose rate of electron showers.  To achieve complete
polymerization of monomer in the space A which occupies a larger
portion of reaction space in wood, there is no other alternative
other than the use of monomer-polymer or monomer-prepolymer systems
to take advantage of the gel effect which prevents the prompt col-
lision of irradiated monomers and thus favors the growth of polymeric chains.  In other words, the homopolymers are primarily produced in the space A while the graft polymers are produced in the
space B (12,14,18).

     Figure 1 shows the relation between $E_L'$, $E_T'$ and polymer fractions
for various WPCs.  It is revealed that the polymers in the space A
either being polymerized or impregnated can increase the dynamic
modulus ($E_T'$) along the tangential direction at the measureing temperature below their Tg and thus reinforce the mechanical strength of
veneer with a significant dissolution of the mechanical anisotropy
specific to the wood structure simultaneously.  On the other hand,
those polymers produced in the space B are obviously indifferent from
any changes in the mechanical properties of wood regardless of the
tensile stress from their actions in bulking the wood system.  It
is also a well-known fact that they can considerably improve the

dimensional stability of wood by supressing the swelling of the
system due to moisture in an extent to which the bulking of size
conforms.

Figure 2 shows the multiple features in the temperature dispersion of $E_L'$ and $E_T''$ for various WPCs which were prepared by the
impregnation of polymer solutions and by the radiation induced polymerization of impregnated monomers or monomer-polymer systems under
electron beam respectively.

It is revealed that a variety of distinctive rises of $E_T''$ peaks appear
at temperatures ca. 20°C higher than the Tg of respective polymer and
also multiple features in the lower temperature shift of the $E_T''$ peak
at around 230°C conforming to the depression of $E_T'$. It was assumed
that the former change came from the interaction between the surface
cell walls and the polymer which existed in the space A. The latter
changes came authentically from the collapse of the fibril structure
of genuine wood due to the enhanced thermal shears which were promoted by polymers in the space B, depending on their state in wood.

As a result, it may be presumed that overall, the state of
the art of polymers in WPCs can be classified into the following
three types:

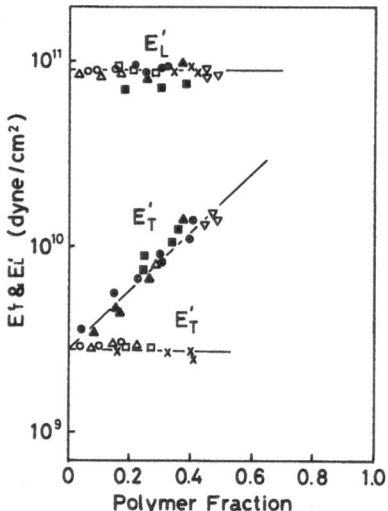

Figure 1.  Effect of polymer fraction on dynamic modulus for longitudinal
           direction ($E_L'$) and tangential direction ($E_T'$) of WPC by various
           polymer systems at room temperature:
           O St(irradiation); △ MMA(irradiation); □ AN(irradiation);
           X BA(irradiation); ● PSt(impregnation); ▲ PMMA(impregnation);
           ■ AN(irradiation)/PSt(impregnation); ◇ Parraffine(impregnation);
           ▽ AA(irradiation).

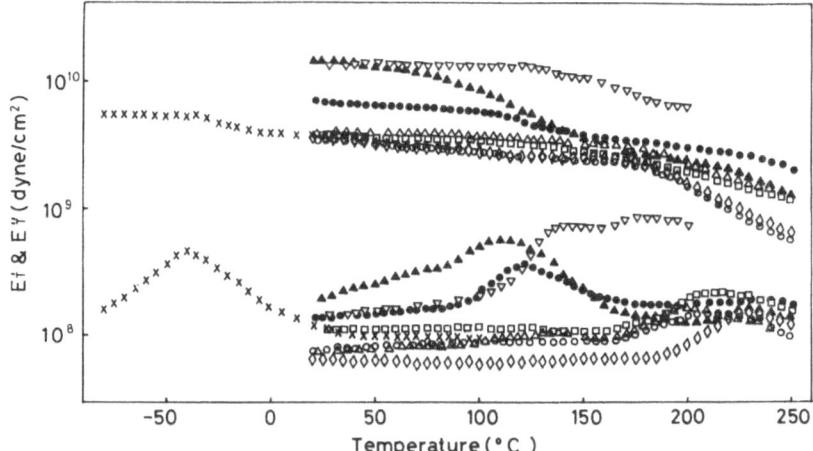

Figure 2. Temperature dependence of the dynamic modulus $E_T'$ and the loss modulus $E_T''$ at 110 Hz for WPC by various polymer systems: o St(irradiation; Δ MMA(irradiation); X BA(irradiation); ● PSt(impregnation); ▲ PMMA(impregnation; ◇ control wood; □ AN(irradiation); ▽ AA(irradiation).

(1) Type-I WPC has polymers locating genuinely in the space A.

Type-I includes the polymers in those WPCs (PSt-, PMMA-, and Paraffin - WPC etc.) which were prepared by the impregnation of polymers dissolved in organic solvents and the successive removal of the solvents.

(2) Type-II WPC has polymers in the space B.

Type-II includes the polymers in those WPCs (St-, MMA-, and AN-WPC etc) which were prepared first by impregnating those monomers with swelling capabilities by themselves to oven-dried woods or the non-polar monomers (St, MMA etc.) to moisture preswollen woods and then, irradiated under electron beam and finally, extracted by organic solvent to remove the residual parts in the space A.

(3) Type-III WPC has polymers coexisting in both the space A and B.

Type-III includes the polymers in those WPCs (AA-WPCs, BA, and St-AN WPCs, etc by irradiation and monomer (AN, AA etc)-polymer (PSt, PMMA etc) WPCs etc) which were prepared first by impregnating those monomers to oven-dried or moisture preswollen woods, then irradiated under electron-beam, post-polymerized under heat cure at 80°C and then, extracted by organic solvents to remove homopolymers in the space A. Those WPCs can leave a large portion of polymers in the space A even after the solvent extraction. Those monomers in monomer-polymer systems with swelling capabilities by themselves originated in the space B and combined with other reactive polymer chains in the space A, being not terminated on lignins which are coating cell walls.

In Type-I WPC, polymers can interact with the surface of cell walls and thus increase $E_T'$ and $E_T''$ of WPC systems. The $E_T''$ peak at Tg of polymers shifts to higher temperature. While in Type-II WPC, polymers interact with the cellulosic system inside of cell walls and thus, do induce a remarkable shift to the lower temperature of $E_T''$ peak at around 230°C due to the promotion of the thermal shear. It leads to the decrystallization of the crystalline parts which bundle the fibril system of wood (18,20). In Type-III WPC, polymers own both capabilities and thus, accelerate the decrystallization by the promoted thermal shear with a specific indication of a remarkable $E_T''$ at around 180°C conforming to the increase in the thermal motion of amorphous cellulosic chains as observed in AA-WPC etc.(20). However, as for PBA with $T_g$ being located far below the room temperature, there was no indication of any recognizable rise of $E_T''$ as well as any increase of $E_T'$ in the ordinary range of measuring tempera- ture. But it revealed an obvious rise of $E_T''$ besides an increase of $E_T'$ below -40°C.

Figure 3 shows the temperature dependent variation of the relation between $E_T'$ and the polymer fraction for a series of the PMMA-WPCs which were prepared by the impregnation of different amounts of PMMA dissolved in benzene to the oven-dried beech veneers and then, by the removal of the solvent at 80°C in vacuo successively. As had been observed for the PBA-WPC, there was also no recognizable increase of $E_T'$ with the polymer fraction above a temperature ca. 20°C higher than the $T_g$ of PMMA.

Irrespective of the tensile stress from the initial shear due to the bulking effect of those polymers produced in the space B, they do not alter $E_T'$ and $E_T''$ of wood and hence, are superficially indifferent from any change in the macroscopic mechanical structure of wood at an ordinary temperature range. However, they are capable to interact with wood system in the low and high temperature range in hindering the rotational motion of methylol groups on glucosidic rings of cellulosic chains at a submolecular level and also in endorsing the thermal shear to promote the earlier corruption of the fibril system in wood at a macromolecular level.

Eventually, it may be allowable to presume that anyhow, the fibril system which supports the mechanical structure of wood owns some mechanically sensitive sites around the microcrystalline parts which bundle the tails of microfibrils consisting of the lamella and must be located on the surface of cell walls being exposed to polymers in the space A. In this context, the tensile stress from the internal shear along the T-direction can be dispersed to the L-direction in a sense as predicted by the scissor-rod model for the mechanical structure of wood. It was also found that the elongation in the oven-dried size of WPCs ($\Delta L_{bulking}$) occured due to the strain from the bulk volume of polymers produced in the space B and that the elongation in wet size of WPCs due to swelling by moisture was

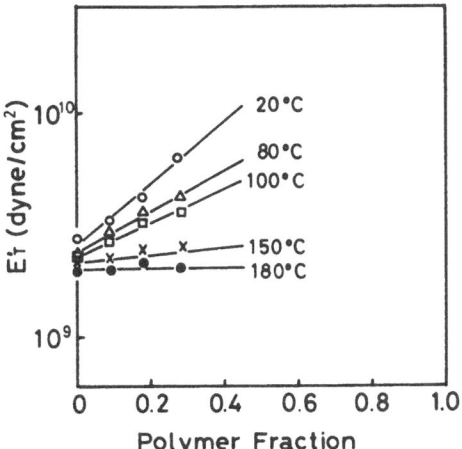

Figure 3. Effect of polymer fraction of PMMA (impregnation) system on $E_T'$ of WPC at various temperature.

approximately equal to 12.9  $L_{bulking}$ assuming the swelling limit of ordinary beech veneer towards T-direction as 12.9.

    Figure 4 compares the relations between the ASE values and the bulking degree in % for various WPCs which were produced at a total irradiation dosage of 6 Mrad under electron beam.  Figure 5 also compares the relations between the ASE values and the polymer contents for the above WPCs.  The overview of the results in both figures indicates that the PSt from the polymerization of monomers impregnate in the space B is utmost effective in rejecting the moisture with the least bulking of wood system at an equivalent level of polymer content, next comes the PBA and then, PMMA prepared by impregnation of monomers.  It should be noted hereby that this turn in the utility of monomers in rejecting the moisture effectively is valid irrespective of the impregnation methods, either in terms of the preswelling by moisture or by liq. $NH_3$ (21).  Thereby, the plots of both PSt and PMMA WPCs from the impregnation of monomers to the acetone-replaced veneer which was preswollen by liquid $NH_3$ and then, replaced by acetone at low temperature (ca. $-75°C$) were found to be located approximately on extension of the lines from the plots of both WPCs which were prepared by the impregnation of monomers to the moisture preswollen veneers.  The polymerization of AN behaves abnormally in that they gave rather lower ASE values irrespective of the greater bulking degree.  This is probably due to the reason that PAN can not be dissolved in AN unlike other monomers. A considerable part of PAN stays in the system as microcrystalline

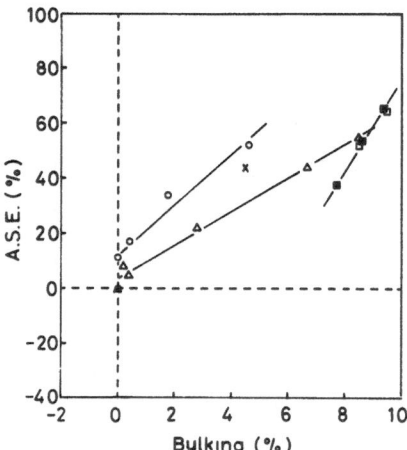

Figure 4. Relation between A.S.E. and Bulking of base wood by Polymers.
   O St(irradiation); △ MMA(irradiation); □ AN(irradiation);
   Χ BA(irradiation); ● PSt(impregnation); ▲ PMMA(impregnation);
   ■ AN(irradiation)/PSt(impregnation).

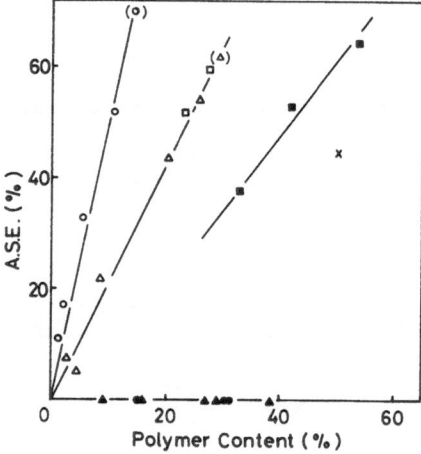

Figure 5. Effect of Polymer Content on A.S.E. of WPC: The symbols
   are the same as shown in Fig.4.

powders which are stuck to cell walls as needles and hence, must be poorly coordinated with the cellulosic lamella (12). This is endorsed by the fact that after the treatment of those WPC by di-methylformamide (DMF) solution, they showed an increase in $E_T'$ and scarce traces of white powders in the newly cut out cross section of the veneer. Thereby, the increase of $E_T'$ must conform to the portion of the polymer which was extracted by DMF in the space A being left as polymeric layers on the surface of cell walls after the removal of the solvent in cacuo. It suggests that the wider mol-ecular cross-section and the lower the polarity of the monomer, the higher the ASE of the WPC system. The performance of PAN is par-ticular in disclosing the new adsorption sites for moisture in accord-ance with its strong capability to bulk the wood system due to CN groups besides the aforementioned peculiarity in making crevice being stuck as needles on polymerization. It is quite obvious that those WPCs (Ps- and PMMA-WPCs) prepared by the impregnation of polymer solution can not show any gains in the ASE values as well as the bulking % of the system and that the AN (irradiation)/PSt (impreg-nation)-WPC which was prepared first by polymerizing impregnated AN under electron beam and then by the impregnation of PSt solution successively gave approximately the same values as those of AN-WPC for both ASE and bulking degree.

Figure 5 also indicates that St and MMA can hardly be polymer-ized in the space A under the electron beam irradiation whereas, as for BA, the polymerization proceeds considerably in the space A to an extent to which the lower value of ASE vs the higher polymer content of that polymer conforms.

It is also observed that the plots of both PSt- and PMMA-WPC prepared by liquid $NH_3$ swelling technique are located on the ex-tension of the line which consisted of the plots of those WPCs from the moisture preswollen woods.

Although polymers in the space A were tightly adhered to the surface of cell walls bringing about a specific increase in the $E_T'$ of the system, those polymers must be peeled off after repeated cycles of wet and drying processes.

Figure 6 shows the irreversible depression of $E_T'$ of WPC systems after one cycle of wet and drying processes for AN-PSt systems which were prepared first by polymerizing impregnated AN under electron beam and secondly by the impregnation of a thick solution of PSt. It is obvious that the higher the internal dynamic modulus $E_0'$ and the lower the ASE value of WPC systems, the greater the depression of $E_0'$.

It shows that being accompanied by some rupture of the surface fractions of the cell walls themselves from the fatigue due to the internal shear by the swelling and contraction of the wood system.

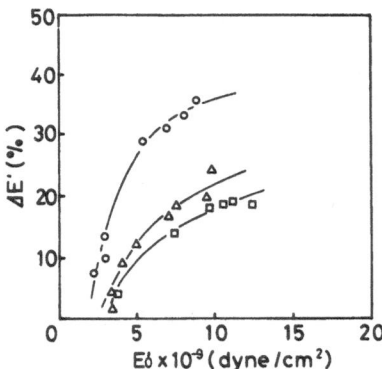

Figure 6. Relation between E' and $E_O'$ of WPC by AN (irradiation)/
PSt (impregnation) systems.
$$E' = (E_O' - E_S')/ E' \times 100 \ (\%)$$
where $E_O'$; $E_T'$ of WPC before swelling.
$E_S'$; $E_T'$ of WPC after swelling.
○ A.S.E. = 0%; △ A.S.E. = 37.7%; □ A.S.E. = 53.3%.


The magnitude of the imposed internal work which induces the
peeling of polymers with a rupture of the surface of cell walls can
be estimated by the magnitude of the internal work imposed on the
system during one wet and dry cycle, namely by the area which is surrounded by a hysteresis loop in the variation of a strain force $E_O' \times$
$\Delta L/L$ vs the displacement $\Delta L/L$. Figure 7 shows the relation between
S and $E_O'$. It reveals the presence of proportionality relations among
slopes depending on the ASE values of WPC systems. Figure 8 also
shows the presence of the proportional relation between S and $(\Delta L/L)^2$
with slopes depending on $E_O'$ of WPC systems. As a result, the following relation is obtained.

$$S \propto E_O' \times (\Delta L/L)^2$$

This means that the swelling degree of the system in % or $\Delta L/L$
must be dimished by inverse root square to the increase in $E_O'$ which
conforms to the increase of polymer fraction in the space A on a
semilogarithmic scale.

The deterioration of the overcoat laquer of clear finished
plywood due to fatigue on exposure to the repeating cycles of wet
and dry processes was remarkably prevented when the surface veneer
was replaced by its AN-WPC product with an improved dimensional
stability against moisture. Therein, the tensile stress which was
imposed on the laquer film due to the swelling and contraction of the
wood system during each run of the cycle must be considerably
diminished depending on the ASE values of AN-WPCs.

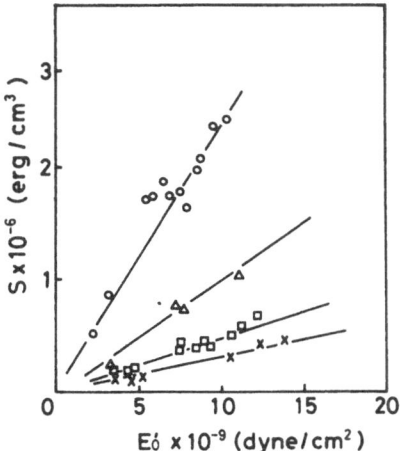

Figure 7. Relation between the induced opperant internal work per
unit volume (S) and $E_0'$ of WPC by AN(irradiation)/PSt(impregnation) systems.
○ A.S.E.= 0%; △ A.S.E.=37.7%; □ A.S.E.=53.3%; ✗ A.S.E.=65.0%

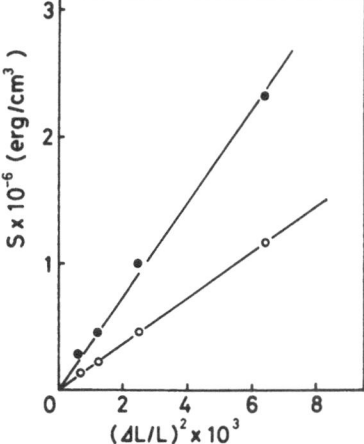

Figure 8. Relation between S and $(\Delta L/L)^2$ of WPC by AN(irradiation)/
PSt(impregnation) systems. ● $E_0'$=10x10⁹ dyne/cm² ; ○ $E_0'$=5x10⁹
dyne/cm²

Figure 9 shows the relation between the ASE values of AN-WPCs and the number of repeated cycles in the least square mean for the indication of a crack or a streak of microcracks on the nitrocellulose lacquer coat during the test. Thereby, the sample was first wet through by immersion in water for 12 hrs, kept in dry ice for 12 hrs successively and then, dried at 80°C for 24 hrs in each run of the test cycle and the base plywood was composed of a 0.6 mm thick AN-WPC product of beech veneer (surface), a 4 mm thick lauwan core and 0.6 mm thick beech veneer (back). It should be noted that they were especially plied in a parallel configuration along L-direction of wood to augument the displacement along T-direction due to swelling and contraction for the exaggeration of difference among the test results.

Irrespective of that the internal work imposed on the system during one cycle of wet and dry processes must basically be proportional to $( L/L)^2$ or the square power of the ASE value of the system. It is revealed that in reality, the critical number for the indication of any cracks increases in a higher power of ASE value than the predicted square power. This is presumably due to the reinforcing effect of the glue joint which may diminish the actual displacement as the ASE value of the surface veneer increases.

Therefore, although the perfect prevention of any swelling of wood at a ASE value close to 100 is our ultimate goal, it can be said that the ASE value higher than 60 will be practically good enough

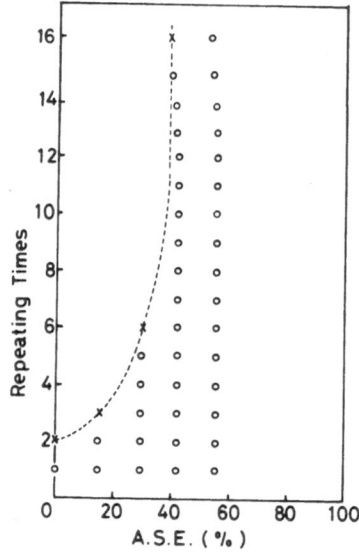

Figure 9. Relation between repeating cycles of wet and dry and A.S.E. of WPC by AN.

to prevent the weather aging of surface coat of plywood. It should be noted that the knowledge obtained so far will give a valuable insight in the selection of monomers for the water-proof surface coating of plywood flooring materials by polyester resin.


Prepolymer-monomer system

With regard to another kind of WPCs which were prepared by the electron curing of the prepolymer-monomer systems such as polyester resin and the like which were impregnated in wood, it is likely that extra cross-linking reactions occur between the residual reactive sites of polyester network in the space A and the reactive tails of encapsulated polymers creeping from the space B into the space A. Those cross-linking reactions will issue a tight binding among the polyester networks in the space A and the polymers encapsulated in the space B through polymer joints which were created by the recombination type termination of propagating radicals from both spaces. The termination of the free ends of residual cross-linking chains in the polyester network on the reactive component in the lignin layers must also occur reinforcing the performance of lignin network which cement the cellulosic fibrils in wood. These particular cross-linking bonds must induce multiple contractions in the wood structure of polyester WPCs depending on the prepolymer/monomer ratio whereas the encapsulation of polymers inside of cell walls induces the bulking of wood.

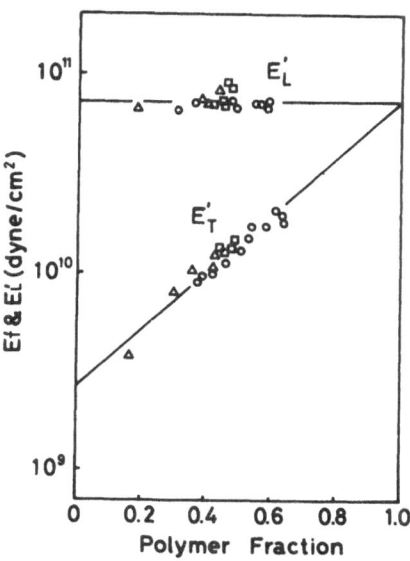

Figure 10. Effect of polymer fraction on $E'_L$ and $E'_T$ of WPC by various PE systems. ○ PE/St; △ PE/St/AN; □ PE/St/AA.

Figure 10 shows a very similar feature to that in Fig. 1 with
an indication of a perfect fit of the plots to the theory irrespective of a considerable presence of encapsulated polymers which
should basically bring an obvious deviation depending on the extent
of bulking to which the amount of polymer encapsulated in the space
B should conform. The strong interaction between the wood system
and a dense polyester network must be reflected in the temperature
dispersions of $E_T'$ and $E_T''$ and particularly of $E_L'$ and $E_L''$ of the systems.
Irrespective of the kinds of polymer in WPCs, the indicated peak
temperature of $E_T'$ bands of polymers in the space A were always
located ca 20-30°C higher than the genuine Tg of the corresponding
polymers being affected by the interaction with the wood system.
The lower temperature shifts of $E_T''$ peak at around 230°C were induced
by the uncrystallization of quasi-crystalline portion of fibrils and
the corruption of fibril structure due to the enhanced thermal shear
which conforms first to the strength of the interaction between the
wood system and encapsulated polymers inside of cell walls and
secondly to the density of polyester network in the space A.

Figure 11 shows the temperature dispersions of $E_T'$ and $E_T''$ for
PE/St-WPCs in various prepolymer/monomer ratios. It is revealed that
the shift of $E_T''$ peak temperature occurs conforming to the density
of cross-linking bonds which was governed by the prepolymer/monomer
ratio. The $E_T''$ peak at around 200°C for PE/St-WPCs was observed as
the shift of $E_T''$ peak of wood at 230°C based on the structure change
of wood. It is supported that PE/St-WPCs have a strong interaction
between the wood system and the polyester network.

Figure 12 and Figure 13 show the temperature dispersions of $E_T'$
and $E_T''$ of PE/St/AN-WPCs and PE/St/AA-WPCs, respectively. The low
temperature shift of $E_T''$ peak at 230°C of wood was also observed in
a similar manner as the temperature feature of $E_T''$ for PE/St-WPCs.
With regard to PE/St/An-WPCs, $E_T''$ peak at 230°C of wood was observed
at around 180°C. This must be due to the particular enhancement of
thermal shear endorsed by AA which interacts with cellulosic system
specifically (19).

PE/St/AN-WPCs show a rise of $E_T''$ peak at around 50°C in proportion to the AN content in the prepolymer-monomer systems. It
is mentioned that polyester resins in space A are composed of the
networks of loosely cross-linked polyester with PAN because formed
PAN is insoluble in AN monomer and hardly corss-linked to polyester.
On the other hand, the $E_T''$ peak at around 110°C for PE/St/AAWPC comes
from the glass transition of polyester resin to the higher temperature with the increase of the AA ratio.

The distinctive feature of $E_T''$ dispersions in the higher temperature range suggests that although both AN and AA are capable of
penetration to cell walls by themselves to be distributed in both
spaces, their reaction patterns must differ in making the tight

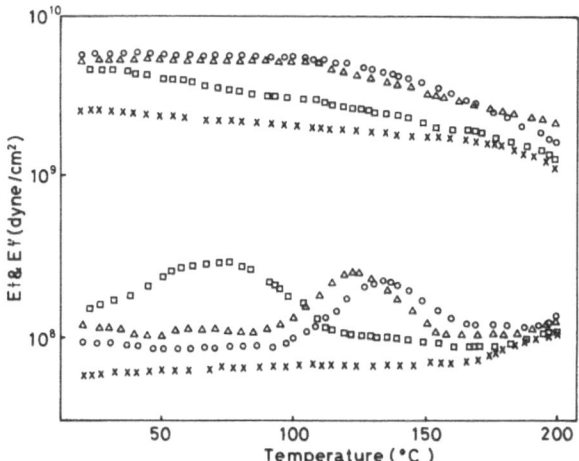

Figure 11. Temperature dependence of the dynamic modulus $E_T'$ and the loss modulus $E_T''$ at 110 Hz for WPC by PE/St systems. O P/M =75/25; ▲ P/M =52/48; ☐ P/M =32/68; Ⅹ P/M =0/100. Polymer contents of the samples are 98.8, 82.3, 57.6, and 19.2 (%), respectively.

cross-bond connecting the encapsulated polymers in the space B with the polyester network in the space A. This is presumably due to the reason that as the copolymerization rate of St with AN is faster than the cross-linking reaction rate of St with PE under electron beam irradiation. Secondly, the reactive tails of encapsulated PAN in the space B tend to couple with lignin layer on the surface of cell walls rather than with reactive residues of the PE/St network in the space A. Thirdly, as PAA chains are still, the reactive tails of encapsulated PAA in the space B tend to grow in the space A and couple with PE/St network in the space A rather than to be terminated on lignin. Besides, the genuine cross-linking reaction of AA with PE/St may occur rather than the copolymerization with St in the space A.

So far the state of the art of cross-linking bonds and their mechanical performances in PE/monomer WPCs have been discussed in correlation with the assignment of the distinctive $E_T''$-band or shoulder in the temperature dispersion of $E_T'$ and $E_T''$ of those WPCs. This is primarily because of the reason that they include basically the contribution from wood as well as that from polymer with E' locating between $E_L'$ and $E_T'$ of wood. It is also well known that the value of $E_L'$ and $E_L''$ of wood are located in one figure higher than those of $E_T'$ and $E_T''$. Accordingly, the authentic character of the strong bonding between polyester and wood will appear distinctively

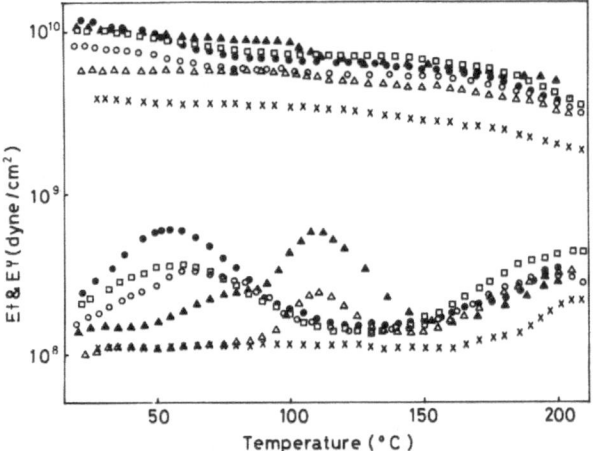

Figure 12. Temperature dependence of the dynamic modulus $E'_T$ and
the loss modulus $E''_T$ at 110 Hz for WPC by PE/St/AN systems.
✗ AN; △ PE/St/AN=13/12/75; ○ PE/St/AN =17/16/67; □ PE/St/AN=
25/25/50;● PE/St/AN = 38/37/25; ▲ PE/St = 50/50.  Polymer
contents of samples are 11.9, 33.2, 61.3, 63.5, 76.4 and
70.1 (%), respectively.

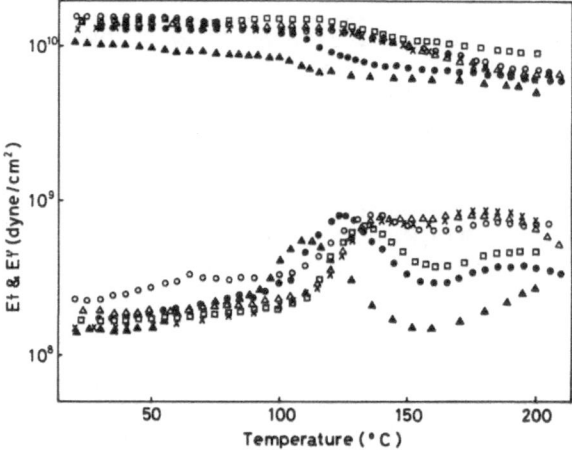

Figure 13. Temperature dependence of the dynamic modulus $E'_T$ and the
loss modulus $E''_T$ at 110 Hz for WPC by PE/St/AA systems.
The symbols are the same as shown in Fig. 12.  Polymer
contents of samples are 83.6, 73.6, 83.9, 83.0, 98.4, and
70.1, respectively.

in the temperature dependence of $E_L'$ and $E_L''$ which specifically reflect the change in wood fibrils themselves.

Figure 14 shows the temperature dependence of $E_L'$ and $E_L''$ of wood, AA- and PE/St-WPCs. The distinctive $E_L'$ band with its obvious peak at 255°C is observed for wood itself corresponding to the broad $E_T''$ band which includes the in-artifact contribution parallel to the genuine fibril direction besides that perpendicular to the fibril. This band was previously assigned to come from the collapse of fibril structure due to the spiral dislocation of lamella system from the uncrystallization of microcrystalline celluloses which bundle the microfibrils (18,19). As for AA- and PE/St-WPCs, obvious rise of $E_L''$ band is observed for the latter with the indication its peak at 185°C and its tail extending to the higher temperature range in which the former $E_L''$ band of wood itself is located. Thereby, the indicated peak temperature of the former $E_L''$ band corresponds approximately to the iso-dynamic modulus temperature above which the depressions proceeds with the order of $E_L'$ levels from wood, AA- and PE/St-WPCs being in the reverse turn of those below that temperature. The $E_L''$ band at around 185°C must reflect the genuine change in the structure of wood itself and hence, is assigned to come from the motion of amorphous cellulose due to the collapse of fibril structure. It should be noted hereby that, there was not any indication of such an iso-dynamic modulus point neither in the dispersion of $E_T'$ of any kind of WPC nor in the dispersion of $E_L'$ of the foregoing monomer and polymer-monomer WPCs. This agrees with the former presumption that in an extreme saying, every fraction of polymers in the genuine polyester WPC must be tied together and combined with wood system tightly by cross-linking bonds including unsaturated prepolymer-monomer networks in the space A and encapsulated polymers from the monomer distributed in the space B. In this respect, the polyester WPC is qualified to be not a mere composite of wood and polymer but is essentially a wood-polymer complex provided that the monomer in the prepolymer- monomer system of the space A can be well partitioned to the space B in an extent to which the upper extreme of polymer bulking conforms.

In this regard, PE/St-WPCs must own a distinguished property in confronting the abrasion shear as well as in rejecting moisture. Table 1 shows the result of the abrasion and the wet and dry tests of PE/St (38/62) WPC veneer which was prepared by the impregnation of St diluted PE (prepolymer) to the moisture preswollen veneers and by polymerizing the impregnated system at a total dosage of 6 Mrad under I.C.T. type accelerator (300 kV, 50 mA). Thereby, the test pieces were adhered on the plywood in a parallel configuration and the wet and dry cycle test was conducted in the way as described previously. The abrasion test was conducted using #180 sand paper adhered on the rotating disk at 500 rpm under load of 1kg.

Figure 15 shows the relation between ASE and bulking with

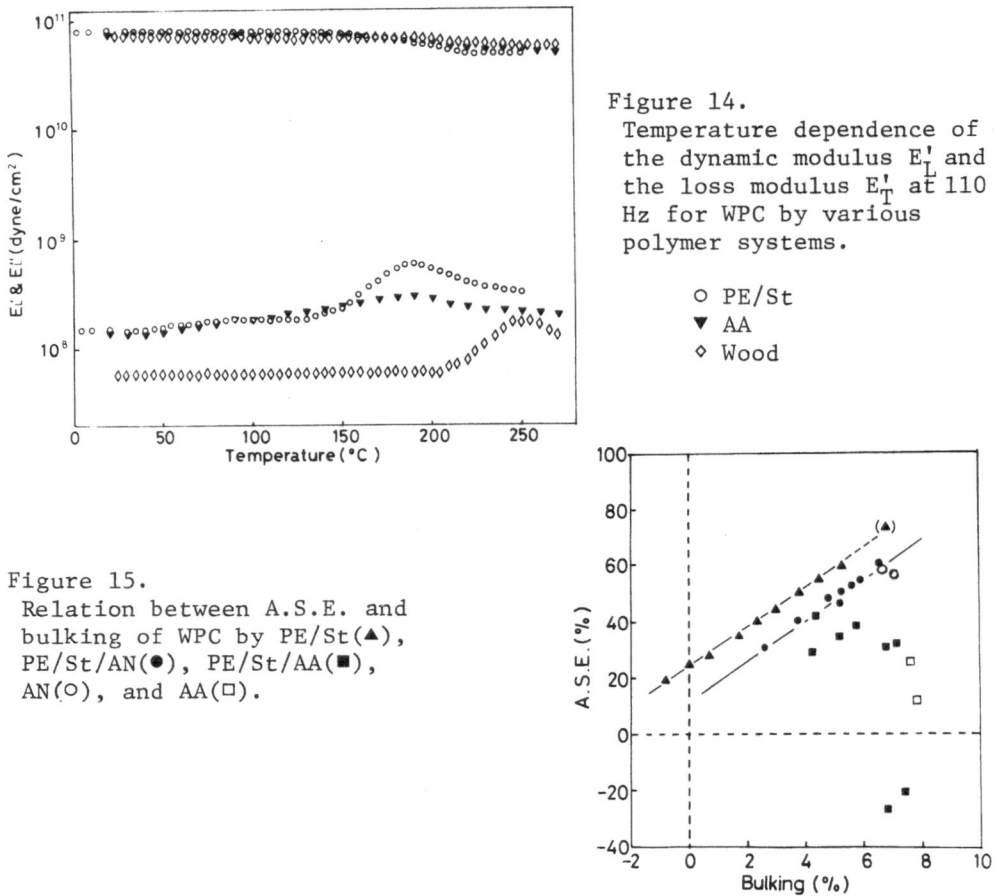

Figure 14.
Temperature dependence of
the dynamic modulus $E_L'$ and
the loss modulus $E_T'$ at 110
Hz for WPC by various
polymer systems.

○ PE/St
▼ AA
◇ Wood

Figure 15.
Relation between A.S.E. and
bulking of WPC by PE/St(▲),
PE/St/AN(●), PE/St/AA(■),
AN(○), and AA(□).

regard to various polyester-WPCs. This figure shows clearly the super-
ior capability of PE/St-WPCs in rejecting moisture with an indication
of ASE 20% even at zero bulking. First, this must come from the authentic capability of St injecting moisture and secondly, from the effect
of the dense cross-linking network which brings out the contraction of
size to reject the penetration of moisture inside of the cell walls.

We have reached so far ca ASE 70% for St-WPC by the first swelling
of the beech veneer in liq. $NH_3$ at $-75°C$, replacing $NH_3$ with acetone
at $-75°C$ successively then, impregnating St to the acetone replaced
veneer at room temperature and finally, polymerizing St by the irradiation of electron beam (21). As the application of this technique,
it is allowable to impregnate St to the acetone-replaced veneer in the
first step and then, impregnate PE/St system to that veneer in the second step. The plot of the expected ASE value corresponding to the
observed elongation of St-WPC is located on the extension of the plot
for PE/St-WPC being surrounded by bracket.

TABLE I

Results of the Wet and Dry Cycle Test and the Anti-
Abrasion Test of WPC by Two-Step Irradiation Process

| | P.C. (%) | Bulking (%) | A.S.E. (%) | Wet and dry test[*1] (cyles) | Abrasion Resis-[*2] tance(mg) |
|---|---|---|---|---|---|
| Blank | -- | - | 0 | 2 | 133.9 |
| PE/St (38/62) | 47.3 | 4.86 | 49.1 | 12 | 38 |

*1  The wet and dry cycle test;
     In water (20 25°C), 16 hrs, 80°C dry oven 5 hrs per 1 cycle.
*2  The abrasion test;
     The reduced amount of the test piece per 100 rpm = $W_0-W/5$
     The abrasion test condition;
     The load 1 kg, # 180 sand paper, The rotation = 500 rpm
     where $W_0$= the weight of a test piece before the abrasion test
            W = the weight of a test piece after the abrasion test
     Test pieces were adhered on the plywood placed parallel


     In respect of the technical feasibility in bulking the wood
system in a wider extent to which the elevation of ASE conforms, AN
was mixed with PE/St taking account of the greater capability in
bulking the wood system by itself.  PE/St/AN-WPCs are capable to
reach the ASE value close to 60% at the maximal bulking of ca. 7%.

     However, irrespective of the higher capability of AA in bulking
the wood system and the higher reactivity of that monomer in the
copolymerization with PE/St system than those of AN, PE/St/AN WPCs
show an abnormal relation between ASE and bulking in that they give
lower values of ASE against bulking degree and a negative value of
ASE at a high polymer content.  This is probably due to the creation
of new adsorption sites for moisture by the cleavage from the extreme
swelling of the hydrophobic polymers side of the cell walls.

     The evaluation on the utility of the WPC products has been based
so far on the criteria primarily with regard to the contradictive
balance between mechanical strength and ASE and secondly, with regard
to the performance and the state of art of polymer in WPCs depending
on the function specific to each monomer and prepolymer and their
reactivities.  In this regard, the peculiar quality of the compact
network of cross-linking bonds in the polyester WPC gave some solution by binding polymers in the space A, grafting polymers in the
space B and binding tightly to wood.  However, the balance between
the cross-linking in the network and the ASE value of the system is
still a matter of discussion.  When we decrease PE/St ratio to im-

Figure 16.
  Relation between A.S.E. and
  polymer content of WPC by
  PE/St systems.

     ○ P/M = 75/25
     △ P/M = 52/48
     □ P/M = 32/68

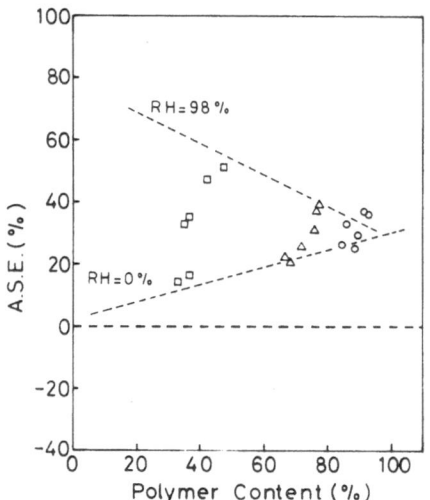

prove the ASE value by distributing St into the space B of the
moisture preswollen veneer as much as possible, the density of cross-
linking bonds in the network decreases with a simultaneous decrease
of polymer content in the spcace A.

     Figure 16 shows the relation between ASE and polymer content
depending the PE/St ratio of PE/St-WPCs.  The upper broken line
shows the attainable limit of ASE values of various PE/St WPCs which
were prepared by the impregnation of PE/St to the maximally preswollen veneers cured at R.H 98% whereas the lower broken line shows
the lower limit of ASE values of various PE/St WPCs which were prepared by the impregnation of PE/St systems to the oven-dried veneers
stored at R.H 0%.  The plots locating between both extremes belong
to those WPC with base veneers cured at various R.H.  The slopes of
both lines indicate reverse signs.  The ASE values of PE/St-WPCs prepared from the oven-dried veneers increase with polymer content due
to the effect of the dense network in rejecting moisture.  The ASE
values of PE/St-WPC prepared from the maximally preswollen veneer
decrease with polymer content because of the lower bulking of the
system.  The poverty of monomer into the space B conforms to the increase of PE/St ratio.

     In a previous study (12), we measured the E' and E" of the
genuine PE/St films with the PE/St ratios corresponding to those of
the impregnated systems in the foregoing PE/St-WPCs.  It is obvious
that the indicated peak temperature of the E" band of each film corresponds exactly to its $T_g$ which depends on the cross-linking density
of polyester network.  Thus, it is revealed that the higher the PE/
St ratio, the stronger the cross-linking bonds.  It can be made in
PE/St-WPCs in accordance with the increase of the polymer content
in the system.  As shown in Fig. 16, the upper and lower broken lines
cross each other at a point corresponding to the polymer content of

TABLE II

A.S.E. and A.A.E. of WPC by Two-Step Irradiation Process

| First Process | | | Second Process | | | |
|---|---|---|---|---|---|---|
| | P.C. (%) | Bulking(%) | | P.C. (%) | Bulking(%) | A.S.E.(%) |
| PE /St (38/62) | 53.3 | 2.2 | PE /St (75/25) | 88.9 | 1.65 | 49.8 |
| St | 13.2 | 5.17 | PE /St (75/25) | 99.3 | 4.09 | 54.9 |
| AN | 24.5 | 10.09 | PE /St (75/25) | 129.3 | 9.03 | 63.6 |
| PE /St (38/62) | 47.3 | 4.86 | --- | --- | --- | 49.1 |
| PE /St (75/25) | 91.8 | 0.90 | --- | --- | --- | 33.8 |
| AN | 27.1 | 10.1 | --- | --- | --- | 65.9 |
| St | 10.2 | 4.53 | --- | --- | --- | 51.7 |

100% and the ASE value of ca 30%. Those values must be the extremes of PE/St-WPCs for both items in the evaluation of the practical utility of the products.

Anyhow, unless we rely on other means, this must be an essential barrier to the goal which demands a high ASE value more than 60% for the wet and dry tests as referred to previously as well as a high polymer content close to 120% as the packing limit of polymer in wood and a compact polyester network with $T_g$ close to 120°C (the $T_g$ of the highest quality product in Japan) for the abrasion test.

In this respect, we explored an impregnation method of unsaturated polyester systems in terms of two steps. As the monomer in the first step, we used AN and St. Those monomers were impregnated to the moisture preswollen veneers and then, the systems were irradiated at a total dose of 6 Mrad (300 kV, 50 mA). After the first irradiation process, PE/St (75/25) was impregnated to the irradiated veneers in situ and polymerized at a dose of 6 Mrad (300 kV, 50 mA) additionally and the, heat cured at 80°C for 3 hrs or directly plied to a plywood with a phenol-melamine glue under hot press. The polymer content (%) and the bulking degree (%) of the products after the first and the second processes are described in Table 2 with the ASE of the final products. The abrasion resistance of PE/St (75/25) WPC surface veneers prepared by this method exceeded the previous one described in Table 1.

CONCLUSION

It can be concluded that the properties of PE/St WPCs fancy
veneers in terms of two steps impregnation method will suffice the
present goal in providing a tough surface veneer for the preparation
of fancy plywoods available for the flooring material and outside
sidings.

REFERENCES

1.  D. G. Adams, E. T. Choong and R. C. McIlhenny, Forest Prod. J.,
    20 (1970) 25.
2.  D. L. Kenaga and V.T. Stannet, Forest Prod. J., 12 (1962) 161.
3.  E. J. Gibson, R. A. Laidlaw and G. A. Smith, J. Appl. Chem., 16
    (1966) 58.
4.  R. A. Laidlaw, L. C. Pinion and G. A. Smith, Holzforschung 21
    (1967) 97.
5.  K. Taneda and I. Hasegawa, Bulletin of the Forest Product Insti-
    tute of Hokkaido, Jan., (1971) 12.
6.  J. K. Miettinen, Polym. Prep., 10 (1969) 494.
7.  T. Autio and J. K. Miettinen, Forest Prod. J., 20 (1970) 36.
8.  J. D. Nordstrom and J. E. Hinsch, Polym. Prep., 10 (1969) 473.
9.  A. S. Hoffman, J.T.Jameson, W. A. Salmon, D. E. Smith and D. A.
    Trageser, Polym. Prep., 10 (1969) 486.
10. J. Masuda, 9th Japan Conference and Radioisofopes May 13-15
    (1969) 486 Tokyo.
11. T. Handa, I. Seo, H. Akimoto, M. Saito and Y. Ikeda, Proc. 15th
    Japan Congress on Material Research. (1971) 158.
12. T. Handa, S. Yoshizawa, H. Kaido, I. Seo, N. Otsuka, M. Saito
    and Y. Ikeda, Proc. 18th Japan. Congr. Mater. Res., (1975) 126.
13. T. Handa, S. Yoshizawa, M. Fukuoka and M. Suzuki, Polym. Sci.
    and Technol., vol. II "Polymer Alloys II" D. H. Klempner and K.
    C. Frisch (Eds.), Plenum Pub., 1980, p.263.
14. T. Handa, N. Otsuka, H. Akimoto, Y. Ikeda and H. Saito, Mokuzai
    Gakkaishi 19 (1973) 493.
15. T. Handa, S. Yoshizawa and K. Hatakeyama, J. Appl. Polym. Sci.,
    23 (1979) 1527.
16. T. Handa, M. Fukuoka, S. Yoshizawa and Y. Hashizume, Rep. Prop.
    Polym. Phys. Jap. 22 (1971) 367.
17. T. Handa, S. Yoshizawa, M. Fukuoka, Y. Hashizume and I. Seo,
    Proc. 21st. Japan Congr. Mater. Res., (1978) 304.
18. T. Handa, S. Yoshizawa, Y. Ikeda and M. Saito, Kobunshi Ronbun-
    shu 33 (1976) 147.
19. T. Handa, S. Yoshizawa, M. Suzuki and T. Kanamoto, Kobunshi
    Ronbunshu 35 (1978) 117.
20. T. Arima, Mokuzai Gakkaishi 19 (1973) 435.
21. T. Handa, Y. Hashizume, T. Ishii, T. Kanamoto, S. Yoshizawa and
    T. Nakamura , A.C.S. Series, submitted.

# PHASE DOMAIN SIZE AND CONTINUITY IN SEQUENTIAL IPN's:  A REVIEW

L.H. Sperling, J.M. Widmaier, J.K. Yeo and J. Michel*

Materials Research Center
Coxe Laboratory 32
Lehigh University
Bethlehem, Pennsylvania  18015

## INTRODUCTION

An interpenetrating polymer network, IPN, is defined as a combination of two polymers in network form, at least one of which is synthesized and/or crosslinked in the immediate presence of the other.  While an IPN is closely related to polymer blends, grafts, and blocks, it can be distinguished in two ways:  (1) An IPN swells, but does not dissolve in solvents, and (2) creep and flow are suppressed.

The name "interpenetrating polymer networks" dates back to the work of John Millar in 1960 (1). However, more than 20 years later, the definition is still changing as ideas about the system evolve.  Millar's research involved a study of polystyrene/polystyrene IPN's, where both networks were identical.  In 1967, academic research began with the work of Frisch, et al. (2), and Sperling and Friedman (3), and is now taking place in many laboratories (4-6).

Sperling and Friedman began research on what is now called sequential IPN's, where polymer network I is synthesized first, then monomer II plus crosslinker and initiator is swollen in and polymerized.  Using poly(ethyl acrylate)/polystyrene IPN's, these workers noted the presence of two glass transitions, and that

* The authors are pleased to acknowledge the financial support of the National Science Foundation through Grant No. DMR-8106892, Polymers Program.

their samples were milky.  They concluded that the two networks
were phase separated, but had only a rough idea as to the phase
domain size.

One qualitative experiment performed by Sperling and Friedman
(7) to estimate the phase domain size, however, is worth reporting
here.  When the samples were held so that white light was trans-
mitted through the samples into the eye, they appeared reddish.
When the samples were held so that they were viewed by reflected
light, however, they appeared bluish.  Such optical effects indicated that the domains had to be smaller than the wavelength of
light.  From the intensity of the colors, a domain size of between
200 and 2000 Å appeared most likely.  In any event, these workers
had no idea as to the domain shapes, or size distribution.

In 1972, Huelck, et al. (8) reported transmission electron
microscopy studies on this system.  They found domains of about
500-1000 Å, depending on the composition.  The phase domain size
also depended on the compatibility of the polymer pair.  For the
system poly(ethyl acrylate)/poly(styrene-co-methyl methacrylate),
compositions richer in methyl methacrylate yielded smaller domains
than systems richer in styrene.  It should be noted that poly(ethyl
acrylate) and poly(methyl methacrylate) are chemically isomeric.

The domains were nearly uniform in size, in the form of a cellular structure.  Polymer network I made up the comb, and network II
always made up the "honey".  However, there was some suggestion, even
at that date, of dual phase continuity in some compositions.

After several more papers devoted to the characterization of
IPN's via dynamic mechanical spectroscopy and transmission
electron microscopy (4), Donatelli, et al. derived a semi-
empirical equation which expressed the domain size of sequential
IPN's in terms of the crosslink density of network I, the volume
fractions of the two networks, and their interfacial tension (9).
While the derivation contained some rather important assumptions,
it correctly predicted the domain size of full IPN's (both polymers
crosslinked), and semi-1 IPN's (only polymer I crosslinked).

More recently, Yeo, et al. (10) derived a new equation, using
more fundamental concepts, and removing some of the assumptions.
This derivation will be summarized below.

Both Donatelli and Yeo assumed the presence of isolated
spheres, in spite of growing evidence that at least some compositions exhibited dual phase continuity.  One problem was that
the required equations did not come out in a manageable form for
the assumption of interconnected cylinders, for example.

During all this time, there was growing evidence supporting the notion of dual phase continuity in sequential IPN's. However, it remained for the work of Widmaier and Sperling (11, 12) to establish this important point. These experiments will also be summarized here. The objective of this paper is to review the understanding of phase domain size and continuity in sequential IPN's, updating a recent monograph (4) in this important area.


THE BASIC EQUATION FOR SEQUENTIAL IPN DOMAIN DIAMETERS

Physical Model

A working model of the domain structure in IPN's and related materials is depicted in Figure 1. Polymer II constitutes a spherical core and is in a contracted (deformed) state, while polymer I surrounds the core and is in an expanded (deformed) state While the model in Figure 1 is oversimplified, it describes the physical situation of a spherical domain. Related to this model, several essential relationships are evolved:  Let $\phi_1$ and $\phi_2$ be the volume fractions of polymer I and II, and $\alpha$ the linear deformation ratio of polymer network I from swelling.

A spherical Domain of Polymer II
surrounded by a shell of Polymer I

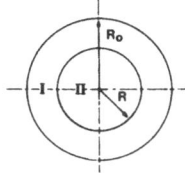

$R_0$ . Radius of an imaginary spherical
region containg both polymers
I & II

R    Radius of a polymer II domain

Figure 1:  A working model of a domain.

The following simple geometric constraints will be noted:

$$\phi_1 + \phi_2 = 1 \tag{1}$$

$$R^3 = (1 - \phi_1) R_o^3 = \phi_2 R_o^3 \tag{2}$$

$$\alpha = \phi_1^{-1/3} \tag{3}$$

Subscripts 1 and 2 denote polymer (or monomer) I and polymer (monomer) II, respectively, throughout, and R is the radius of a polymer II domain.

Process Path

Several assumptions are made for the derivation:

1)  Thermodynamic equilibrium processes exist throughout the development of the domain formation.
2)  The domains have identical diameters with a spherical shape.
3)  The polymer networks obey Gaussian statistics.
4)  A sharp interfacial boundary exists between the two phases.

The process path of domain formation is illustrated in Figure 2.  Initially, in state 1, network I is completely separated from monomer II (plus crosslinker).  In State 2, polymer network I is swollen with the monomer II mixture.  The path from State 1 to State 2 is accompanied by the mixing (dilution) between polymer I and monomer II, and mutual concomitant expansion of polymer I caused by swelling with monomer II mixture.  The free energy of polymerization on going from State 2 to State 3 will be ignored, as it is not of interest to this problem.  Also, the enthalpic 1,2 contact energies between monomer II and polymer I will be assumed to be the same as the polymer II-polymer I enthalpic contact energies.

State 3 is the hypothetical, mutually mixed state, where polymers I and II are mixed and mutually diluted.  Network I is stretched in the Flory-Rehner mode (13), although maximum swelling (with excess monomer) is not assumed.  Demixing (phase separation) between polymer I and polymer II, with concomitant deformation of polymer II with further deformation of polymer I into a shell leads to State 4.

State 4 shows a phase-separated state, with a spherical domain of polymer II forming as the core, surrounded by polymer

I, deformed into a spherical shell.  In reality, however, the
core and shell are not sharply demarked, with some chains
mechanically trapped in the wrong phase.  Since the distance of
diffusion required is less than 500 Å in most cases, this is of
the same order as the dimensions of the chains themselves.

Referring to Figure 2, the molecular rearrangements taking
place on transforming State 3 to State 4 requires amplification.
Certainly no covalent bonds are broken during the process, as
would be required from a literal interpretation of the model.
Instead, phase separation ensues at an early stage of the polymerization of monomer II, when the free energy of mixing becomes

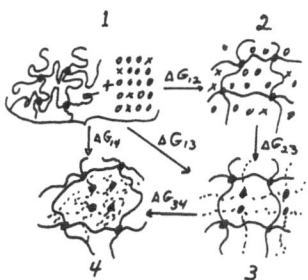

Figure 2.  A simplified path of domain formation.

positive.  This is probably at or before the gel stage for many
IPN systems.  Thus, the molecular migration begins earlier than
illustrated in the model, and hence State 3 is hypothetical, for
calculation purposes only.

Thermodynamics of Phase Separation

For a closed system at constant pressure and temperature,
the Gibbs free energy (hereafter free energy) change, $\Delta G$, is
given by

$$\Delta G = \Sigma(\Delta H_{i,i+1}) - T\Sigma(\Delta S_{i,i+1}) \qquad (i = 1,2,3) \qquad (4)$$

where $\Delta H_{i,i+1}$, $\Delta S_{i,i+1}$ represent the enthalpy and entropy changes involved in the process from state i to state i+1, respectively; and T is the absolute temperature.

Extending equation (4) to the domain formation process, the free energy change for polymer II domain formation, $\Delta G_d$, can be expressed as

$$\Delta G_d = \sum_{i=1}^{3} (\Delta H_{i,i+1}) - T \sum_{i=1}^{3} (\Delta S_{i,i+1}) + \Delta G_i \qquad (5)$$

where $\Delta G_i$ represents the interfacial free energy change for domain formation.

The quantity $\Delta G_i$ is a thermodynamic property related to the process taking place from State 3 to State 4. Figure 2 indicates that the path from State 1 to State 3 via State 2 can be replaced by the direct path from State 1 to 3. Therefore, eq. (5) reduces to

$$\Delta G_d = \Delta H_{13} + \Delta H_{34} - T(\Delta S_{13} + \Delta S_{34}) + \Delta G_i \qquad (6)$$

In fact, $\Delta H_{13}$ and $\Delta H_{34}$ are the heat of mixing and the heat of demixing (the negative heat of mixing) between polymers I and II, respectively, such that the sum of these two terms can be assumed to be zero. This, of course, is an approximation, which assumes that the contact heats between monomer II and polymer I do not change on polymerization. The phase separation from State 3 to State 4 involves configurational and conformational re-arrangements which are not necessarily balanced by the trans-formation from State 1 to State 3. There will be an entropic component to each rearrangement that will not necessarily cancel each other. (The free energies of polymerization are not con-sidered.)

The quantity $\Delta S_{13}$ is equal to the sum of the entropy of mixing, $\Delta S_m$, and the entropy change for the elastic deformation of polymer I being swollen with polymer II, $\Delta S_{sw}^{I}$. The quantity $\Delta S_{34}$ is equal to the sum of the entropy change of demixing, $\Delta S_{dm}$; the rearrangement entropy change for elastic deformation (con-traction of the polymer II network upon deswelling), $\Delta S_{dsw}^{II}$; and the entropy change for elastic deformation (biaxial inflation) of polymer I network, $\Delta S_{df}^{I}$. Again, $\Delta S_m$ and $\Delta S_{dm}$ cancel each other, ending up zero. The interfacial free energy change, $\Delta G_i$ consists of the interfacial free energy change for quiescent domain formation, $\Delta G_i^o$ A term for the entropy change on placing polymer I and II molecules in each domain, $\Delta S_p$, must also be added to $\Delta S_{34}$. In summary, the free energy change for polymer II domain formation can be expressed as follows:

$$\Delta G_d = -T(\Delta S^I_{sw} + \Delta S_p + \Delta S^{II}_{dsw} + \Delta S^I_{df}) + \Delta G_1 \qquad (7)$$

The superscripts I and II indicate changes for polymers I and II only.

In the following sections, each of the thermodynamic quantities in eq. (7) will be developed.

### Entropy Change for the Elastic Deformation of Polymer I by Swelling with Polymer II

Polymer network I undergoes an isotropic deformation by being swollen with polymer II, which gives rise to an entropy change. The entropy change upon affine elastic deformation is given by (13-17),

$$\Delta S^I_{sw} = - \frac{\nu'_I}{2} \mathcal{R} (3\alpha^2 - 3 - \ln \alpha^3) \qquad (8)$$

where $\nu'$ is the number of moles of effective network chains, $\alpha$ represents the linear deformation ratio upon isotropic swelling, and $\mathcal{R}$ represents the gas constant.

Then, from eq. (2) and Figure 1,

$$\nu'_1 = \frac{4}{3} \pi R_o^3 \phi_1 \nu_1 = \frac{\pi}{6} (\frac{\phi_1}{\phi_2}) \nu_1 D_2^3 \qquad (9)$$

where $D_2$ is the diameter of the polymer II domain and $\nu_1$ is the number of effective network chains (in moles) of polymer I per unit volume of dry polymer I.

The quantity $\alpha$ is equal to $\phi_1^{-1/3}$, as defined in eq. (3). Therefore, $\Delta S^I_{sw}$ is finally expressed as

$$\Delta S^I_{sw} = - \frac{\pi}{12} (\frac{\phi_1}{\phi_2}) \nu_1 \mathcal{R} (3\phi_1^{-2/3} - 3 + \ln \phi_1) D_2^3 \qquad (10)$$

### Rearrangement Entropy Change

The thermodynamic probability, $\Omega$, of placing $N_1$ polymer I linear molecules and $N_2$ polymer II linear molecules into the shell domain and its core could be expressed as

$$\Omega = \phi_1^{N_1} \phi_2^{N_2} \qquad (11)$$

where $\phi_1$ and $\phi_2$ represent the volume fraction of the shell and the core, respectively.

The entropy change, $\Delta S$, and $\Omega$ are related by the Boltzmann equation

$$\Delta S = -k\ell n\ \Omega \tag{12}$$

where k is Boltzmann's constant.

Combining eq. (11) and (12), the rearrangement entropy change $\Delta S_p$, of placing the polymers I and II in domains and expressing $N_1$ and $N_2$ in moles yields

$$\Delta S_p = -\ \mathcal{R}\ (N_1\ \ell n\ \phi_1 + N_2\ \ell n\ \phi_2) \tag{13}$$

Then, referring to Figure 1 and equation (2),

$$N_1\ =\ \frac{4}{3}\pi\ (R_o^3 - R^3)\ \frac{\rho_1}{M_1}\ =\ \frac{\pi}{6}\ (\frac{\phi_1}{\phi_2})\ (\frac{\rho_1}{M_1})\ D_2^3 \tag{14}$$

and

$$N_2\ =\ \frac{4}{3}\pi R^3 (\frac{\rho_2}{M_2})\ =\ \frac{\pi}{6}\ (\frac{\rho_2}{M_2})\ D_2^3 \tag{15}$$

where $\rho_1$ and $\rho_2$, and $M_1$ and $M_2$ represent the densities and molecular weights of polymers I and II, respectively.

Substituting eq. (14) and eq. (15) into eq. (13), the rearrangement entropy is finally expressed as

$$\Delta S_p = -\ \frac{\pi}{6}\ \mathcal{R}\ (\frac{\phi_1}{\phi_2}\ \frac{\rho_1}{M_1}\ \ell n\ \phi_1 + \frac{\rho_2}{M_2}\ \ell n\ \phi_2)\ D_2^3 \tag{16}$$

Entropy Change for Elastic Deformation of Polymer II

Polymer II is polymerized in the presence of polymer I such that polymer II is in a swollen state. The swollen polymer II then deswells into the spherical core, which constitutes an isotropic contraction type of deformation. Therefore, again from eq. (9) and eq. (2),

$$\nu_2' = \frac{4}{3}\pi\ R_o^3\ \phi_2\ \nu_2 = \frac{\pi}{6}\ \nu_2\ D_2^3 \tag{17}$$

where $\nu_2'$ is the number of moles of elastically effective polymer II network chains, not considering the total volume. The polymer II linear contraction is given by equation (8), and

$$\alpha = (\frac{\phi_2}{\phi_1 + \phi_2})^{1/3} = \phi_2^{1/3} \tag{18}$$

such that

$$\Delta S_{dsw}^{II} = - \frac{\pi}{12} \, v_2 \, \boldsymbol{\rho} \, (3 \, \phi_2^{2/3} - 3 - \ln \phi_2) D_2^3 \tag{19}$$

## Entropy Change for Biaxial Elastic Deformation (Inflation of Polymer I)

As the domain model in Figure 1 indicates, polymer I is biaxially deformed into a spherical shell. The deformation entropy change, $\Delta S_{df}^{I}$, is related to different strains in different directions and hence could be represented in terms of $\lambda_x$, $\lambda_y$, and $\lambda_z$. Evaluation of $\Delta S_{df}^{I}$ is done by working out the geometry of biaxial deformation. As depicted in Figure 3, an element in a sphere of the swollen network ends up strained and in a new location in the spherical shell. The volume of the isotropically swollen element, $\Delta V_1$, equals $\Delta X_1 \Delta Y_1 \Delta Z_1$.

Upon demixing and moving to $r_2$, its volume becomes $V_2 = \phi_1 \, \Delta X_2 \, \Delta Y_2 \, \Delta Z_2 = \phi_1 \, \Delta X_2^2 \, \Delta Z_2$ (i.e. biaxially deformed).

The entire shell at $r_1$ moves to $r_2$ upon demixing and is strained such that

$$\phi_1 = \frac{4 \pi \, r_2^2 \, dr_2}{4 \pi r_1^2 \, dr_1} \tag{20}$$

i.e.

$$dr_2 = \phi_1 \, (\frac{r_1^2}{r_2^2}) \, dr_1 \tag{21}$$

Let $\alpha_{zs}$ be the strain ratio in radial direction, then

$$\alpha_{zs} = \frac{dr_2}{dr_1} = \phi_1 \, \frac{r_1^2}{r_2^2} \tag{22}$$

The area ratio of the shells at $r_2$ and $r_1$ is $4 \pi r_2^2 / 4 \pi r_1^2 = r_2^2 / r_1^2$ such that the strain ratio in x direction is expressed as

$$\alpha_{xs} = (\frac{r_2^2}{r_1^2})^{1/2} = \frac{r_2}{r_1} \tag{23}$$

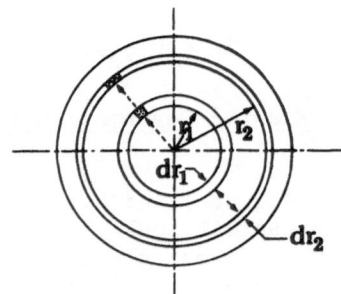

Figure 3,   Working geometry for biaxial deformation -- I.

Since the deformation is biaxial,

$$\alpha_{ys} = (\frac{r_2^2}{r_1^2})^{1/2} = \frac{r_2}{r_1} \tag{24}$$

Now it is necessary to find $r_2$ in terms of $r_1$ and $\phi_1$.  After phase separation, the material which was inside $r_1$ resides in a spherical shell between R (Figure 1) and $r_2$, as deformed (see Figure 4), where R is the final value for $r_2$ after phase domain formation is complete.  Then:

$$\frac{4}{3} \pi r_1^3 \phi_1 = \frac{4}{3} \pi (r_2^3 - R^3) \tag{25}$$

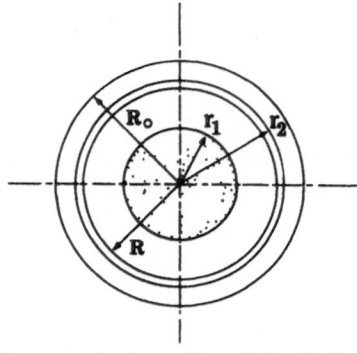

Figure 4.   Working geometry for biaxial deformation - II.

such that

$$r_2 = (R^3 + \phi_1 \; r_1^3)^{1/3} \tag{26}$$

Substituting eq. (2) into (26),

$$r_2 = \{R_o^3(1-\phi_1) + \phi_1 \; r_1^3\}^{1/3} \tag{27}$$

Recognizing that $\lambda_x = \alpha \; \alpha_{xs}$, eqs. (3), (23), and (27) can be combined to give

$$\lambda_x = \phi_1^{-1/3}\{R_o^3(1-\phi_1) + \phi_1 \; r_1^3\}^{1/3}/r_1 \tag{28}$$

Let $r_1$ equal $fR_o$, such that f represents a fractional radius position going from 0 to 1 over the sphere. Then,

$$\lambda_x = \lambda_y = \phi_1^{-1/3}\{R_o^3(1-\phi_1) + \phi_1 f^3 \; R_o^3\}^{1/3}(fR_o)^{-1} \tag{29}$$

$$\lambda_x = \lambda_y = \{1 + \frac{\phi_2}{\phi_1} \; f^{-3}\}^{1/3} \tag{30}$$

Likewise, from $\lambda_z$ equalling $\alpha$ times $\alpha_{zs}$, and eqs. (3), (22) and (27),

$$\lambda_z = \phi_1^{-1/3} \; \phi_1 r_1^2 \{R_o^3(1-\phi_1) + \phi_1 \; r_1^3\}^{-2/3} \tag{31}$$

$$\lambda_z = \{1 + (\frac{\phi_2}{\phi_1})f^{-3}\}^{-2/3} \tag{31a}$$

For a spherical shell of radius $r_1$ and thickness $dr_1$, its volume is $4\pi r_1^2 \; dr_1$ which contains $\nu_1 \; \phi_1$ $(4\pi r_1^2)dr_1$ elastically effective network chains.

From a consideration of the entropy change on affine deformation via swelling (13-16),

$$d(\Delta S_{df}^I) = \frac{\nu_1}{2}\mathcal{R}\phi_1(4\pi r_1^2) \; (\lambda_x^2+\lambda_y^2+\lambda_z^2 - 3 - \ln \lambda_x\lambda_y\lambda_z)dr_1 \tag{32}$$

The quantity $\ln \lambda_x\lambda_y\lambda_z$ is seen to equal zero.

$$d(\Delta S_{df}^I) = \frac{\pi}{4} \; \nu_1(\frac{\phi_1}{\phi_2})\mathcal{R} \; D_2^3\{2(1 + \frac{\phi_2}{\phi_1} \; f^{-3})^{2/3} +$$

$$(1 + \frac{\phi_2}{\phi_1} \; f^{-3})^{-4/3}-3\}f^2df \tag{32a}$$

$$d(\Delta S^I_{df}) = \frac{\pi}{4} \nu_1 (\frac{\phi_1}{\phi_2}) \mathcal{R} \, D_2^3 \{2f^2(1 + \frac{\phi_2}{\phi_1}f^{-3})^{2/3}$$

$$+ f^2(1 + \frac{\phi_2}{\phi_1}f^{-3})^{-4/3} - 3f^2\}df \qquad (32b)$$

Integrating equation (32b) from $f = 0$ to $f = 1$ (17), the entropy change for biaxial elastic deformation of polymer I is finally expressed as

$$\Delta S^I_{df} = \frac{\pi}{4} \nu_1 \mathcal{R} (\frac{1}{\phi_2}) (2\phi_1^{1/3} - \phi_1^{4/3} - \phi_1) D_2^3 \qquad (33)$$

## Interfacial Free Energy Change for Domain Formation

The interfacial free energy change for domain formation is brought about from the intrinsic interfacial tension between the two polymers, $\gamma^o$. This can be expressed as

$$\Delta G_i^o = \pi \gamma^o D_2^2 \qquad (34)$$

for spheres.

Now, the free energy change for domain formation, $\Delta G_d$, can be expressed by substituting equation (10), (16), (19), (33), and (34) into equation (7). After appropriate manipulation and rearrangement, $\Delta G$ is given by

$$\Delta G_d = -\frac{\pi}{6} \mathcal{R} T [-(\frac{\phi_1}{\phi_2} \frac{\rho_1}{M_1} \ln \phi_1 + \frac{\rho_2}{M_2} \ln \phi_2) +$$

$$\frac{\nu_1}{2} (\frac{1}{\phi_2}) (3\phi_1^{1/3} - 3\phi_1^{4/3} - \phi_1 \ln \phi_1) -$$

$$\frac{\nu_2}{2} (3\phi_2^{2/3} - 3 - \ln \phi_2)] D_2^3 + \pi \gamma^o D_2^2 \qquad (35)$$

In order to determine the domain size, $D_2$, which gives a minimum in the free energy, the first partial derivative of equation (36) with respect to $D_2$, $\partial(\Delta G_d)/\partial D_2$, is equated to zero and solved for $D_2$, i.e.

$$\frac{\partial(\Delta G_d)}{\partial D_2} = -\frac{\pi}{2}\mathcal{R}T[-(\frac{\phi_1}{\phi_2}\frac{\rho_1}{M_1}\ln\phi_1 + \frac{\rho_2}{M_2}\ln\phi_2) +$$

$$\frac{\nu_1}{2}(\frac{1}{\phi_2})(3\phi_1^{1/3} - 3\phi_1^{4/3} - \phi_1\ln\phi_1) -$$

$$\frac{\nu_2}{2}(3\phi_2^{2/3} - 3 - \ln\phi_2)]D_2^2 + 2\pi\gamma^\circ D_2 = 0 \qquad (36)$$

Thus solving for $D_2$

$$D_2 = 4\gamma^\circ[\mathcal{R}T(A\nu_1 + B\nu_2 - C)]^{-1} \qquad (37)$$

where

$$A = \frac{1}{2}(\frac{1}{\phi_2})(3\phi_1^{1/3} - 3\phi_1^{4/3} - \phi_1\ln\phi_1) \qquad (38)$$

$$B = \frac{1}{2}(\ln\phi_2 - 3\phi_2^{2/3} + 3) \qquad (39)$$

$$C = \frac{\phi_1}{\phi_2}\frac{\rho_1}{M_1}\ln\phi_1 + \frac{\rho_2}{M_2}\ln\phi_2 \qquad (40)$$

Thus, eq. (37) provides $D_2$ in terms of the volume fraction and crosslink density of each component, and the interfacial tension. Terms for molecular weight are provided, if applicable. Specific forms of eq. (37) for individual cases will be derived in the next section.

SPECIFIC FORMS FOR EQUATION 37

Characteristics

The experimental variables required for equation (37) are the volume fraction, density, crosslink density, molecular weight, interfacial tension and temperature.

All of these variables are measurable experimentally or ob-tainable by calculation from experiment.

Equation (37) can be applied to various IPN's cases in specific forms. A few of the most important follow. Others are developed in ref. (30).

(1) The Case of $\nu_1 \neq 0$ and $\nu_2 \neq 0$ for sequential IPN's:

With $M_1 = M_2 = \infty$, eq. (37) is simplified to

$$D_2 = 4\gamma^\circ [\mathcal{R} T(A\nu_1 + B\nu_2)]^{-1} \tag{41}$$

Equation (41) predicts that the domain diameter of polymer II depends on the interfacial tension as well as on the crosslink densities of both networks I and II. However, an evaluation of the relative magnitudes of the constants A and B indicate that $\nu_1$ is about 10 times as important as $\nu_2$.

(2)  The Case of $\nu_1 \neq 0$ and $\nu_2 = 0$:

This is a semi-IPN of the first kind. With $M_1 = \infty$, eq. (37) reduces to

$$D_2 = 4\gamma^\circ [\mathcal{R} T(A\nu_1 - \frac{\rho_2}{M_2} \ln \phi_2)]^{-1} \tag{42}$$

(3)  The Case of $\nu_1 = 0$ and $\nu_2 \neq 0$:

This is the case of semi-IPN of the second kind. With $M_2 = \infty$, eq. (37) gives

$$D_2 = 4\gamma^\circ \{ \mathcal{R} T(B\nu_2 - \frac{\phi_1}{\phi_2} \frac{\rho_1}{M_1} \ln \phi_1) \}^{-1} \tag{43}$$

## The Donatelli Equation

Before proceeding to the experimental part, the Donatelli equation corresponding to equation (37) will be reviewed. The theory of Donatelli, et al. (9,18) indicates a linear form:

$$D_2 = 2\gamma^\circ \phi_2 / \mathcal{R} T\nu_1 \phi_1 [(\frac{1}{\phi_1})^{2/3} + (\frac{\phi_2}{\phi_1}) \frac{1}{\nu_1 M_2} - \frac{1}{2}] \tag{44}$$

and a cubic form:

$$\frac{\nu_1^2 D_2^3}{c^2 K^2} \phi_1 + (\frac{\phi_2}{M_2} - \frac{\phi_1 \nu_1}{2}) D_2 = \frac{2\gamma^\circ \phi_2}{\mathcal{R} T} \tag{45}$$

where c, K $(= r_0/M^{1/2})$ are constants, and $\phi$ represents the volume fraction. In the calculation of the domain sizes in Table I. column C [Eq. (45)], c = $\sqrt{2}$ and K = 8 x $10^{-9}$ cm(mole/gm)$^{1/2}$, respectively. The quantity $\nu_1$ is defined for the dry network I state (25).

Several special solutions of equation (45) are of interest.

(1)   The Case of $\phi_2 \cong 0$ or $\gamma \cong 0$

When $\gamma = 0$, equation (45), which is a cubic with three solutions, reduces to

$$D_2^2 = c^2 K^2 / 2 \nu_1 \tag{46}$$

and

$$D_2 = 0 \tag{47}$$

For $K = 8 \times 10^{-9}$ cm(mole/gm)$^{1/2}$ and $\nu_1 = 3.39 \times 10^{-5}$ mole/gm., then $D_2 = \pm 100\text{Å}$, which agrees well with experiments on polystyrene/polystyrene homo IPN's (19). Similar results are obtained as $\phi_2$ approaches zero.

(2)   The Case of $\nu_1 = 0$

As the crosslink density of polymer I approaches zero, the structure and morphology of a graft type copolymer is approached. In this case, eq. (45) yields

$$D_2 = 2\gamma M_2 / RT \tag{48}$$

For $\gamma = 3$ dyn/cm and $M_2 = 2 \times 10^5$ g/mole, a domain diameter of the order of 4000-5000 Å is obtained. Note that eq. (48) does not have a dependence on $\phi_2$. This value compares well with the cellular structures within the rubber particles in HIPS, for example. (21)

Equation (48) may be of interest to the field of colloid science, setting a lower limit on the domain size that can be obtained with oligomeric and polymeric dispersions.

(3)   The Case of $\phi_2 = 1$

When $\phi_2$ approaches unity (100% polymer II), eq. (45) predicts that $D_2$ goes to infinity. (As the pure phase, polymer II must have infinite bounds.)

Interestingly, the case of $\nu_1 = 0$, $\phi_2$ near unity and polymer II crosslinked (a semi-II IPN of nearly all polymer II) was studied by Donatelli et al. They confirmed the size of the polymer II domain increases as $\phi_2$ approached unity.(20)

(4) On the "Constant" C

Originally, C was taken as the square root of 2, which was based on the best estimate of the average number of crosslinked chain segments needed to circumnavigate a domain. For a final equation without arbitrary constants, this assumption was the most important.

Instead of taking $C = \sqrt{2}$, a variable quantity may be assumed. For example,

$$C = 1/(1-\phi_2)^{1/2} \qquad\qquad (49)$$

tends to modify eq. (45) for an IPN. (Note that for $\phi_2 = 0.5$, $C = \sqrt{2}$, still.) Equation (49) follows from an analysis that C probably increases as $\phi_2$ increases, but is otherwise arbitrary.

If $\gamma = 0$, eq. (45) yields

$$D_2^2 = \frac{C^2 K^2 M_{c,1}}{2\rho_1} \qquad\qquad (50)$$

Substituting eq. (49) into eq. (46) yields

$$D_2^2 = C^2 K^2 M_{c,1}/2\,\rho_1 \qquad\qquad (51)$$

Then, from eqs. (9) and (11);

$$D_2 = K \left[ \frac{M_{c,1}}{2(1 - \phi_2)\rho_1} \right]^{1/2} \qquad\qquad (52)$$

which predicts that $D_2$ goes to infinity as $\phi_2$ goes to unity. For $\phi_2$ approaching zero, a finite value of the order of 100 Å is obtained for $D_2$, depending on $M_{c,1}$. If C is taken as a constant ($\sqrt{2}$), then $D_2$ is a constant if $\gamma$ is zero. Thus, for a homo-IPN, $D_2$ may be either a constant or a variable, depending upon the exact assumptions.

POLY(N-BUTYL ACRYLATE)/POLYSTYRENE IPN'S AND RELATED MATERIALS

Yeo, et al. (22-24) have done morphology studies using TEM on poly(n-butyl acrylate)/polystyrene, PnBA/PS, IPN's and related materials.

Table I shows the domain sizes for this system, illustrating a wide range of crosslink levels. Column A illustrates the Yeo theory. Columns B and C of Table I contain the calculated values

Table 1

Experimental and Theoretical Domain Sizes for Poly(n-butyl acrylate)/Polystyrene IPN's and Semi IPN's (22)

| System* | Variables $\nu \times 10^5$ mole/cm$^3$ $M \times 10^{-5}$ gms/mole | Volume Ratio | Experiment | Domain Diameter, $D_2$ (Å) Theory, $\gamma^\circ$ = 3.65 dynes/cm A | B | C |
|---|---|---|---|---|---|---|
| | $\nu_1=3.7$ $\nu_2=21.8$ | 25/75 | 800 | 845 | 1,170 | 447 |
| | | 40/60 | 650 | 644 | 883 | 362 |
| | | 50/50 | 550 | 572 | 725 | 321 |
| IPN | $\nu_1=21.8$ $\nu_2=21.8$ | 25/75 | 200 | 207 | 200 | 141 |
| | | 40/60 | 170 | 169 | 148 | 115 |
| | | 60/40 | 150 | 143 | 98 | 93 |
| | $\nu_2=21.8$:constant | | | | | |
| | $\nu_1=3.7$ | 50/50 | 550 | 572 | 725 | 321 |
| | $\nu_1=14.0$ | | 260 | 224 | 192 | 137 |
| | $\nu_1=21.8$ | | 195 | 154 | 124 | 103 |
| | $\nu_1=25.0$ | | 120 | 136 | 108 | 94 |
| Semi-1 | $\nu_1=3.7$ $M_2=3.0$ | 25/75 | 1,250 | 1,314 | 1,037 | 441 |
| | | 40/60 | 1,000 | 1,084 | 804 | 359 |
| Semi-1 | $\nu_1=21.8$ $M_2=3.0$ | 25/75 | 250 | 227 | 193 | 140 |
| | | 60/40 | 180 | 161 | 97 | 92 |
| Semi-2 | $M_1=2.0$ $\nu_2=10.0$ | 33/67 | 3,330 | 3,260 | -- | -- |

* TEGDM for crosslinker I and DVB for crosslinker II.
A Present theory.
B The linear form of Donatelli equation, eq. (44).
C The cubic form of Donatelli equation, eq. (45).

of the domain diameter according to linear and cubic forms of the Donatelli equation.

Both the Yeo theory and the Donatelli, et al., theory show that the domain sizes are sensitive to the value of interfacial tension.

It should be noted that eqs. (44) and (45) do not provide any values for the semi-2 IPN case.

PHASE DOMAIN CONTINUITY IN IPN'S

Both Donatelli and Yeo assumed isolated spheres in their respective derivations. The experimental data shown in Table I was based on chord lengths, assuming spheres. Via transmission electron microscopy, this appears not to be such a bad approximation, as most systems show a cellular structure of either nearly round or somewhat elongated form. Often, however, there is some indication of phase continuity (20). Modulus studies also indicated some degree of phase continuity (2). The problem of how to get direct experimental evidence remained for more than ten years.

The approach selected for the present study involves the selected decrosslinking and dissolution of one network or the other (11,12). This involved the use of acrylic acid anhydride, AAA, as the crosslinker. This crosslinker quantitatively hydrolyzes after exposure to warm ammonia water overnight. The now linear polymer component was leached out in a Soxhlet extractor for ten days, using either toluene or acetone. Density measurements were made on a picnometer, and scanning electron microscopy was carried out.

In Table II, experimental values of $M_c$ estimated from swelling measurements and shear modulus at 10 seconds, are compared with the theoretical values calculated from the monomer/crosslinker ratio. The data show differences in effectiveness between divinyl-benzene and acrylic acid anhydride as crosslinkers. When DVB is used as the crosslinker, the experimental values agree well with the theoretical ones. The close agreement may be fortuitous however, since one would expect physical crosslinks to contribute to the experimentally measured corsslink level, expecially when modulus was used. Apparently, network defects and physical corsslinks nearly cancel each other out in this system.

For AAA on the other hand, the $M_c$ values are much higher than expected, indicating lower crosslink formation. Partial hydrolysis of AAA could not explain the results since less than 5% free acrylic acid was found in the AAA by infra-red analysis. An alternative explanation for the lower than expected crosslink densities for AAA crosslinked networks may be the possibility of ring formation. It is well known that the crosslinking in diacrylate or dimethacrylate systems is not complete, due to competition between

Table II: Homopolymer Network Characterization Data (11)

| Polymer Network | Average molecular weight between crosslinks, $M_c$ | | | | | | |
|---|---|---|---|---|---|---|---|
| | theoretical | from modulus data | from swelling data, solvent: | | | | |
| | | | toluene | THF | benzene | MEK | acetone |
| PnBA (DVB) | 13,000 | 14,400[a] | 12,200 | 12,700 | 12,600 | 13,200 | 12,800 |
| PnBA (AAA) | 13,000 | 38,100[a] | 36,200 | 35,900 | 37,400 | 41,300 | 42,600 |
| PS (DVB) | 10,500 | 10,800[b] | 8,300 | ----- | ----- | 11,000 | ----- |
| PS (AAA) | 10,500 | 27,700[b] | 21,900 | ----- | ----- | 25,800 | ----- |

(a) at room temperature

(b) at 160°C

intermolecular crosslinking and intra-molecular cyclization. In
AAA, the distance between the two vinyl end-groups favors intra-
molecular cyclization. This data is helpful in understanding the
domain sizes reported below.

After the decrosslinking step, the soluble polymer thus obtained was found to have a similar molecular weight to that of the
corresponding linear polymer, prepared in the same experimental
conditions, but without addition of crosslinker. As reported in
Ref. (11), the average molecular weight is around $3.5 \ 10^5$ g/mole for
polystyrene, which is in accordance with previous results from this
laboratory. The difference in molecular weight between linear and
decrosslinked poly(n-butyl acrylate) is around 10% when 0.3% chain
transfer agent was added. The GPC chromatograms show a minor
broadening of the molecular weight distribution for the decrosslinked
polymer. This indicates that the major point of attack was a chemical degradation of the three-dimensional junction sites.

## Sequential PnBA/PS IPN's

Sequential PnBA/PS IPN's were prepared with various compositions,
similar to the work of Yeo, et al. (22-24). Thus, the material was
rigid when polystyrene was the major component, and soft when poly
(n-butyl acrylate) dominated, confirming previous studies by Yeo
et al, (7,8). The exact compositions are shown in Tables III and IV,
and ranged from 30/70 to 90/10.

In the range 30/70 to 70/30, all the samples were opaque whereas for the 80/20 and 90/10 PnBA/PS ratios, they were more transparent but still opalescent. This indicates a phase separated,
heterogeneous system.

## A-type IPN's:   PnBA (AAA)/PS(DVB)

When soaked in ammonium hydroxide, network I is decrosslinked
and can be extracted, see Table III. The amount of PnBA extracted
using acetone roughly equals the amount in the IPN. By using
toluene, the amount of extractable material is higher than expected,
especially for high polystyrene content samples.

Some PS (from imperfect crosslinking of the polymer network II)
is also extracted. For example, analysis by U.V. spectroscopy at 262
mm reveals 11.0% of PS in the sol fraction of a 50/50 decrosslinked
IPN, which explains the high values obtained when toluene was employed as the extracting fluid.

Using acetone, which is a good solvent for the PnBA phase but
a bad one for the PS, the extraction data in Table III show that at
each composition the PnBA phase was quantitatively removed from the
IPN. This means that a) the decrosslinking reaction occurs even

Table III:   Results of Soxhlet Extraction for Decrosslinked
             PnBA (AAA)/PS (DVB) IPN's with Various Compositions
             (11)

| IPN Weight Composition PnBA/PS | Sol Fraction (%) | |
|---|---|---|
| | Toluene as solvent | Acetone as solvent |
| 0/100 | 6.1 | ---- |
| 30/70 | 37.8 | 29.5 |
| 40/60 | 46.4 | 38.3 |
| 45/55 | 48.8 | 43.8 |
| 50/50 | 53.2 | 49.6 |
| 55/45 | 62.7 | 56.1 |
| 60/40 | 69.2 | 61.6 |
| 70/30 | 75.9 | 71.2 |
| 80/20 | 85.1 | 83.0 |
| 90/10 | 94.4 | 94.6 |
| 100/0 | 100 | 100 |

when the PnBA network is in an IPN form (effective diffusion of the
ammonium hydroxide in the network) and b) the grafting level between polymer I and polymer II is low, probably less than five
percent.

The optical properties and the macroscopic aspects of the
samples in the dry state after extraction are of certain importance.
Surprisingly, for every composition the decrosslinked and extracted
IPN was almost transparent.  This must be caused by the partial
collapse to a size where light-scattering is reduced (see below).
Note also that when the IPN had more than 30% PS, a monolithic sample
remained in the thimble; at 30%, after drying, the material was very
brittle.  For the lower compositions studies (10 and 20% PS), small
grains were obtained.  This means that for mid-range compositions,
polymer II was also continuous.  Roughly speaking, the limit of
polymer II continuity is between 70/30 and 80/20 PnBA/PS.

The remaining polystyrene network was characterized by swelling
studies.  For each composition, when possible, the equilibrium
swelling degree in toluene was calculated, and are listed in Table
IV.

The change in $q_v$ with percent PnBA needs explanation.  One can
not consider the extracted and decrosslinked IPN, i.e. the remaining
polymer II as a conventional PS network.  The decrease in density
(see Table IV) together with electron micrographs presented below

Table IV:   Density measurements and Swelling Behavior for
            PnBA (AAA)/PS (DVB) IPN's and Decrosslinked IPN's
            with Various Compositions (11)

| IPN weight composition PnBA/PS | IPN density (g/cm$^3$) | Decrosslinked and extracted IPN $\rho$ (g/cm$^3$) | $q_v$(a) |
|---|---|---|---|
| 0/100  | -----  | -----  | 4.70  |
| 30/70  | 1.053  | 0.984  | 5.78  |
| 40/60  | 1.022  | 0.781  | 6.43  |
| 45/55  | 1.079  | 0.845  | 6.33  |
| 50/50  | 1.067  | 0.795  | 7.04  |
| 55/45  | 1.059  | 0.818  | 10.37 |
| 60/40  | 1.052  | 0.744  | 15.35 |
| 70/30  | 1.023  | 0.965  | 18.96 |

(a)   True swelling plus interstitial solvent held by
      capillary forces.

suggest a porous material.  Some liquid is imbibed in the pores
during swelling as a more or less pure phase.  The materials that had
a higher concentration of PnBA, on swelling have more solvent imbibed,
and hence an apparently larger $q_v$.

   As mentioned above, the density of the decrosslinked, extracted,
A-type IPN's is in the range of 0.8 - 0.9 gms/cm$^3$, clearly lower than
fully dense, and indicating a partial collapse of the structure.  The
invariance of the numbers is probably a consequence of the partial
collapse of the porous structure, also.

   It can be seen that for samples containing over 50% of PnBA,
i.e., when the PS phase is the swollen component in the IPN, $q_v$
deviates from the homopolystyrene equilibrium swelling volume, thus
indicating more and more effect of the pore volume.

Electron Microscopy

   In general, all scanning electron micrographs of decrosslinked
and extracted IPN's revealed the internal appearance of a sponge with
pores of 500Å to one micron.  The decrease in density (see Table IV)
although less than theoretical, confirms a porous structure.  Figures
5 and 6 show micrographs for a mid-range composition.  In Figure 5, the
the lighter regions represent the remaining polymer, which is polystyrene and the dark zones are voids where network I was located.
Undoubtedly, the PS phase is continuous.  The voids seem to continuous too, and hence the poly(n-butyl acrylate) phase must have been
continuous before extraction.  At a higher magnification, Figure 6,

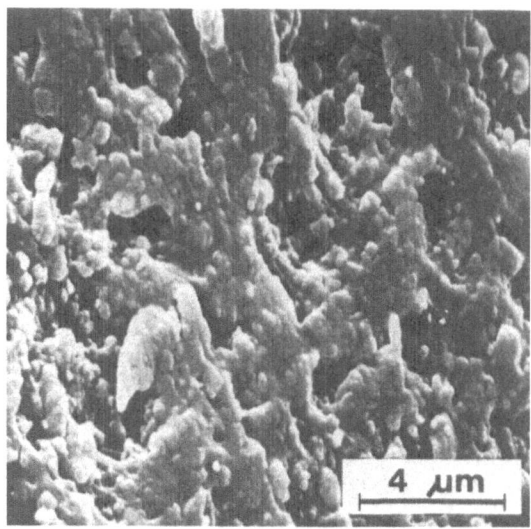

Figure 5:   Scanning electron micrograph of decrosslinked and
            extracted A-type IPN [PnBA(AAA)/PS(DVB), 50/50],
            low magnification. (11)

it can be seen that the continuous phase is formed by an agglomerate
of spheres.  The diameter of these spheres is approximately 100 nm.
This is in accordance with Table I.

For IPN's with only 30% PS, the decrosslinked material collapsed
during the drying step after extraction of polymer I and reveals,
Figure 7, a cracked structure with grooves.  This was suggested by
the value of the density found for that sample, see Table IV.

It is of particular interest to examine the morphology of the
remaining phase in the case of obvious macroscopic discontinuity
(less than 20% PS in the IPN).  Each individual grain exhibited a
porous structure even when resulting from a 90/10 PnBA/PS IPN,
Figure 8.  At high magnification, for the 80/20 ratio, there is no
doubt the voids are continuous to a certain extent, Figure 9.  But
for 90/10, Figure 10, one observes individual particles (300-400 nm
wide) formed by the aggregation of a few spheres.

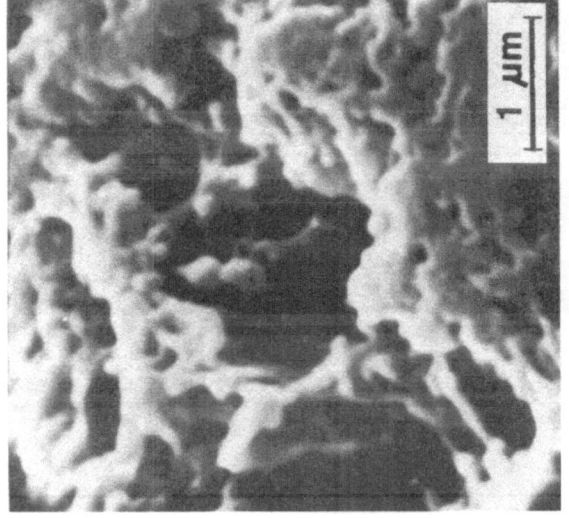

Figure 9: Scanning electron micrograph of decrosslinked and extracted A-type IPN [PnBA(AAA)/PS(DVB), 80/20], high magnification.

Figure 8: Scanning electron micrograph of the remaining polymer II network granular powder after decrosslinking and extraction of (90/10 PnBA(AAA)/PS(DVB) IPN. Low magnification.

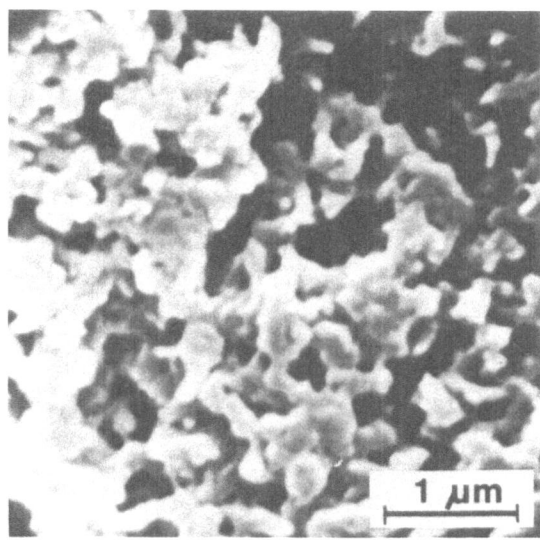

Figure 10:   Scanning electron micrograph of decrosslinked and ex-
             tracted A-type IPN [PnBA(AAA)/PS(DVB), 90/10], high
             magnification.

AN OVERALL COMPARISON OF TECHNIQUES

     Transmission and scanning electron microscopy give dif-
ferent appearances of what is nominally the same sample.  The theoretical equations, of course, have their own limitations and approximations.  It was of interest to compare all three at once, see
Figure 11 (12).  As above, the theoretical value of $\gamma$ was taken to
be 3.65 dynes/cm  for equation (41).  The IPN was based on poly(n-
butyl acrylate)/polystyrene compositions.

     The two experimental techniques and theory agree reasonably
well.  The domain sizes via SEM were about twice as large as those
obtained via TEM.  This is attributed to internal cyclization
reactions when the AAA is used as the crosslinker.  The lower actual
crosslinking level was supported by a higher degree of equilibrium
swelling.

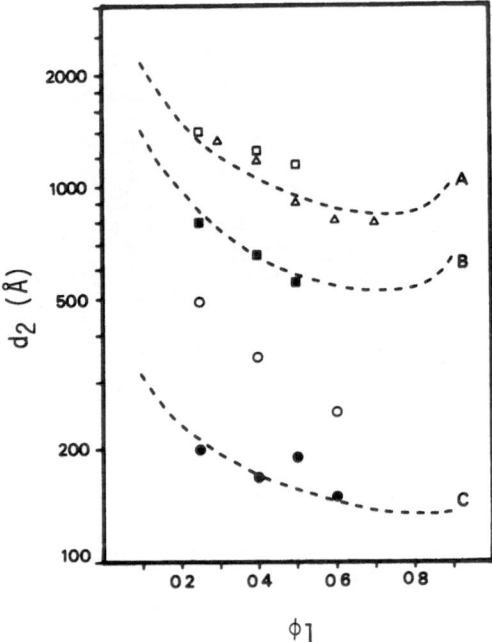

Figure 11.  Polystyrene domain size in poly(n-butyl acrylate)/
            polystyrene IPN's.  Open circles, squares, and
            triangles, domain size for three crosslink levels
            as determined by SEM.  Solid circles and squares,
            corresponding data by TEM.  Dotted lines, equation
            (41), $\gamma$ = 3.65 dynes/cm., based on stoichiometry.

CONCLUSIONS

     The use of a labile crosslinking agent allows a different ap-
proach to the study of interpenetrating polymer networks.  With
acrylic acid anhydride as the crosslinker, hydrolysis leads to a
linear polymer, easy to extract and characterize, and to a pure homopolymer network whose characteristics can be compared with similar networks prepared by classical methods.

     When AAA was used to crosslink PnBA, i.e. polymer I, and DVB
to crosslink PS, i.e., polymer II, nearly the whole amount of decrosslinked polymer can be extracted, indicating a low level of
chemical grafting between polymer I and polymer II.

Interesting morphological details heretofore not available were found for PnBA(AAA)/PS(DVB) IPN's: the remaining network II presents a porous structure as indicated by density measurements and scanning electron microscopy. The voided regions, which correspond to the location of the poly(n-butyl acrylate) phase before extraction, are inter-connected in the composition range studied.

Most interestingly, the extent of continuity of the remaining polystyrene depends on its concentration in the IPN. For mid-range compositions, high magnification scanning electron micrographs showed dual phase continuity. When the amount of polymer II in the IPN was decreased below 30%, the sample crumbled by itself indicating a macro-scopically discontinuous second phase. However, at a microscopic level some of the spheres were still interconnected. It is only when polymer II is around 10% that the discontinuity is evident and only a few individual particles are connected.

The major conclusions from this paper related to the dual phase continuity of PnBA/PS in sequential IPN's are as follows:

(1) Above 20% of polymer network II, its phase domain structure was continuous.
(2) Throughout the composition range studied, polymer network I was continuous.

In this paper, two theories of phase domain size in sequential IPN's were developed and compared, using results based on poly(n-butyl acrylate)/polystyrene. While the earlier theory of Donatelli yielded good results as far as it went, the newer theory of Yeo is much more comprehensive. Both theories assume the existence of isolated spheres, a major drawback.

Scanning electron microscopy and density measurements on de-crosslinked and extracted materials indicate that midrange composi-tions have dual phase continuity. However, assuming equivalent spheres for the domains, both transmission and scanning electron microscopy yield good comparisons with theory.

On a broader scale, this research program has now progressed from the first step of determining that two phases existed in sequential IPN's, to determining phase domain sizes via TEM, to developing theories to express the domain sizes in terms of cross-link densities, volume fractions, and interfacial energies, to the examination of aspects of dual phase continuity. At the present time the work is continuing with a examination of the actual polymer coil dimensions in IPN's, via small-angle neutron scattering of partially deuterated samples (25).

REFERENCES

1.  J.R. Millar, J. Chem. Soc., 1311 (1960).
2.  H.L. Frisch, D. Klempner, and K.C. Frisch, Polym. Lett. $\underline{7}$, 775 (1969).
3.  L.H. Sperling and D.W. Friedman, J. Polym. Sci., $\underline{A-2}$ $\underline{7}$, 425(1969).
4.  L.H. Sperling, "Interpenetrating Polymer Networks and Related Materials", Plenum, New York, 1981.
5.  Yu.S. Lipatov and L.M. Sergeeva, "Interpenetrating Polymeric Networks", Naukova Dumka, Kiev, 1979.
6.  D. Klempner, Angew. Chem., $\underline{90}$, 104(1978).
7.  L.H. Sperling and D.W. Friedman, unpublished.
8.  V. Huelek, D.A. Thomas, and L.H. Sperling, Macromolecules, $\underline{5}$, 340,348(1972).
9.  A.A. Donatelli, L.H. Sperling, and D.A. Thomas, J. Appl. Polym. Sci., $\underline{21}$, 1189(1977).
10. J.K. Yeo, L.H. Sperling, and D.A. Thomas, submitted, Polymer.
11. J.M. Widmaier and L.H. Sperling, accepted, Macromolecules.
12. J.M. Widmaier and L.H. Sperling, unpublished.
13. P.J. Flory, "Principles of Polymer Chemistry," Cornell, Ithaca. 1953.
14. K. Dusek and W. Prins. Adv. Polymer Sci., $\underline{6}$, 1 (1969).
15. F.T. Wall and P.J. Flory, J. Chem. Phys. $\underline{19}$, 1435 (1951).
16. L.R.G. Treloar, "The Physics of Rubber Elasticity," 3rd ed., Clarendon Press, Oxford, 1975.
17. J.K. Yeo, "Morphology and Behavior of Poly(n-butyl acrylate)/ Polystyrene Interpenetrating POlymer Networks and Related Materials," Ph.D. Thesis, Lehigh University, 1981.
18. J. Michel, S.C. Hargest, and L.H. Sperling, J. Appl. Polym. Sci., $\underline{26}$, 743 (1981).
19. D.L. Siegfried, J.A. Manson, and L.H. Sperling, J. Polym. Sci., Polym. Phys. Ed., $\underline{16}$, 583(1978).
20. A.A. Donatelli, L.H. Sperling, and D.A. Thomas, Macromolecules, $\underline{9}$, 671, 676 (1967)
21. J.A. Manson and L.H. Sperling, "Polymer Blends and Composites," Plenum, 1976.
22. J.K. Yeo, L.H. Sperling, and D.A. Thomas, accepted, Polym. Eng. Sci.
23. J.K. Yeo, L.H. Sperling, and D.A. Thomas, Polym. Eng. Sci., $\underline{21}$, 696 (1981).
24. J.K. Yeo, L.H. Sperling, and D.A. Thomas, J. Appl. Polym. Sci., $\underline{26}$, 3283 (1981).
25. L.H. Sperling, G.D. Wignall, A.M. Fernandez, and J.M. Widmaier, in preparation.

# RIM SYSTEMS FROM INTERPENETRATING POLYMER NETWORKS

R. Pernice*, K.C. Frisch, and R. Navare

Polymer Institute, University of Detroit
4001 W. McNichols Road
Detroit, Michigan 48221

## ABSTRACT

Simultaneous interpenetrating polymer networks (SIN-IPN's) based on polyurethane and polyepoxides were synthesized according to the processing conditions of RIM technology. Catalytic systems were selected in order to minimize or avoid formation of any covalent bonds between the epoxy and the polyurethane chains. In this way topological polymer alloys (IPN's) were obtained which exhibited improved physical properties. Both low modulus and high modulus urethane elastomers were used in combination with epoxy resins derived from glycidyl ethers of bisphenol A. In addition, combinations of the latter with novolac epoxy resins were employed together with high modulus urethane elastomers. Maxima in ultimate tensile strength occurred at the ratio of 20/80 of epoxy resin/polyurethane systems. These maxima were interpreted as evidence of a permanent entanglement between the two different types of polymer networks. A thermoanalytical study using a Rheovibron and DSC was carried out to give further support for the formation of IPN's. A comparison of the above SIN-IPN's with reinforced high modulus polyurethane RIM systems (RRIM) was made. Glass reinforced IPN-RIM systems were also prepared with low modulus urethane elastomers.

*R. Pernice, Montepolimeri, Unita' di Ricerca, P. Marghera-Venezia, Italy.

INTRODUCTION

Interpenetrating polymer networks (IPN's) are polymer alloys consisting of two or more distinct crosslinked polymer networks held together by permanent entanglements with only accidental covalent bonds between the polymers, i.e. they are polymeric catenanes (1-9). Previous investigations have demonstrated that IPN's have improved physical properties over those of their component networks. It was therefore of interest to investigate the use of IPN's in reaction injection molding (RIM) technology. The spectacular rise of RIM and more recently of reinforced RIM (RRIM) in the automotive and other industries has focused attention to relatively high modulus molded parts since they can compete successfully with processes such as conventional injection molding and SMC (10-11). On a volume basis RIM and RRIM are by far the most energy efficient materials and processes (12). There are several methods of increasing the modulus of urethane-based RIM parts. They are: 1. increasing the hard segments in the urethane polymer; 2. introducing isocyanurate groups into the urethane polymer (13,14); 3. using reinforcing materials e.g. glass fibers, mica, wollastonite, etc. in urethane polymers (RRIM).

The disadvantages of RRIM are the relatively high viscosities of the liquid components containing the reinforcing fillers, abrasiveness of the fillers and mechanical problems with unstable suspensions of solid reinforcements in liquid streams in most currently used RIM machines.

Another method to achieve higher modulus RIM parts is the use of PU-based IPN's in combination with a glassy polymer, e.g. a polyepoxide or unsaturated polyester. This study describes the use of SIN-IPN's (i.e. simultaneous formation of the polymeric networks forming the IPN) consisting of polyurethanes and epoxy or unsaturated polyesters. The alloying of the glassy epoxy or unsaturated polyesters (relatively high Tg's), respectively with elastomeric PU's should result in the formation of IPN's exhibiting some properties of reinforced RIM's without the necessity of including solids during the RIM processing.

EXPERIMENTAL

The materials used and their descriptions are listed in Table I. The polyols and the epoxy resins were demoisturized at 80°C under vacuum (1 mm Hg) for about 24 hours. The catalysts and the isocyanate were used as received. All the samples were prepared according to the normal processing conditions of RIM technology. The IPN's based on Epon 828 were post-cured in an oven at 121°C for 1 hour. In the case of the epoxy novolac

resin, however, complete polymerization of the RIM plaques required a post-curing time of up to three hours instead of 1 hr.
at 121°C, although studies are underway to shorten the post-
cure by the use of higher temperatures.

RESULTS AND DISCUSSION

The formulations, processing conditions and physical properties of the IPN's consisting of a combination of urethane elastomers
with bisglycidyl ether of bisphenol A (Epon 828) are shown in
Tables II and III.  The respective reference urethane elastomers
are included in these tables for comparison purposes.  The
catalyst combination of dibutyltin dilaurate (T-12), $BF_3$ etherate,
$BCl_3$ amine complex and DMP-30 was selected in order to achieve
rapid and complete polymerization of the individual networks.
Other effective epoxy hardeners containing active hydrogen atoms
were avoided in order to prevent or minimize copolymerization
between the epoxy and urethane networks.  It is apparent from
Tables II and III and Fig. 1, the maximum in tensile strength
occurred in IPN's containing 20% of Epon 828.  The flexural modulus increased while the elongation and split tear tended to decrease with increasing epoxy content.  IPN's based on high modulus
urethane elastomer and a combination of epoxy resins (bisglycidyl
ether of bisphenol A (Epon 828) and epoxy novolac (D.E.N. 431))
are shown in Table IV.  The same trends in physical properties
were noted as for the above described IPN's.  A maximum in
tensile strength was also observed at 20% of epoxy content (Fig. 1).

Glass reinforced high modulus urethane RIM elastomers (70/
30 polymer polyol/short chain diols) were also prepared for comparison, using silane-treated 1/16" milled glass fibers (10-30%
concentration).  The resulting properties of the RRIM urethanes
are shown in Table V.  Comparing Tables II and V, it can be observed that the tensile strength and the other physical properties of the RRIM urethane containing 10% glass fibers are very
similar to the IPN based on 80% urethane/20% epoxy.  The relatively low elongations at break shown in Tables II and V are due to the
high concentrations of hard segments in the urethane network.  The
elongation can be improved by reducing the content of short diols
in the formulation as can be seen in Table III.  Tables VI and
VII show the effect of glass fiber reinforcement on the physical
properties of IPN's based on low modulus urethane elastomer and
10% or 20% or Epon 828 respectively.  Both combinations exhibited
improvement of the tensile strength and heat sag with increasing
the glass fiber (1/16" milled glass fibers) concentration in the
range 10-30% as illustrated in Fig. 2.  The flexural modulus also
increased with increasing glass fiber content while the elongation
and split tear tended to decrease.

TABLE I

MATERIALS

| Designation | Chemical Composition | Eq. Wt. | Supplier |
|---|---|---|---|
| Niax 31-28 | Poly(oxypropylene) triol, ethylene oxide capped and grafted with 21% acrylonitrile | 2003 | Union Carbide |
| Niax 50-1180 | Mixture of short chain diols | 47.6 | Union Carbide |
| Isonate 191 | Modified MDI | 138.9 | The Upjohn Co. |
| T-12 | Dibutyltin dilaurate | | M & T Corp. |
| $BF_3(C_2H_5)_2O$ | Lewis acid etherate complex | | Eastman Chem. |
| XU 213-$BCl_3$ (amine)$_m$ | Lewis acid amine complex | | Ciba-Geigy |
| DMP-30 | 2,4,6-Tris(dimethylaminomethyl) phenol | | Rohm & Haas |
| Epon 828 | Bis(glycidyl ether) of bisphenol A | 185 - 192* | Shell Chemical Co. |
| D.E.N. 431 | Epoxy novolac resin | 172 - 179* | Dow Chemical |
| Hysol AC 44368 | Mold release agent | | Dexter Co. |
| Glass fibers P 117 | 1/16" Milled glass fibers silane treated | | Owens Corning Fiberglas |

(*) Grams of resin containing one gram-equivalent of epoxide (ASTM D 1652)

TABLE II

HIGH MODULUS URETHANE-EPOXY IPN'S

| | | | | | |
|---|---|---|---|---|---|
| Ratio Diol/Polyol | 30/70 | 3/70 | 30/70 | 30/70 | 30/70 |
| Niax 31-28 | 35.3 | 31.8 | 28.2 | 24.7 | 21.2 |
| Niax 50-1180 | 15.1 | 13.6 | 12.1 | 10.6 | 9.1 |
| Isonate 191 | 49.5 | 44.5 | 39.6 | 34.6 | 29.7 |
| T-12 | 0.012 | 0.012 | 0.012 | 0.012 | 0.012 |
| $BF_3$ etherate | -- | 0.44 | 0.66 | 0.66 | 0.66 |
| $BCl_3$ amine complex | -- | 1.0 | 2.0 | 3.0 | 4.0 |
| DMP-30 | -- | 0.21 | 0.21 | 0.21 | 0.21 |
| Epon 828 | -- | 10 | 20 | 30 | 40 |
| Pot life, sec. | 60 | 50 | 60 | 60 | 55 |
| Cured in mold, min/$^{\circ}$C | | | 5/100 | | |
| Post-cured, min$^{\circ}$C | | | 60/121 | | |
| Shore hardness D | 78 | 85 | 85 | 85 | 85 |
| Tensile strength, MPa at break, 70$^{\circ}$F, 21$^{\circ}$C | 44.8 | 44.5 | 49.8 | 37.5 | 23.9 |
| Elongation at break, % 70$^{\circ}$F, 21$^{\circ}$C | 30 | 20 | 20 | 10 | 10 |
| Split tear, KN/m, 70$^{\circ}$F, 21$^{\circ}$C | 61.2 | 22.0 | 19.8 | brittle | brittle |
| Flexural modulus, MPa: -20$^{\circ}$F, -29$^{\circ}$C 70$^{\circ}$F, 21$^{\circ}$C 158$^{\circ}$F, 70$^{\circ}$C | 1920 1007 435 | 2238 1311 483 | 2543 1593 580 | 2736 1815 779 | 2746 1702 1061 |
| Izod impact, ft. lb/in$^2$ 70$^{\circ}$F, 21$^{\circ}$C | 6.0 | 3 | 5 | 2.5 | 2.5 |
| Heat sag, in. (250$^{\circ}$F, 121$^{\circ}$C, 1 hr.) | 0.5 | 0.7 | 0.8 | 0.5 | 1.4 |
| Modulus ratio, -29$^{\circ}$C/70$^{\circ}$C | 4.4 | 4.6 | 4.4 | 3.5 | 2.6 |

TABLE III

LOW MODULUS URETHANE-EPOXY IPN'S

| | | | | | |
|---|---|---|---|---|---|
| Ratio Diol/Polyol | 15/85 | 15/85 | 15/85 | 15/85 | 15/85 |
| Niax 31-28 | 55.9 | 50.3 | 44.7 | 39.1 | 33.5 |
| Niax 50-1180 | 9.9 | 8.9 | 7.9 | 6.9 | 5.9 |
| Isonate 191 | 34.2 | 30.8 | 27.9 | 24.0 | 20.5 |
| T-12 | 0.012 | 0.012 | 0.012 | 0.012 | 0.012 |
| $BF_3$ etherate | -- | 0.44 | 0.66 | 0.66 | 0.66 |
| $BCl_3$ amine complex | -- | 1.0 | 2.0 | 3.0 | 4.0 |
| DMP-30 | -- | 0.21 | 0.21 | 0.21 | 0.21 |
| Epon 828 | -- | 10 | 20 | 30 | 40 |
| Pot life, sec. | 60 | 60 | 90 | 170 | 240 |
| Cured in mold, min/$^{\circ}$C | | | 5/100 | | |
| Post-cured, min/$^{\circ}$C | | | 60/121 | | |
| Shore hardness D | 68 | 70 | 70 | 73 | 75 |
| Tensile strength, MPa at break, 70$^{\circ}$F, 21$^{\circ}$C | 24.4 | 21.8 | 28.9 | 25.5 | 17.8 |
| Elongation at break, % 70$^{\circ}$F, 21$^{\circ}$C | 80 | 50 | 30 | 10 | 5 |
| Split tear, KN/m, 70$^{\circ}$F, 21$^{\circ}$C | 24.5 | 22.8 | 17.5 | 10.8 | brittle |
| Flexural modulus, MPa: | | | | | |
| -20$^{\circ}$F, -29$^{\circ}$C | 1029 | 1086 | 1557 | 1149 | 1890 |
| 70$^{\circ}$F, 21$^{\circ}$C | 356 | 362 | 680 | 895 | 920 |
| 158$^{\circ}$F, 70$^{\circ}$C | 48 | 61 | 118 | 168 | 266 |
| Heat sag, in. (250$^{\circ}$F, 121$^{\circ}$C, 1 hr.) | 1.8 | 1.7 | 1.5 | 1.5 | 1.5 |
| Modulus ratio, -29$^{\circ}$C/70$^{\circ}$C | 21.3 | 17.7 | 13.2 | 11.1 | 7.1 |

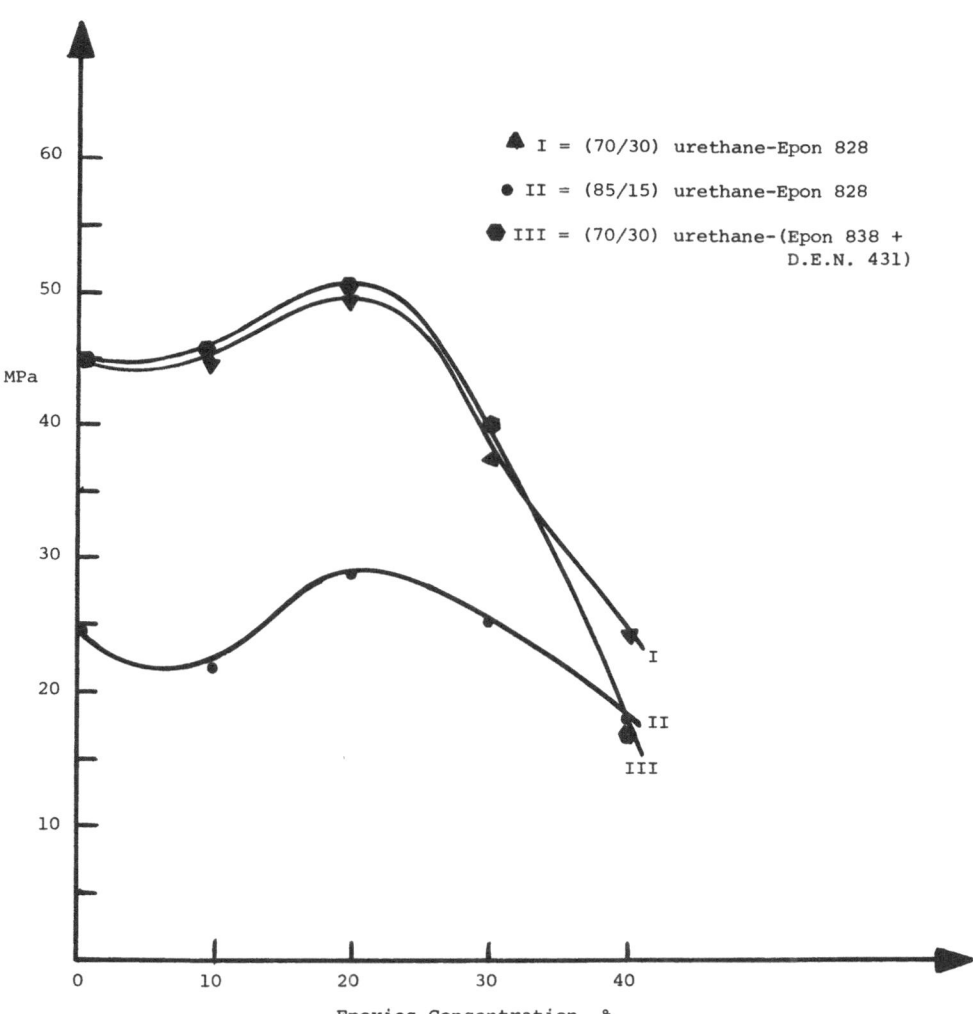

Figure 1.   Tensile Strength of IPN's Urethane-Epoxies

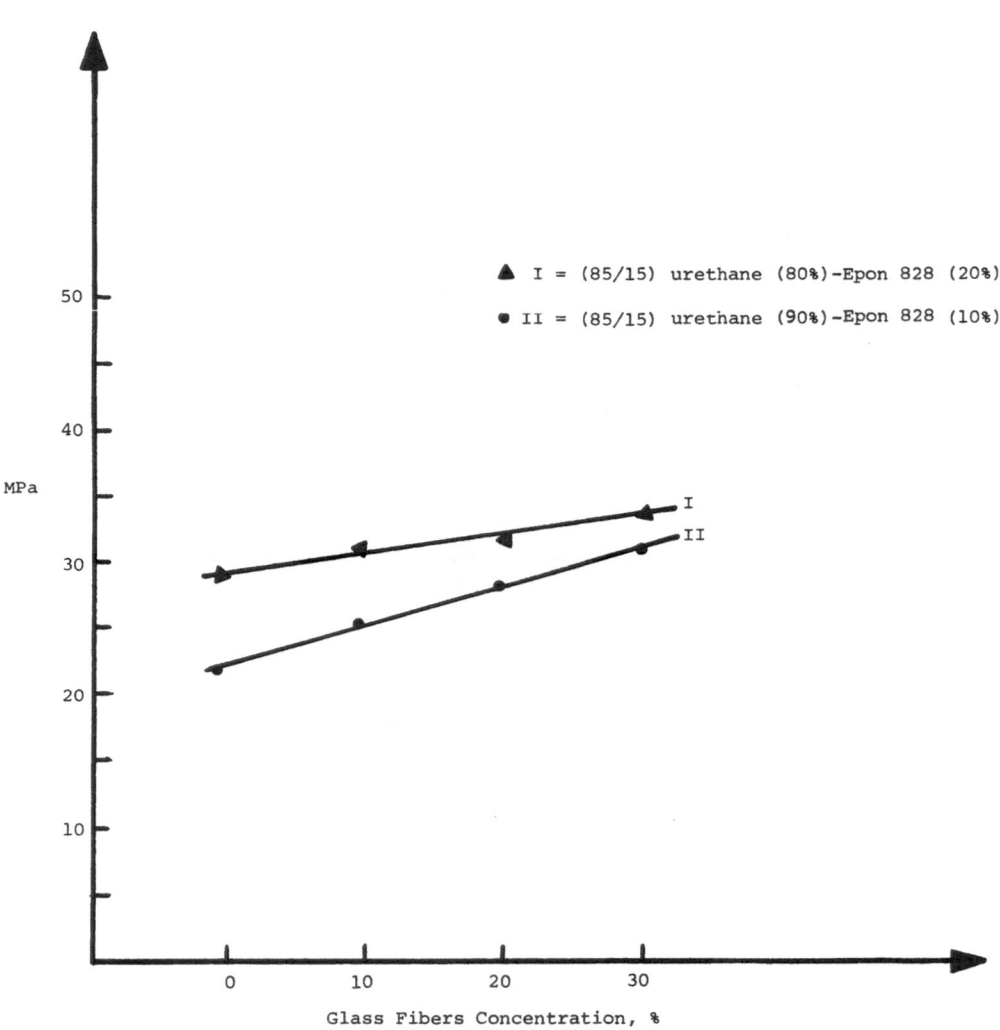

Figure 2.  Tensile Strength of Glass-Reinforced IPN's

TABLE IV

HIGH MODULUS URETHANE-EPOXY IPN'S
(Combination Novolac-Glycidyl Ether of Bisphenol A Epoxies)

| | | | | |
|---|---|---|---|---|
| Ratio Diol/Polyol | 30/70 | 30/70 | 30/70 | 30/70 |
| Niax 31-28 | 31.8 | 28.2 | 24.7 | 21.2 |
| Niax 50-1180 | 13.6 | 12.1 | 10.6 | 9.1 |
| Isonate 191 | 44.5 | 39.6 | 34.6 | 29.7 |
| T-12 | 0.012 | 0.012 | 0.012 | 0.012 |
| $BF_3$ etherate | 0.44 | 0.66 | 0.66 | 0.66 |
| $BCl_3$ amine complex | 1.0 | 2.0 | 3.0 | 4.0 |
| DMP-30 | 0.21 | 0.21 | 0.21 | 0.21 |
| Epon 828 | 5 | 10 | 15 | 20 |
| D.E.N. 431 | 5 | 10 | 15 | 20 |
| Pot life, sec. | 65 | 90 | 180 | 300 |
| Cured in mole, min/$^\circ$C | | 5/100 | | |
| Post-cured, min/$^\circ$C | | 180/100 | | |
| Shore hardness D | 82 | 83 | 85 | 85 |
| Tensile strength, MPa at break, 70$^\circ$F, 21$^\circ$C | 45.0 | 51.1 | 40.2 | 17.4 |
| Elongation at break, % 70$^\circ$F, 21$^\circ$C | 15 | 10 | 10 | 5 |
| Split tear, KN/m, 70$^\circ$F, 21$^\circ$C | 14.7 | 10.2 | brittle | brittle |
| Flexural modulus, MPa: | | | | |
| -20$^\circ$F, -29$^\circ$C | 2223 | 2459 | 2666 | 2678 |
| 70$^\circ$F, 21$^\circ$C | 1378 | 1681 | 1877 | 1778 |
| 158$^\circ$F, 70$^\circ$C | 783 | 1013 | 1179 | 1125 |
| Heat Sag, in. (250$^\circ$F, 121$^\circ$C, 1 hr.) | 0.8 | 1.1 | 1 | 1 |
| Modulus ratio, -29$^\circ$C/70$^\circ$C | 2.8 | 2.3 | 2.3 | 2.4 |

TABLE V

GLASS REINFORCED URETHANE ELASTOMERS

| | | | |
|---|---|---|---|
| Ratio Diol/Polyol | 30/70 | 30/70 | 30/70 |
| Niax 31-28 | 32 | 28.6 | 24.9 |
| Niax 50-1180 | 13.7 | 12.2 | 10.6 |
| Isonate 191 | 44.3 | 39.4 | 34.5 |
| T-12 | 0.010 | 0.010 | 0.010 |
| 1/16" Glass Fibers | 10 | 20 | 30 |
| $BCl_3$ amine complex | --- | --- | --- |
| DMP-30 | --- | --- | --- |
| Epon 828 | --- | --- | --- |
| Pot life, sec. | 50 | 50 | 50 |
| Cured in mold, min/$^{\circ}$C | | 5/100 | |
| Post-cured, min/$^{\circ}$C | | 60/100 | |
| Shore hardness D | 83 | 85 | 88 |
| Tensile strength, MPa at break, 70$^{\circ}$F, 21$^{\circ}$C | 50.5 | 56.4 | 59.9 |
| Elongation at break, % 70$^{\circ}$F, 21$^{\circ}$C | 15 | 15 | 5 |
| Split tear, KN/m, 70$^{\circ}$F, 21$^{\circ}$C | 36 | 36 | 26 |
| Flexural modulus, MPa: | | | |
| -20$^{\circ}$F, -29$^{\circ}$C | 2461 | 3376 | 4382 |
| 70$^{\circ}$F, 21$^{\circ}$C | 1610 | 2159 | 3067 |
| 158$^{\circ}$F, 70$^{\circ}$C | 651 | 884 | 1185 |
| Heat sag, in. (250$^{\circ}$F, 121$^{\circ}$C, 1 hr.) | 0.6 | 0.5 | 0.3 |
| Modulus ratio, -29$^{\circ}$C/70$^{\circ}$C | 3.8 | 3.8 | 3.7 |

TABLE VI

GLASS REINFORCED URETHANE-EPOXY IPN'S

| | | | |
|---|---|---|---|
| Ratio Diol/Polyol | 15/85 | 15/85 | 15/85 |
| Niax 31-28 | 50.3 | 50.3 | 50.3 |
| Niax 50-1180 | 8.9 | 8.9 | 8.9 |
| Isonate 191 | 30.9 | 30.9 | 30.9 |
| T-12 | 0.012 | 0.012 | 0.012 |
| $BF_3$ etherate | 0.44 | 0.44 | 0.44 |
| $BCl_3$ amine complex | 1.0 | 1.0 | 1.0 |
| DMP-30 | 0.21 | 0.21 | 0.21 |
| Epon 828 | 10 | 10 | 10 |
| 1/16" Glass fibers | 10 | 20 | 30 |
| Pot life, sec. | 150 | 180 | 270 |
| Cured in mold, min/$^\circ$C | | 5/100 | |
| Post-cured, min/$^\circ$C | | 60/121 | |
| Shore hardness D | 75 | 75 | 75 |
| Tensile Strength, MPa at break, 70$^\circ$F, 21$^\circ$C | 25.1 | 28.1 | 30.9 |
| Elongation at break, % 70$^\circ$F, 21$^\circ$C | 40 | 20 | 20 |
| Split tear, KN/m, 70$^\circ$F, 21$^\circ$C | 25.2 | 23.6 | 22.2 |
| Flexural modulus, MPa: | | | |
| -20$^\circ$F, -29$^\circ$C | 1506 | 1793 | 2237 |
| 70$^\circ$F, 21$^\circ$C | 600 | 780 | 819 |
| 158$^\circ$F, 70$^\circ$C | 129 | 205 | 264 |
| Heat sag, in. (250$^\circ$F, 121$^\circ$C, 1 hr.) | 1.0 | 0.8 | 0.7 |
| Modulus ratio, -29$^\circ$C/70$^\circ$C | 11.7 | 8.7 | 8.5 |

TABLE VII

GLASS REINFORCED URETHANE-EPOXY IPN'S

| | | | |
|---|---|---|---|
| Ratio Diol/Polyol | 15/85 | 15/85 | 15/85 |
| Niax 31-28 | 44.7 | 44.7 | 44.7 |
| Niax 50-1180 | 7.9 | 7.9 | 7.9 |
| Isonate 191 | 27.4 | 27.4 | 27.4 |
| T-12 | 0.012 | 0.012 | 0.012 |
| $BF_3$ etherate | 0.66 | 0.66 | 0.66 |
| $BCl_3$ amine complex | 2.0 | 2.0 | 2.0 |
| DMP-30 | 0.21 | 0.21 | 0.21 |
| Epon 828 | 20 | 20 | 20 |
| 1/16" Glass fibers | 10 | 20 | 30 |
| Pot life, sec. | 120 | 150 | 180 |
| Cured in mold, min/$^{\circ}$C | | 5/100 | |
| Post-cured, min/$^{\circ}$C | | 60/100 | |
| Shore hardness D | 80 | 80 | 80 |
| Tensile strength, MPa at break, 70$^{\circ}$F, 21$^{\circ}$C | 30.9 | 31.5 | 33.4 |
| Elongation at break, % 70$^{\circ}$F, 21$^{\circ}$C | 10 | 10 | 10 |
| Split tear, KN/m, 70$^{\circ}$F 21$^{\circ}$C | 23.2 | 21.4 | 18.4 |
| Flexural modulus, MPa: | | | |
| -20$^{\circ}$F, -29$^{\circ}$C | 1761 | 1971 | 2284 |
| 70$^{\circ}$F, 21$^{\circ}$C | 965 | 1004 | 1005 |
| 158$^{\circ}$F, 70$^{\circ}$C | 168 | 193 | 349 |
| Heat sag, in. (250$^{\circ}$F, 121$^{\circ}$C, 1 hr.) | 1.2 | 1.1 | 1.0 |
| Modulus ratio, -29$^{\circ}$C/70$^{\circ}$C | 10.5 | 10.2 | 6.5 |

The phase relations in the above IPN-RIM systems are being investigated using a Rheovibron and Differential Scanning Calorimeter.

ACKNOWLEDGEMENT

The authors wish to express their gratitude to Montedison Polimeri, S.p.A. Italy, for the establishment of a research grant which made this investigation possible.

REFERENCES

1.  H.L. Frisch, D. Klempner and K.C. Frisch, J. Polym. Sci. A-2, 8, 921 (1970).

2.  K.C. Frisch, D. Klempner and H.L. Frisch, J. Appl. Polym. Sci., 18, 689 (1974).

3.  S.C. Kim, D. Klempner and K.C. Frisch, J. Appl. Polym. Sci., 21, 1289 (1977).

4.  A.J. Curtius, M.J. Covitch, D.A. Thomas and L.H. Sperling, Polym. Eng. & Sci., 12, 101 (1972).

5.  V. Huelck, D.A. Thomas and L.H. Sperling, Macromolecules 5, 340 (1972).

6.  R.L. Touhsaent, D.A. Thomas and L.H. Sperling, J. Polym. Sci., 46, 175 (1974).

7.  D. Klempner, H.K. Yoon, K.C. Frisch and H.L. Frisch in "Polymer Alloys" II, edited by D. Klempner and K.C. Frisch, Plenum Publishing Co., New York, 1980, p. 185-201.

8.  H.L. Frisch, D. Klempner, H.K. Yoon and K.C. Frisch, Macromolecules, 13, 1016 (1980).

9.  J.A. Munson and H. Sperling "Polymer Blends and Composites", Plenum Publishing Co., New York (1976).

10. K.C. Frisch, Rubber Chem. and Technol., 53, 126 (1980).

11. L.J. Lee, Rubber Chem. and Technol., 53, 543 (1980).

12. M.J. Mikulec, SPI-FSK International Conference, Strasbourg, France, June 9-13, 1980, Carl Hanser, Munich-Vienna, p. 137-148.

13. H. Ulrich, J. Cell. Plastics, 17, No. 1, 31 (1981).

14. K.C. Frisch and D. Klempner, Unpublished results.

# MICROPHASE SEGREGATION IN SEGMENTED

# AMINE-CURED POLYURETHANES

C. R. Desper and N. S. Schneider

Army Materials and Mechanics Research Center
Organic Materials Laboratory
Watertown, Massachusetts 02172

## INTRODUCTION

The influence of chemical composition on microphase segregation in segmented polyurethanes based on 2,4 and 2,6 toluene diisocyanate (TDI) was investigated earlier by Schneider, Sung and co-workers (1-3). In these studies, polymers of varying degrees of phase segregation were prepared from the two types of TDI coupled to polyether or polyester soft segments and chain extended with butanediol. The extent of phase mixing was judged from infared estimates of interurethane hydrogen bonding as well as from thermal properties, particularly the soft segment glass transition temperature Tg. In general, for 2,4TDI-butanediol, which forms an amorphous hard segment structure, phase segregation was weak in samples with a 1000 molecular weight soft segment, but enhanced with increasing soft segment molecular weight, for the polyether as compared to the polyester soft segment, and with the use of 2,6TDI which forms a crystalline hard segment structure.

More recent studies by Sung and co-workers (4-5) on 2,4 TDI polyurethanes, prepared by solution polymerization and cure with ethylene diamine, have shown that strong phase segregation occurs in these materials. Since the hard segment structure is amorphous, the improved phase segregation must be the result of the increased hydrogen bonding of the urea groups formed in the amine cure. The present work is directed toward understanding the structure and properties of polyurethanes based on 2,4 TDI and 80/20 mixtures of 2,4 and 2,6 TDI, cured with MOCA, which represent one of the most important commercial classes of polyurethane elastomers. These melt polymerized, amine cured polyurethanes are thermoset materials due to the formation of some biuret crosslinks. Although such

233

crosslinking is usually thought to play a major role in the properties, it will be seen that these materials, like those studied by
Sung, are highly phase segregated.  Comparisons are also made with
polyurethanes cured with a non-carcinogenic replacement for MOCA.

    The most direct measure of microphase segregation is offered
by small angle X-ray scattering (SAXS).  The quantitative application of SAXS has been recently described by several workers (6-8).
The parameters used in the present work are the interface thickness
$a_I$ (assuming a sigmoidal density profile), the squared electron
density fluctuation $(\Delta\rho)^2$, and the average inhomogeneity lengths
$\ell_H$ and $\ell_S$ for hard and soft segments.

THEORETICAL

    In the treatment of SAXS data obtained using a slit-collimated
instrument, such as the Kratky camera used in the present work, one
must distinguish between the experimental scattering curve $\tilde{I}(s)$ and
the true scattering curve $I(s)$, where the independent variable $s$ is
defined in terms of the X-ray wavelength $\lambda$ and the scattering angle
$2\theta$:

$$s = (2/\lambda) \sin \theta. \tag{1}$$

In the present work, copper radiation with $\lambda=0.15418$nm was used.
The data may be treated theoretically either in terms of the experimental (or slit-smeared) intensity function $\tilde{I}(s)$ or the true (or
desmeared) intensity function $I(s)$.  In the present instance the
true intensity function $I(s)$ was obtained from the experimental intensity $\tilde{I}(s)$ using the iterative method of Glatter (9) to correct
for distortion of the data arising from the finite length and width
of X-ray beam and of the receiving slit.  The correction process,
known as "desmearing", can have a profound effect on the shape of
the scattering curve, as shown in Figure 1.  For this reason, theoretical methods have been devised to deal with the data in terms of
the slit-smeared intensity function as well as in terms of the true
intensity function.  With the exception of the interphase thickness
determination, the methods based on the true (desmeared) intensity
function were used in the present work.

Thickness of Interphase Boundaries

    For a system of two phases of finely divided particles having
sharp boundaries, the SAXS intensity becomes inversely proportional
to the fourth power of the angle, or of s, at higher angles:

$$\lim_{s \to \infty} [I(s)]= K_p/s^4. \tag{2}$$

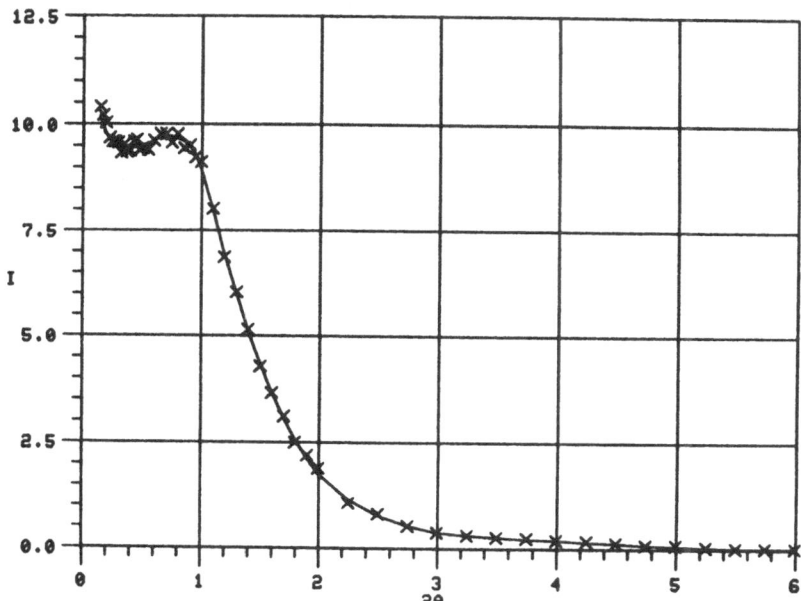

Figure 1a.   SAXS intensity curve smeared by
             slit system.

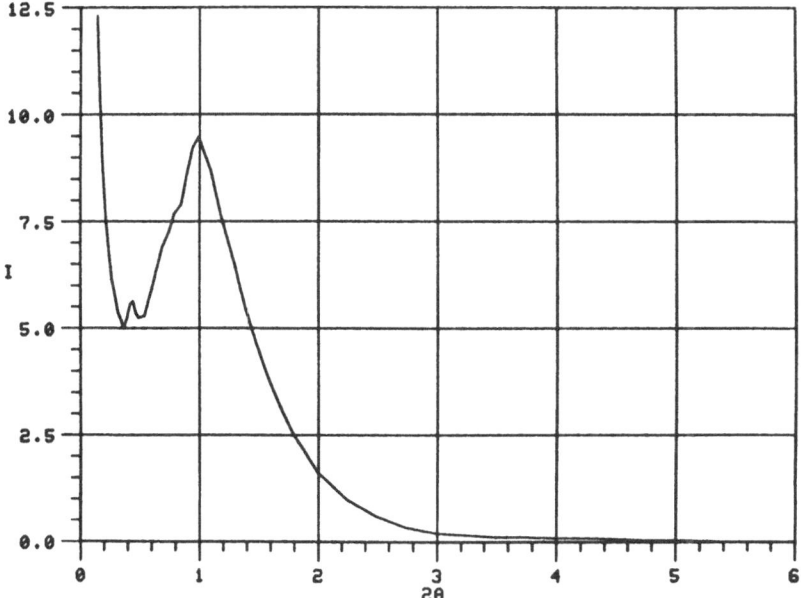

Figure 1b.   SAXS intensity curve desmeared.

This relationship is known as Porod's law, and the constant $K_p$, known as the Porod law constant. Departures from Porod's law can arise from either thermal density fluctuations within phases and from finite thickness of the interphase boundaries. Koberstein, Morra, and Stein (6) have recently reviewed various theoretical treatments of these departures in terms of material models. The thermal density fluctuation is not of interest per se in the present context, but must be taken into account before the data may be used for characterizing the interphase thickness. In the present work a diffuse scattering background was determined by extrapolation from the wide angle region (above $2\theta=8°$), using an empirical equation:

$$\ln[I_B (s)] = K_1 + K_2 s^4. \tag{3}$$

The constants $K_1$ and $K_2$ were determined by a least squares curve fit in the range $2\theta=8°$ to $16°$. Since a wide angle goniometer was used to record data in this range, the intensity values were scaled to the level of the Kratky camera data by reference to a datum common to both instruments at $2\theta=8°$. The diffuse background correction was applied prior to desmearing.

Three basic models of the material structure have been used to treat the diffuse boundary problem [see Koberstein et al]: the sigmoidal model, the thermodynamic model of Helfand (11), and the linear gradient model. The sigmoidal model will be used here. Koberstein has shown that the profile given by the Helfand model is quite similar in shape. The linear gradient model is considered unrealistic in terms of the thermodynamics of the system.

Using the sigmoidal model, the Porod-law relation becomes

$$[I(s)] = [K_p/s^4] \exp (-4\pi^2 \sigma^2 s^2) \tag{4}$$

at high s values, where $\sigma$ is the standard deviation of the Gaussean smoothing function:

$$h(r) = \frac{1}{\sqrt{2\pi\sigma^2}} \exp [\frac{- r^2}{2\sigma^2}]. \tag{5}$$

This has the effect of broadening the interfaces to result in an electron density function $\rho(r)$ given by:

$$[\rho(r) -\rho_B]/[\rho_A -\rho_B]= \int_{\infty}^{r} h(t)dt \tag{6}$$

near the interface. The parameter $\sigma$ is identified with the standard deviation (in length units) of the gradient in the direction normal to the interface. Although $\sigma$ characterizes the thickness of the interface, it has a very specialized definition, and its identification as an "interphase thickness" is misleading. The smoothing function h(r) falls to 60.6% of its maximum value in a distance $\sigma$; thus the density gradient is 60.6% of its maximum values at a distance $\sigma$ on either side of the central plane. Thus identification of $\sigma$ as the "interface thickness" is misleading by at least a factor of two.

A more meaningful interface thickness may be defined by:

$$a_I \ h(o) = \int_{-\infty}^{\infty} h(r) \ dr. \tag{7}$$

This defines an integral thickness $a_I$ as the breadth of a rectangle of height h(o) which yields the same area as the integral under the smoothing function. Bonart (12, 13) uses such an integral breadth formulation for the sigmoidal profile, while Bates (14) has introduced an integral breadth using the Helfand profile. From (5) and (7) one may obtain:

$$a_I = \sqrt{2\pi} \ \ \sigma = 2.5066 \ \sigma \tag{8}$$

In the interval the fractional density change ranges from 0.105 to 0.895. Thus the distance $a_I$ spans 79% of the total density change between the two microphases.

The empirical method of Koberstein, Morra, and Stein (6) was used to determine $\sigma$ values from smeared intensity using the relationship:

$$s^3 \tilde{I}(s) = K \exp [-38(\sigma s)^{1.81}] \tag{9}$$

assuming infinite slit length. To calculate $\sigma$, a least squares straight line was fit to the plot of $\ell n[s^3\tilde{I}(s)]$ versus $s^{1.81}$ in the range $2\theta = 3°$ to $7°$. All other parameters were determined from desmeared data. A typical Koberstein plot is shown in Figure 2.

The integral thickness $a_I$ defined here is not to be confused with the thickness parameter E of the linear gradient model. Koberstein et al (6) have shown that, with certain mathematical approximations, the same SAXS data may be interpreted in terms of a linear gradient model with an interface thickness which compares to the sigmoidal result by:

$$E = \sqrt{12} \ \sigma = 3.464 \ \sigma \tag{10}$$

The linear gradient thickness E will not be reported here, since it does not approximate the interface in a real system. For the

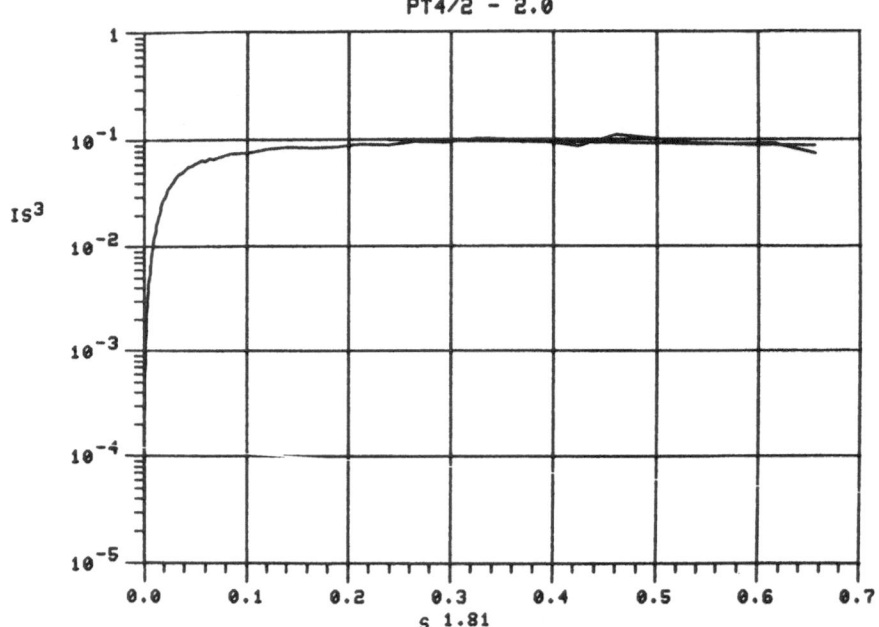

Figure 2.   Koberstein plot for determination of
interphase thickness.  S. is in nm.

purpose of comparison, however, an existing result quoted in terms
of linear gradient thickness E may be converted to an integral
thickness basis by:

$$a_I = E \sqrt{(\pi/6)} = 0.7236E \tag{11}$$

## Density Fluctuation

The average squared electron density fluctuation $\overline{(\Delta\rho)}^2$ is
obtained by the well-known integration of the SAXS scattering
curve (8):

$$2\pi^2 I_e \overline{(\Delta\rho)}^2 V = \int_0^\infty s^2 I(s)\ ds. \tag{12}$$

The intensities were placed on an absolute basis for this purpose
by reference to a standard Lupolen sample as described by Kratky
(15).

Ophir and Wilkes (7) describe the effect of internal phase
mixing and of diffuse boundaries on the electron density fluctuation.  Using their notation, we shall designate the electron
densities of the pure hard and soft segment phases as $\rho_{HP}$ and
$\rho_{SP}$, while the electron densities in the presence of internal
mixing are designated $\rho_{HM}$ and $\rho_{SM}$ respectively.  While Ophir

and Wilkes designate the volume fractions of the two phases as $\phi_H$ and $\phi_S$, we shall distinguish between the pure phase and the mixed phase cases: the volume fractions are $\phi_{HP}$ and $\phi_{SP}$ in the former case, $\phi_{HM}$ and $\phi_{SM}$ in the latter case. Ophir and Wilkes further define Model A, corresponding to perfect phase separation (no mixing of hard and soft segments) and sharp boundaries; Model B, allowing some phase mixing but maintaining sharp boundaries; and Model C, allowing both phase mixing and diffuse boundaries. In these cases the electron density fluctuations are given by:

$$\overline{(\Delta\rho)}^2_A = (\rho_{HP} - \rho_{SP})^2 \, \phi_{HP} \, \phi_{SP} \qquad (13)$$

$$\overline{(\Delta\rho)}^2_B = (\rho_{HM} - \rho_{SM})^2 \, \phi_{HM} \, \phi_{SM} \qquad (14)$$

$$\overline{(\Delta\rho)}^2_C = (\rho_{HM} - \rho_{SM})^2 \, (\phi_{HM} \, \phi_{SM} - \phi_3/6) \qquad (15)$$

where $\phi_3$ is the volume fraction of the interphase zones. Identifying the Model C electron density fluctuation with the experimental variance, the correction factor for interphase zones may be written:

$$\overline{(\Delta\rho)}^2_B / \overline{(\Delta\rho)}^2_C = \rho_{HM} \, \rho_{SM} / (\rho_{HM} \, \rho_{SM} - \phi_3/6) \qquad (16)$$

This correction, which is treated by Vonk (16), will be discussed in the Appendix.

To evaluate the effect of internal segment mixing we shall use the variance ratio V defined by:

$$V = \overline{(\Delta\rho)}^2_B / \overline{(\Delta\rho)}^2_A \qquad (17)$$

This ratio may be readily expressed in terms of dimensionless quantities:

$$V = \left[ \frac{\phi_{HM}(1-\phi_{HM})}{\phi_{HP}(1-\phi_{HP})} \right] \left[ \frac{\rho_{HM}}{\rho_{HP}} \right]^2 \left[ \frac{(1-\rho_{SM}/\rho_{HM})}{(1-\rho_{SP}/\rho_{HP})} \right]^2 \qquad (18)$$

Van Bogart (17) has proposed the use of $\sqrt{V}$ as a degree of phase separation. Use of the square root rests on the premise that $\phi_{HM} = \phi_{HP}$ (volume fraction of hard segment is the ideal case value) so the first factor in (18) is unity. In this case:

$$\sqrt{V} \; (\phi_{HM} = \phi_{HP}) = (\rho_{HM} - \rho_{SM})/(\rho_{HP} - \rho_{SP}) \qquad (19)$$

is linear with the Model B density difference, and thus with
the degree of phase separation. In the present instance there
is reason to believe, as will be discussed later, that a better
premise is $\rho_{HM} = \rho_{HP}$; i.e. that the second factor in (18) is unity.
The assumption here is that segmental mixing consists predominantly
of hard segments dissolved in the soft segment phase and not
vice versa. The variance ratio then depends upon three independent variables: the electron fraction hard segment $W_H$; the ratio
$(W_{HD}/W_H)$ of dissolved to total hard segment; and the electron
density ratio $(\rho_{SP}/\rho_{HP})$ of pure phases. The derivation of V in
terms of these variables is given in the Appendix. The functional
dependence of V on $(W_{HD}/W_H)$, using values of $W_H$ and $(\rho_{SP}/\rho_{HP})$
appropriate for the present samples, is given in Figure 3. The
graphs are much closer to a linear function than a parabolic
function for sample compositions in the 0-40% hard segment
range. Thus the square root parameter is not used.

Inhomogeneity Length

The inhomogeneity length $\ell c$, or correlation distance, is
obtained (8,10,17,18) from:

$$\ell_c = \frac{1}{2} \int_0^\infty sI(s)ds / \int_0^\infty s^2 I(s)ds. \qquad (20)$$

From $\ell_c$ one may further identify inhomogeneity lengths $\ell_H$ and $\ell_S$
for the hard segment and soft segment phases, related to $\ell_c$ by:

$$\ell_H = \ell_c / \phi_{SM} \qquad (21)$$

and

$$\ell_S = \ell_c / \phi_{HM} \qquad (22)$$

where $\phi_{HM}$ and $\phi_{SM}$ are the volume fractions of the two phases. The
averages involved are reciprocal averages. For instance, $\ell_c$ in
equation (20) is the reciprocal of the average value of $s$, which
has units of reciprocal distance. Similarly, $\ell_H$ and $\ell_S$ are
average lengths, using such a reciprocal averaging process of
line segments lying within the selected phase when an infinite
number of lines, randomly spaced and randomly oriented, are
hypothetically drawn through the two phase structure.

EXPERIMENTAL

Polymers were synthesized by the two-step procedure at the
compositions indicated in Table I. The code which designates
the polymer composition has the following options: M or P for
the amine chosen, MOCA or POLACURE; T4 or TM for the type of TDI,
2,4 or 80/20, 2,4/2,6 mixture; 1 or 2 for 1000 or 2000 molecular

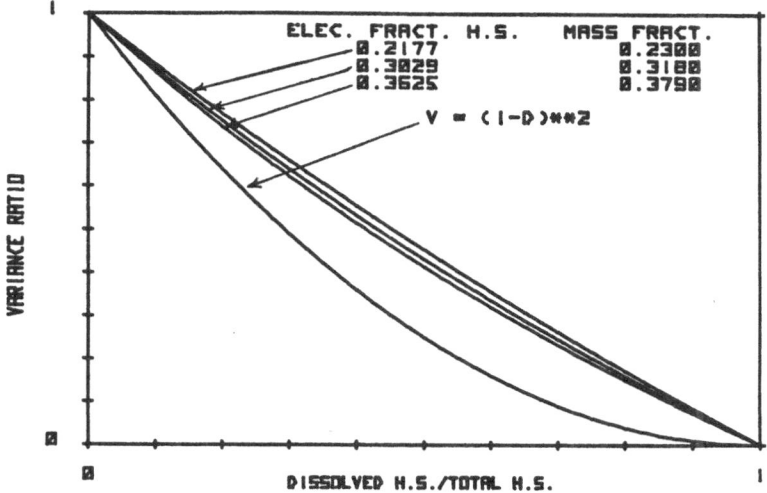

Figure 3a. Variance ratio calculated as a function of $W_{HD}/W_H$, polymers cured with MOCA.

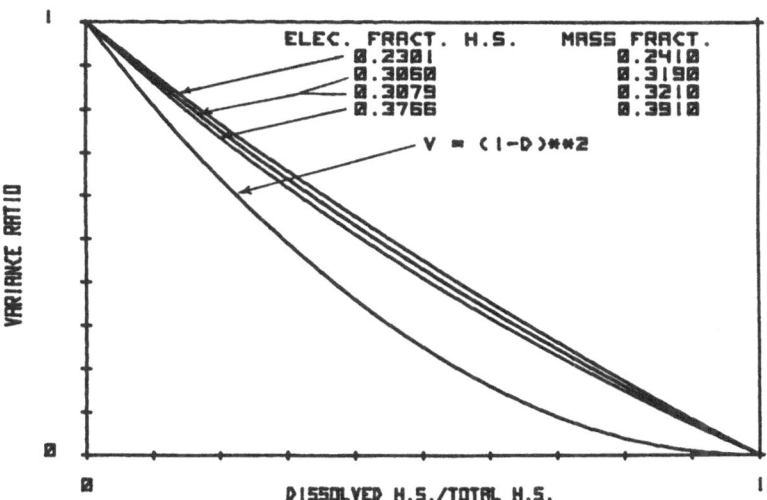

Figure 3b. Variance ratio calculated as a function of $W_{HD}/W_H$, polymers cured with TMAB.

Table I.   Sample Composition

| Designation | TDI[a] | PTMO[b] MW | Prepolymer Excess NCO | Mole Ratio TDI:Amine:PTMO | Hard Segment Content |
|---|---|---|---|---|---|
| MOCA[c] - Based Materials | | | | | |
| MT4/1 - 1.6 | (2,4) | 1000 | 4.43% | 1.68:0.65:1 | 31.8% |
| MT4/1 - 2.0 | (2,4) | 1000 | 6.33% | 2.02:0.97:1 | 37.9% |
| MTM/2 - 2.0 | d | 2000 | 3.55% | 1.99:0.94:1 | 23.0% |
| Polacure[e] - Based Materials | | | | | |
| PT4/1 - 1.6(f) | (2,4) | 1000 | 4.14% | 1.63:0.60:1 | 32.1% |
| PTM/1 - 1.6 | d | 1000 | 4.06% | 1.62:0.59:1 | 31.9% |
| PTM/1 - 2.0 | d | 1000 | 6.17% | 1.99:0.94:1 | 39.1% |
| PT4/2 - 2.0 | (2,4) | 2000 | 3.48% | 1.97:0.92:1 | 24.1% |
| PTM/2 - 2.0 | d | 2000 | 3.48% | 1.97:0.92:1 | 24.1% |

$NCO/NH_2$ Ratio = 1.05 for all materials.

(a) TDI: toluene diisocyanate.
(b) PTMO: poly(tetramethylene)oxide.
(c) MOCA: (4,4')-methylene bis(2-chloroaniline).
(d) Mixture 80% (2,4)-TDI + 20% (2,6)-TDI.
(e) Polacure: trimethylene glycol di-p-aminobenzoate (also called TMAB).
(f) DuPont Adiprene L100 prepolymer used.

Table II.   Mechanical and Thermal Properties

| Designation | 100%E (MPa) (a) | S (MPa) (b) | E (%) (c) | Tear (MPa) (d) | Tg (°C) (e) | T(2) (°C) (f) |
|---|---|---|---|---|---|---|
| MT4/1 - 1.6 | 10.2 | 27.6 | 380 | 3.1 | -51 | (g) |
| MT4/1 - 2.0 | 13.4 | 31.6 | 355 | 3.4 | -57 | (g) |
| MTM/2 - 2.0 | 5.8 | 18.2 | 820 | 2.7 | -69 | (g) |
| PT4/1 - 1.6 | 4.2 | 29.7 | 557 | 1.9 | -55 | 150 |
| PTM/1 - 1.6 | 9.8 | 43.3 | 455 | 3.6 | -53 | 180 |
| PTM/1 - 2.0 | 12.7 | 33.0 | 565 | 3.9 | -55 | 195 |
| PT4/2 - 2.0 | 1.7 | 16.5 | 700 | 0.9 | -74 | 143 |
| PTM/2 - 2.0 | 7.0 | 32.1 | 569 | 3.4 | -73 | 182 |

(a) Tensile modulus at 100% elongation.
(b) Tensile strength.
(c) Ultimate elongation.
(d) Tear strength (Graves).
(e) Soft segment glass transition temperature.
(f) Upper (hard segment) transition temperatures.
(g) Non-repeatable DSC transition at 210°C.

weight polyether; and 1.6 or 2.0 for the TDI/PTMO molar ratio.

The hard segment content listed in Table I corresponds to the total TDI and amine present. However, it is recognized that for the 1.6 TDI/PTMO ratio polymers, the stoichiometry requires that a significant proportion of the TDI used in the end capping step produces urethane interlinkages between macroglycol units. Thus, while the 2.0 ratio polymer may be represented as:

$$[-TDI - AMINE - TDI - PTMO-]_x,$$

the 1.6 ratio polymer may be represented as:

$$\left[ \begin{array}{c} 0.6 \; (-TDI - AMINE - TDI-) \\ + \qquad\qquad -PTMO- \\ 0.4 \; (-TDI-) \end{array} \right]_x.$$

Both of these representations are recognized as idealizations of the more accurate description

$$[(-TDI - AMINE-)_{n-1} - TDI - PTMO]_x,$$

where n assumes a distribution of small integer values whose mean value satisfies the stoichiometry. Nonetheless, even allowing a distribution in n, for the $\bar{n} = 1.6$ case, at least 40% of the linkages between macroglycols consist of a single TDI residue.

SAXS data were obtained using a Kratky camera with a proportional detector at a sample-to-registration plane distance of 22.92 cm. Sample densities were measured by a gas buoyancy method comparing apparent weight in vacuum and in $SF_6$ gas. The gas method was used, rather than liquid buoyancy methods, to avoid polymer swelling effects.

RESULTS AND DISCUSSION

The mechanical property data, Table II, indicate the high property levels typically obtained in MOCA cured materials at the low molar ratios of TDI to PTMO which occur in the commercial elastomers. Comparable property levels are exhibited by POLACURE materials when the 80/20 mixture of 2,4/2,6 TDI (TM) is used. But for reasons which are still not understood the elastomers prepared with 2,4 TDI are distinctly lower in properties, especially 100% modulus, tensile and tear strength.

The soft segment Tg,-50°C in PTMO-1000 and about-70°C in PTMO-2000 samples, indicates that these are strongly phase segregated materials. The improvement in phase segregation with the twofold increase in soft segment molecular weight is typical of both TDI

and MDI polyurethanes.  The Tg of POLACURE samples is comparable to
that of MOCA cured materials and shows no distinction between T4
and TM samples, unlike the mechanical properties.  In the MOCA
cured samples a nonrepeatable DSC transition near 210°C is believed
to be the hard segment transition.  The transition temperature is
close to the temperature at which urethane bond dissociation
occurs.

In the POLACURE samples the upper transition behavior T(2) is
more complex.  But the noteworthy feature, as summarized in Table
I, is that the transitions for the two T4 samples (and others not
discussed here) are much lower than that of the TM samples.  This
indicates that the lower mechanical properties are related in some
manner to the hard segment domain features.

The desmeared SAXS intensity curves are shown in Figure 4.
All samples yielded strong SAXS patterns with diffraction maxima in
the 8-14 nm range.  Electron density variances (see Table III) were
in the $3 \times 10^3$ to $6 \times 10^3$ (moles $e/cm^3$)$^2$.

The characteristic linear dimensions from SAXS measurements
are given in Table IV.  The equivalent Bragg spacings correspond to
an interparticle spacing.  The value is larger in the PTMO 2000
samples than the PTMO 1000 samples, reflecting the higher soft
segment concentration in the former samples.  Differences within
the PTMO 1000 and PTMO 2000 sets are minor.  Detailed interpretation
of the d values would require a more detailed model, taking into
account particle shape and distribution of interparticle distances;
lacking such knowledge an unambiguous interpretation is not possible.

Whereas the Bragg maxima are a measure of interparticle spacings,
the hard segment inhomogeneity length $\ell_H$ is a measure of intraparticle
dimensions.  In contrast to the Bragg spacings, there are strong
differences in the $\ell_H$ values for Polacure TM and T4 samples.  The
PTM $\ell_H$ values are similar to the MOCA results but the PT4 values
are lower by about 40%.  A tentative explanation, consistent with
the differences in mechanical properties, is that in the PT4 samples
the structure consists of isolated hard segment domains, whereas in
PTM samples a more continuous interconnected hard segment structure
occurs, similar to that proposed by Estes, Seymour and Cooper (10)
for MDI polyurethanes.

A comparison of the SAXS intensity curves for a PT4 and a PTM
sample is shown in Figure 5.  In qualitative terms, the Bragg
maxima appear at similar angle values, but the higher $\ell_H$ value for
the PTM sample manifests itself as greater intensity before the
Bragg maximum and more rapid dropoff in intensity after the
maximum.  However, in the Porod tail region well beyond the Bragg
maximum (i.e. above 2° in Figure 5) the two semilog curves are
parallel.  Thus the intensity functions in the tail region are

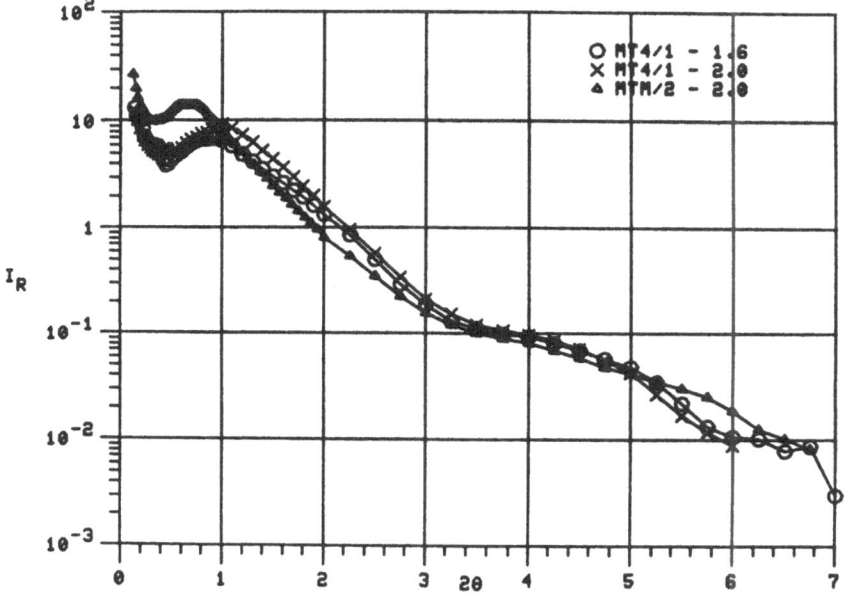

Figure 4a.   Experimental SAXS data, polymers cured
            with MOCA.

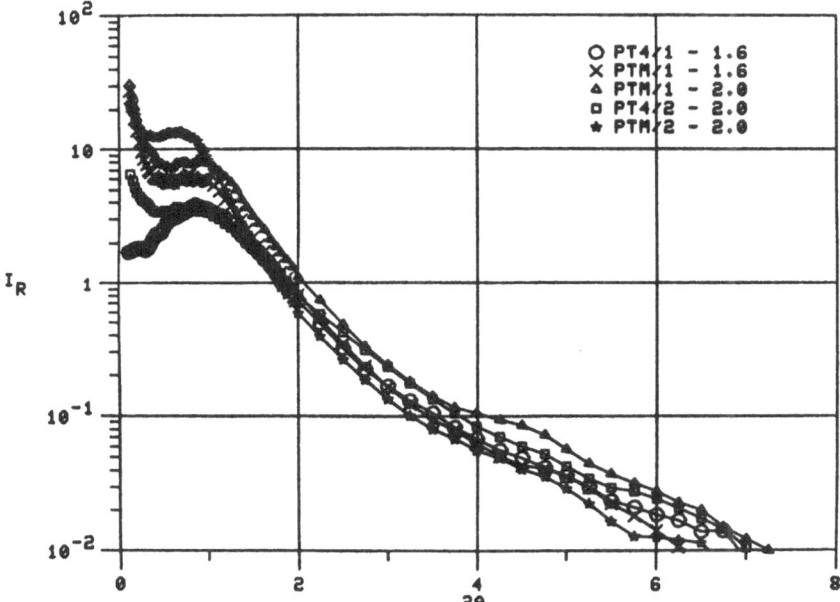

Figure 4b.   Experimental SAXS data, polymers cured
            with TMAB.   Data corrected for specimen
            absorption, thickness, and beam power.

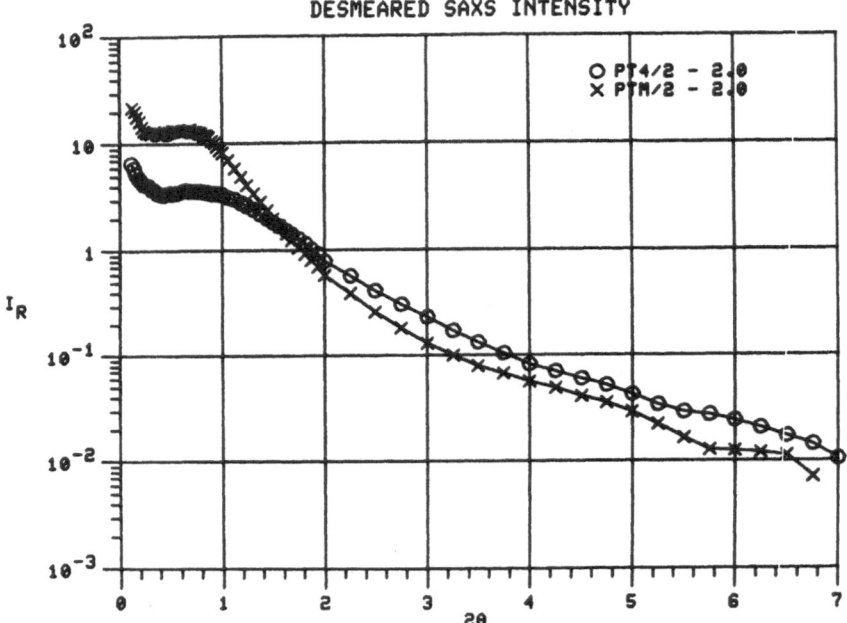

Figure 5.    Comparison of SAXS curves for two samples
             differing only in $l_H$ and $l_S$ values.  Data
             corrected for speciment absorption, thick-
             ness, and beam power.

quite similar in shape, refelecting similar interphase thickness
values of 0.18 and 0.28 nm (see Table IV).

        The σ values Table IV are quite small, comparable in several
instances to the lengths of covalent bonds, e.g. 0.154 nm for the
carbon-carbon single bond.  With this criterion in mind, it is
evident that the interfaces are quite sharp, particularly for the
TMAB cured samples.  These sharp interfaces are attributed to a
high degree of specific interaction with the hard segment phase,
which Sung (4, 5, 19) attributes to three-dimensional hydrogen
bonding between one urea carbonyl and two NH groups in the hard
segment phase.  Amine curing, which leads to urea group formation,
provides for greater driving force for phase segregation than diol
curing, which results in urethane formation with only one NH group
per carbonyl.

        In this regard, we must recall that at a TDI/PTMO ratio of
1.6, 40% of the linkages between macroglycol residues consist of
a single TDI residue.  Such isolated TDI residues will terminate
in urethane linkages and will be less prone to phase segregate than
the larger urea-linked TDI-amine sequences, and may, to a large

Table III.   Experimental Electron Density Variance

| Designation | $M_H$ | $\emptyset_{HP}$ | $(\Delta\rho)_C^2 \times 10^{3*}$ |
|---|---|---|---|
| MT4/1 - 1.6 | 0.318 | 0.255 | 4.14 |
| MT4/1 - 2.0 | 0.379 | 0.310 | 5.16 |
| MTM/2 - 2.0 | 0.230 | 0.180 | 4.42 |
| PT4/1 - 1.6 | 0.321 | 0.263 | 4.55 |
| PTM/1 - 1.6 | 0.319 | 0.261 | 3.30 |
| PTM/1 - 2.0 | 0.391 | 0.326 | 5.43 |
| PT4/2 - 2.0 | 0.241 | 0.193 | 3.87 |
| PTM/2 - 2.0 | 0.241 | 0.193 | 3.77 |

*Electron density ($\rho$) units: moles $e^-/cm^3$

Table IV.   Morphological Dimensions

| Designation | $d*$ | $\ell_H*$ | $\ell_S*$ | $\sigma*$ | $a_I*$ |
|---|---|---|---|---|---|
| MT4/1 - 1.6 | 10 | 3.2 | 15.7 | 0.31 | 0.78 |
| MT4/1 - 2.0 | 9 | 3.5 | 12.3 | 0.30 | 0.75 |
| MTM/2 - 2.0 | 13 | 3.5 | 21.6 | 0.17 | 0.43 |
| PT4/1 - 1.6 | 10 | 2.5 | 13.0 | 0.08 | 0.20 |
| PTM/1 - 1.6 | 11 | 3.3 | 24.1 | 0.16 | 0.40 |
| PTM/1 - 2.0 | 11 | 3.1 | 11.8 | 0.12 | 0.30 |
| PT4/2 - 2.0 | 14 | 2.2 | 16.1 | 0.07 | 0.18 |
| PTM/2 - 2.0 | 13 | 3.7 | 27.1 | 0.11 | 0.28 |

*All values in nanometers (1 nm = 10Å).

Table V.   Estimate of Dissolved Hard Segment

| Designation | $(W_{HX}/W_H)$ (a) | $(W_{HD}/W_H)$ (b) | $(W_{HD}/W_{SM})$ (c) |
|---|---|---|---|
| MT4/1 - 1.6 | 0.169 | 0.28 | 0.11 |
| MT4/1 - 2.0 | 0.022 | 0.23 | 0.12 |
| MTM/2 - 2.0 | 0.027 | 0.15 | 0.04 |
| PT4/1 - 1.6 | 0.185 | 0.36 | 0.14 |
| PTM/1 - 1.6 | 0.199 | 0.48 | 0.18 |
| PTM/1 - 2.0 | 0.029 | 0.32 | 0.16 |
| PT4/2 - 2.0 | 0.034 | 0.34 | 0.09 |
| PTM/2 - 2.0 | 0.034 | 0.34 | 0.09 |

(a) Stochiometric excess hard segment/total hard segment.
(b) Dissolved hard segment/total hard segment (SAXS).
(c) Dissolved hard segment/mixed soft segment phase (SAXS).

extent, be dissolved in the soft segment phase. On the one hand, specific interactions between urea groups will tend to exclude soft segment species from the hard segment phase. On the other hand, the soft segment phase may not be nearly as pure and may contain significant amounts of isolated TDI residue and smaller amounts of longer hard segment sequences. Such dissolved hard accounts for the higher soft segment Tg values in Table II for the samples with the 1.6 molar ratio for TDI/PTMO.

To test this hypothesis, dissolved hard segment content has been calculated from the experimental electrons density variance using the dissolved hard segment model of Figures 3 and 4. The results, given in Table V, show that the soft segment phases may contain as much as an estimated 20% dissolved hard segment. A least squares straight line has been plotted between $(W_{HD}/W_{SM})$, from Table V, and $(1/Tg)$ from Table II. The straight line extrapolation to $(W_{HD}/W_{SM}) = 0$ gives a value Tg = -78°C for the pure soft segment phase, somewhat above the experimental value of -85°C for homopolymer, but in good agreement with the lowest Tg of -78°C reported by Sung (4) for similar amine-cured polyurethanes containing PTMO soft segments.


CONCLUSION

The use of amine curing agents results in a very strong propensity for microphase separation in segmented polyurethanes. The resulting polymers, also denoted as polyurethaneureas, contain urea linkages which promote hard segment aggregation by specific interaction, namely the formation of a three-dimensional hydrogen bonding pattern. The SAXS evidence supports a model of extensive microphase segregation with narrow interphase boundaries, in which the hard segment phases are reasonably pure and the soft segment phases contain varying amounts of dissolved hard segment. The mechanical strength properties are strongly influenced by the state of dispersion of the two phases, and in particular on the relative values of the hard segment phase inhomogeneity length. In a poor material, the hard segments are isolated as microphases with limited lateral dimensions. In better materials the lateral dimensions increase, leading to higher $\ell_H$ values, with little change in the longitudinal period indicated by the Bragg spacing. This increase in lateral dimensions of the hard segment phase results in an overall structure in which the two phases interpenetrate to a greater degree, possibly approaching a structure in which both phases are continuous. Thus two processes - microphase separation per se and possible agglomeration of disperse microphases into a continuous structure - are important factors determining the mechanical properties of a segmented polyurethane.

DISCLAIMER

The contents of this report are not intended as and are not to be construed as an official Department of the Army position or as an endorsement of a particular product by either the author or the Department of the Army.

ACKNOWLEDGMENTS

The authors wish to express their gratitude to Dr. Richard Baron of Polaroid Corporation for making available the polymer samples and mechanical test data. Both sample preparation and testing were carried out at the Polymer Institute of the University of Detroit under the direction of Dr. Kurt C. Frisch. The authors also wish to thank Dr. Gary Hagnauer for providing density data.

REFERENCES

1. N. S. Schneider, C. S. P. Sung, R. W. Matton and J. L. Illinger, Macromolecules, 8, 62 (1975).
2. C. S. P. Sung and N. S. Schneider, Macromolecules, 8, 68 (1975).
3. N. S. Schneider and C. S. P. Sung, Polym. Eng. and Sci., 17, 73 (1977).
4. C. S. P. Sung, C. B. Wu, and C. S. Wu, Macromolecules, 13, 111 (1980).
5. C. S. P. Sung, T. W. Smith, and N. H. Sung, Macromolecules, 13, 117 (1980).
6. J. T. Koberstein, B. Morra, and R. S. Stein, J. Appl. Cryst., 13, 34 (1980).
7. Z. Ophir and G. L. Wilkes, J. Polym. Sci.: Polym. Phys. Edn., 18, 1469 (1980).
8. J. W. C. Van Bogart, A. Lilaonitkul and S. L. Cooper, Adv. in Chem. Series 176, 3 (1979).
9. O. Glatter, J. Appl. Cryst. 7, 147 (1974).
10. G. Porod, Kolloid-Z, 124, 83 (1954).
11. E. Helfand, Acc. Chem. Res. 8, 295 (1975).
12. R. Bonart and E. Muller, J. Macromol. Sci. Phys., B10, 177, (1974).
13. R. Bonart and E. Muller, ibid., 345 (1974).
14. F. Bates, Sc. D. Thesis, Mass. Inst. of Tech., Chem. Eng. Dept., 1982.
15. O. Kratky, Z. Analyt. Chem. 201, 161 (1964).
16. C. G. Vonk, J. Appl. Crystallogr. 8, 340 (1975).
17. J. W. C. Van Bogart, Ph.D. Thesis, Univ. of Wisconsin, Dept. of Chem. Eng., 1981.
18. A. Guinier and G. Fournet, Small Angle Scattering of X-rays, Wiley, New York, p. 158, 1955.
19. C. S. P. Sung, Polymer Sci. and Tech., Vol. 11, (D. Klempner and K. C. Frisch, Eds.), Plenum Press, New York, p. 119, 1980.

APPENDIX

Variance Ratio: $\rho_{HM} = \rho_{HP}$ Case.

Independent Variables

$(\rho_{SP}/\rho_{HP})$ = ratio of pure phase electron densities, soft segment (SS)/hard segment (HS)

$W_H$ = electron fraction HS in a given sample

$(W_{HD}/W_H)$ = ratio dissolved HS/total HS

Convenience Variables (not involved in scattering theory but needed to evaluate electron variables)

$M_H$ = mass fraction HS

$\mu SP$, $\mu HP$ = mass densities (g/cm$^3$) of pure SS and HS

$R_{SP}$, $R_{HP}$ = ratio atomic number/$\Sigma$ atomic weight for pure SS, HS

Relationship of independent variables to convenience variables

$$\rho_{SP} = R_{SP}\,\mu_{SP} \text{ moles } e^-/cm^3, \tag{A-1}$$
and
$$\rho_{HP} = R_{HP}\,\mu_{HP} \text{ moles } e^-/cm^3, \tag{A-2}$$
so
$$\rho_{SP}/\rho_{HP} = (R_{SP}/R_{HP})\,(\mu_{SP}/\mu_{HP}) \tag{A-3}$$
$$W_H = M_H/[M_H + (1-M_H)\,(R_{SP}/R_{HP})] \tag{A-4}$$

Dependent variables

$$\phi_{HP} = W_H/[W_H + (1-W_H)/(\rho_{SP}/\rho_{HP})] \tag{A-5}$$

$$\phi_{HM} = \phi_{HP}(1-W_{HD}/W_H) \tag{A-6}$$

$$\rho_{SM}/\rho_{HP} = 1/[f_{HD} + (1-f_{HD})/(\rho_{SP}/\rho_{HP})] \tag{A-7}$$

where

$$f_{HD} = (W_{HD}/W_H)/[(W_{HD}/W_H)+(1/W_H)-1] \tag{A-8}$$

is the fraction of dissolved hard segment in the soft segment phase.

With the premise $(\rho_{HM}/\rho_{HP})=1$, all variables in eqn. (18) have been expressed in terms of the independent variables, or, if desired, the convenience variable. The resulting equation for shall not be written explicitly. The above equations have been incorporated into a BASIC language program which is available upon request.

13

The parameter values used are listed in Table A-1.  The
density of pure soft segment $\rho_{SP}$ was taken from Bonart and Muller
(13), while $\rho_{HP}$ was obtained by extrapolating the specific volume
of the present samples to $W_H = 1.00$.  The electron density variance
values for Models A, B, and C are listed in Table A-2.

The correction for interphase volume of equation (16) requires
values for $\emptyset_{SM} = 1 - \emptyset_{HM}$, and $\emptyset_3$, the volume fraction interphase
material.  Assuming lamellar structure the latter may be expressed
as:

$$\emptyset_3 = 2\ a_I/d. \tag{A-9}$$

For the purpose of this correction, equation (16) presumes a linear
gradient zone of thickness $a_I$.  The need for a $\emptyset_{HM}$ value for this
correction poses a problem, since $\emptyset_{HM}$ is a dependent variable whose
value is not known until the final solution for dissolved hard
segment is obtained.  This problem was solved by using $\emptyset_{HP}$ as a
first approximation for $\emptyset_{HM}$, applying the correction and calculating a new $\emptyset_{HM}$, and repeating the process until the desired
precision of 0.001 in $\emptyset_{HM}$ was obtained.  Convergence was obtained
in all cases in three to four iterations.

Table A-1.  Material Parameters for Density

Variance Calculations

| System | $R_{SP}$ mole $e^-$/gm | $\mu_{SP}$ gm/cm$^3$ | $R_{HP}$ mole $e^-$/gm | $\mu_{HP}$ gm/cm$^3$ |
|---|---|---|---|---|
| TDI-MOCA-PTMO | 0.5547 | 0.985 | 0.5167 | 1.45 |
| TDI-TMAB-PTMO | 0.5547 | 0.985 | 0.5220 | 1.45 |

Table A-2.  Data for Dissolved Hard

Segment Calculation

| Sample | $\emptyset_3$ | $\emptyset_{HM}$ | $(\Delta\rho)^2_C$ x10$^3$ | $(\Delta\rho)^2_B$ x10$^3$ | $(\Delta\rho)^2_A$ x10$^3$ |
|---|---|---|---|---|---|
| MT4/1-1.6 | 0.156 | 0.17 | 4.14 | 5.07 | 7.71 |
| MT4/1-2.0 | 0.167 | 0.22 | 5.16 | 6.14 | 8.75 |
| MTM/2-2.0 | 0.066 | 0.14 | 4.42 | 4.86 | 5.91 |
| PT4/1-1.6 | 0.040 | 0.16 | 4.55 | 4.79 | 8.35 |
| PTM/1-1.6 | 0.073 | 0.12 | 3.30 | 3.71 | 8.31 |
| PTM/1-2.0 | 0.055 | 0.21 | 5.43 | 5.75 | 9.60 |
| PT4/2-2.0 | 0.026 | 0.12 | 3.87 | 4.04 | 6.62 |
| PTM/2-2.0 | 0.043 | 0.12 | 3.77 | 4.05 | 6.62 |

SYNTHESIS AND PROPERTIES OF SOME COPOLYMERS

OF MALEIC ANHYDRIDE DERIVATIVES

Christel Schneider and Joachim Wolff

Institute of Physical Chemistry
University of Cologne
Cologne, Germany

INTRODUCTION

While the number of papers concerned with the technically important copolymerization of maleic anhydride (MAH) is enormous, only little work has been carried out up to date on the copolymerization of maleic anhydride derivatives.

We were mainly interested in the polymerization behaviour of phenylmaleic anhydride (PhMAH) since the introduction of the bulky phenyl group leads to a very rigid molecule, which should appreciably reduce chain mobility when it is incorporated into a basic polymer like e.g. polystyrene. Besides this sterical effect, the nature of the substituent at the double bond should influence the reactivity of the double bond as well as that of the anhydride ring. To get more insight into these effects and, consequently, into the reaction mechanism, we carried out additional polymerization experiments with monofluoromaleic anhydride (FMAH) and methylmaleic anhydride (MeMAH) and compared the results obtained with data of the well-known MAH/styrene copolymerization.

The very different polymerization behaviour of maleic anhydride and maleimide (MI) has been the subject of a number of investigations. We, therefore, synthesized 2-phenylmaleimide and compared its homo- and copolymerization, especially with styrene, to that of PhMAH.

Since the anhydride group is very attractive for further modification of the polymer by chemical reactions, we carried out the hydrolysis and the esterification of the differently substituted anhydrides and converted the copolymers of PhMAH/styrene to the corresponding imide and hydrazide copolymers.

RESULTS AND DISCUSSION

## Phenylmaleic Anhydride

PhMAH was synthesized according to a prescription by Hill /1/ by reaction of phenylsuccinic acid with $SeO_2$ in an acetanhydride solution.

Surprisingly, inspite of the steric hindrance due to the 1,1- and 1,2-disubstitution, homopolymerization of PhMAH could be achieved in the melt using Co-60-γ-rays as initiator. Elemental analysis, IR and [1]H-NMR spectroscopy clearly showed that polymerization had taken place via the C-C double bond under retention of ring structure (fig. 1). The degree of polymerization, however, is low and was estimated by vapour phase osmometry to be 5 - 10.

Copolymerization of PhMAH with styrene works readily (fig. 2). We have started the reaction with conventional chemical catalysts

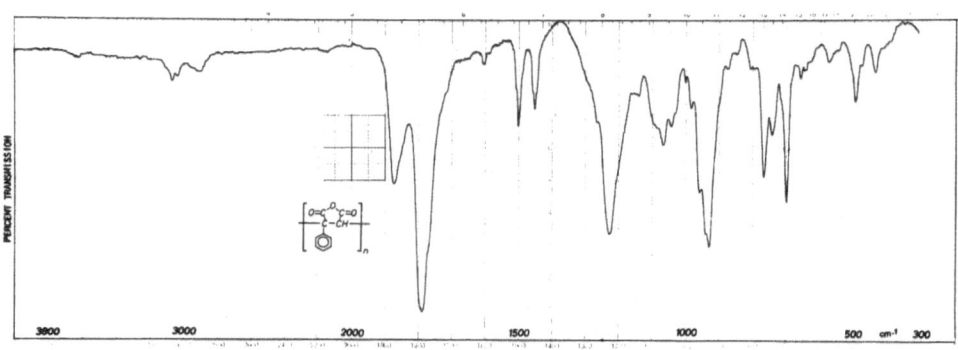

Fig. 1.   IR-spectrum of polyphenylmaleic anhydride

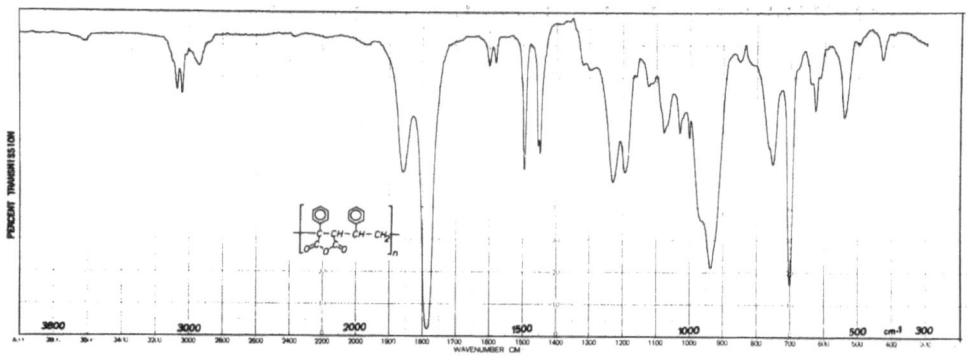

Fig. 2.   IR-spectrum of poly(phenylmaleic anhydride-co-styrene)

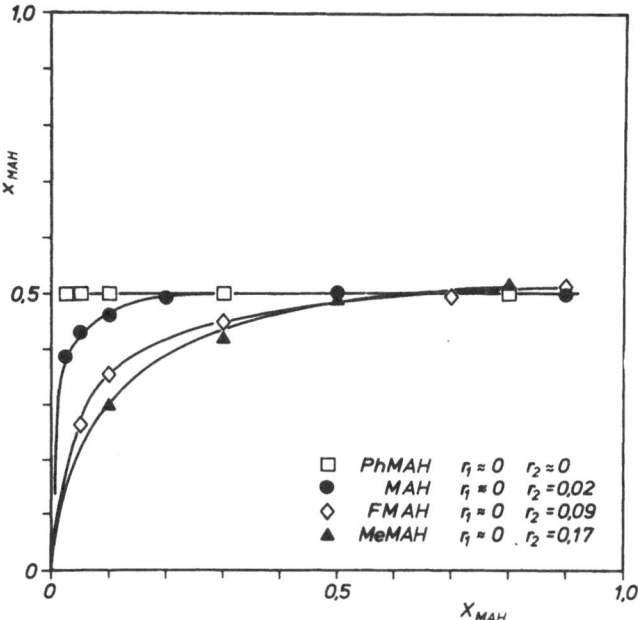

Fig. 3. Copolymerization of XMAH (X = H, F, Me, Ph) with styrene; solvent: benzene; T = 323 K; dose rate: $3.0 \times 10^2$ $Jkg^{-1}h^{-1}$

like AIBN as well as with Co-60-$\gamma$-rays. The polymerization process was carried out in bulk and in various solvents of very different polarity and electron donating strength like acetone, benzene, methylene chloride and THF and we studied the reaction over a wide temperature range (313 K to 398 K). In any case a copolymer with a strictly alternating structure was obtained. As can be seen from fig. 3, the alternation is even more pronounced than in the case of the well-known unsubstituted MAH/styrene copolymer. The reaction rate is strongly dependent on the reaction conditions; the highest rates in solution, which are quite comparable to those of the styrene homopolymerization, were obtained in methylene chloride for a monomer mixture containing about 30 mole% PhMAH (fig. 4). It seems worthwhile to point to the fact that the rate maximum is not attained for an equimolar composition of the monomer mixture but is shifted to lower PhMAH concentrations, the degree depending on the solvent used. We observed similar deviations of the rate maximum from a 1:1 feed in the case of the system MAH/styrene.

As expected, bulk polymerization is characterized by a strong gel effect and is, therefore, very fast. Using $\gamma$-rays with a dose rate of $1.5 \times 10^3$ $Jkg^{-1}h^{-1}$ and working at 339 K, 15% conversion was obtained after 4 min for a monomer mixture consisting of 60 mole%

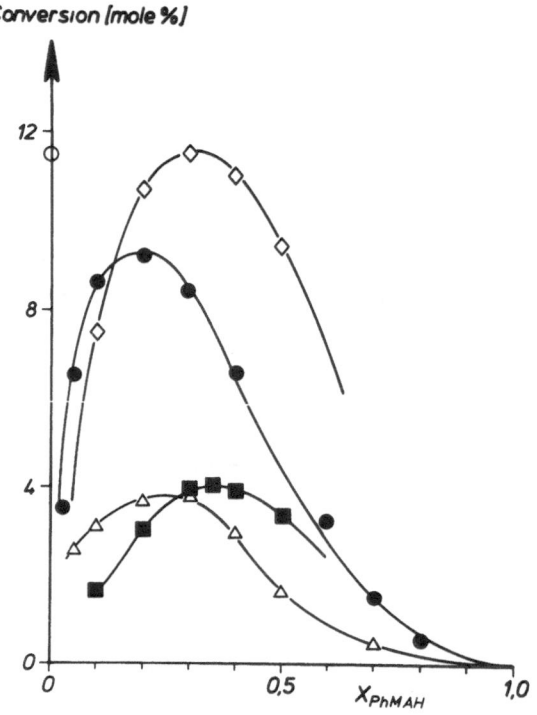

Fig. 4.   Conversion vs. composition of monomer mixture for the copoly-
          merization of PhMAH with styrene in
          ● THF, △ benzene (T = 318 K, t = 23 h)
          ◇ CHCl$_2$ (T = 323 K, t = 5,1 h)
          ■ acetone (T = 323 K, t = 40,5 h)
          ○ homopolym. of styrene in THF
          dose rate: 3,0 x 10$^2$ Jkg$^{-1}$h$^{-1}$

of PhMAH. The rate increased continuously with temperature up to
413 K and was not followed further. Molecular weights up to 20,000
were obtainable depending on the polymerization conditions.

    The copolymers are soluble in polar solvents like methylene
chloride, dioxane, THF or DMF. The softening temperature of the
PhMAH/styrene copolymer is rather high (533 - 553 K depending on chain
length), i.e. about 50 to 70 degrees higher than that of the MAH/
styrene copolymer (table 1). For a technical application it is

Table 1. Softening temperature and thermal decomposition of some copolymers of styrene and isoprene with XMAH and with PhMI

a) Softening temperatures obtained from TMA (in K)

| Poly-styrene | Copolymers of styrene with | | | | Content of cyclic monomers |
| | MAH | MeMAH | PhMAH | PhMI | |
| --- | --- | --- | --- | --- | --- |
| 368-378 | 478 | 493 | 513-533 | 593 | 50 mole% |

| | | Copolymers of isoprene with | | | |
| | | | PhMAH | PhMI | 55 mole% |
| | | | 433 | 463 | |

b) Thermal decomposition obtained by thermogravimetry (in K)

| Poly-styrene | Copolymers of styrene and | | | |
| | MSA | MeMSA | PhMSA | 50 mole% |
| --- | --- | --- | --- | --- |
| 588 | 543 | 543 | 573 | start of decomposition |
| 623 | 583 | 573 | 593 | 5% loss of weight |

considered quite favourable that thermal decomposition begins not until 40 to 50 degrees above the softening temperature.

While we studied the solution copolymerization of the PhMAH/styrene system, a US patent by Dow Chemical Co. on high molecular weight PhMAH copolymers came out in April 1979 /2/. Copolymerization examples given in this patent for the PhMAH/styrene system were mainly carried out in bulk using a 6:1 molar ratio of styrene: PhMAH and reaction temperatures of about 70° to 90° C. The amount of PhMAH in the copolymer was found to depend on the molar ratio of the feed but never to exceed 25 mole%. Molecular weights in the range of 300,000 to 500,000 were obtained and a very peculiar relationship between the molecular weight and the reaction conditions was observed.

Since these results were not in accordance with our data and since the reaction conditions used were quite different from ours, we thoroughly carried out copolymerization experiments according to the reaction conditions described in the patent. However, we cannot confirm the results. In contrast to the patent we always obtain a copolymer of strictly alternating order and find the usual dependencies of rate and molecular weight on the reaction conditions.

If styrene is replaced by p-chlorostyrene, very similar results were obtained. Using α-methyl styrene as comonomer, no polymerization could be achieved, obviously due to steric hindrances. Copolymerization experiments with N-vinyl pyrrolidone failed as well; with acrylic and methacrylic esters only homopolymerization of the ester occurred. However, copolymerization with isoprene and vinyl ethers works readily. The copolymerization of PhMAH with isoprene is less strictly alternating ($r_{PhMAH} \simeq 0$; $r_{IP} = 0.14$) especially at high isoprene concentrations in the feed and the rate is much lower than in the styrene system.

## Monofluoromaleic Anhydride

The synthesis of FMAH was carried out according to a prescription by Raasch, Miegel and Castle /3/; the monomer was carefully purified by fractional distillation.

Attempts to homopolymerize FMAH were not successful; however, copolymerization with styrene works easily (fig. 5). In spite of the electron withdrawing nature of the fluorine atom the copolymer obtained is less alternating than in the case of PhMAH/styrene or even MAH/styrene (fig. 3). The reaction rate is strongly dependent on the nature of the solvent; the highest rates and molecular weights of about 8,000 were obtained in benzene. The copolymers are extremely sensitive to hydrolysis and after a longer period of storage HF is split off, too. Thermal decomposition takes place gradually at about 390 K. Copolymerization of FMAH with methyl methacrylate leads to a copolymer with only small amounts of FMAH ($r_{FMAH} \simeq 0$; $r_{MMA} = 8.1$). Contrary to PhMAH, alternating copolymerization of FMAH with α-methyl styrene can be achieved by using high radiation dose rates.

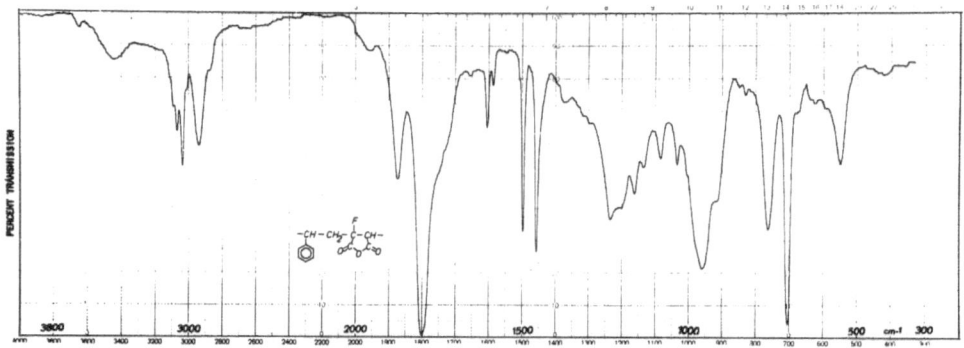

Fig. 5.   IR-spectrum of poly(fluoromaleic anhydride-co-styrene) (slightly hydrolyzed)

## Methylmaleic Anhydride

Copolymerization of MeMAH and styrene initiated by γ-rays was carried out in various solvents. The copolymerization diagram (fig. 3) is in accordance with reactivity ratios given in the literature /4/ for chemical initiation. Alternation is less pronounced than in the case of MAH, FMAH and PhMAH. The conversion versus feed composition curves pass through a maximum but the reaction is obviously affected by chain transfer leading to the formation of allyl radicals as is evident from the rather low molecular weights of 1500-4000.

## Reaction Mechanism

With regard to the copolymerization mechanism, over the past quarter of a century the system maleic anhydride/styrene has been studied in both the bulk phase and in solution by a number of workers because the kinetics of the system were supposed to deviate from the simple terminal model. It has often been assumed that EDA complexes between the monomers were involved in the propagation step of the copolymerization reaction /5/6/7/ or that penultimate groups in the propagating polymer chain played an important role in the polymerization process /8/.

From our results on the phenyl-MAH/styrene system we cannot follow the penultimate model proposed by Dodgson and Ebdon /8/ for the MAH/styrene system. A penultimate effect should be even more pronounced in the case of PhMAH than with the unsubstituted maleic anhydride, but neither do we find any deviation from a 1:1 copolymer composition nor does addition of $ZnCl_2$ lead to a reduction in the amount of PhMAH incorporated in the copolymer.

With regard to the complex model, we carried out series of UV and NMR measurements of EDA complexes in monomer mixtures of styrene with MAH and all derivatives mentioned above. From comparison of the complex constants obtained with the copolymerization data no unequivocal evidence for the operation of a complex model could be obtained.

In the hope of getting a better idea of the copolymerization mechanism of MAH-derivatives with styrene, terpolymerisation of two systems was carried out. In both cases styrene was chosen as the donor monomer while two competitive acceptor monomers were used, one of which was the unsubstituted MAH. When the system PhMAH/MAH/styrene is polymerized, almost no MAH is incorporated into the terpolymer as can be seen from the Slocombe diagram in fig. 6. If, on the other hand, the system FMAH/MAH/styrene is polymerized, MAH is the dominating acceptor unit in the terpolymer. These results are in accordance with the copolymerization data: the higher the order of alternation in the copolymer, the more favoured is the corresponding acceptor monomer in the terpolymerization process.

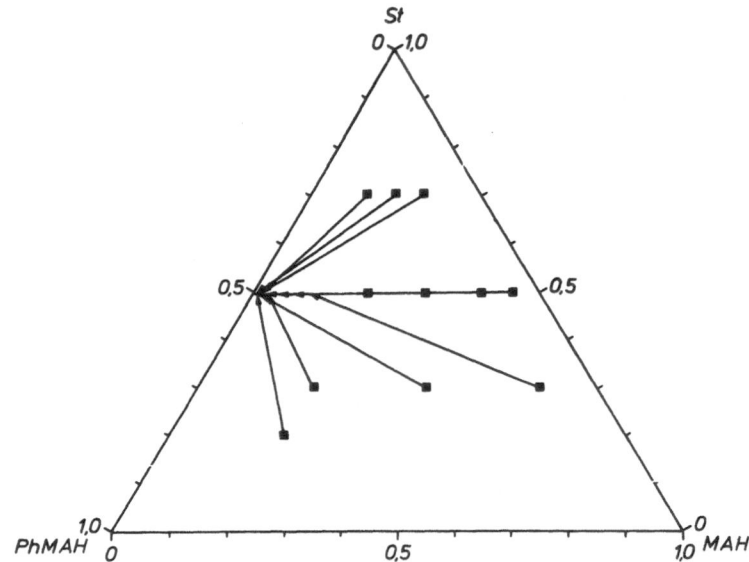

Fig. 6.  Terpolymerization of the system PhMAH/MAH/styrene in
         dioxane;  T = 323 K;  dose rate: $3,0 \times 10^2$ $Jkg^{-1}h^{-1}$

        If the relative rate constants for the addition of a cyclohexyl
radical to different monomers are determined according to a method
of B. Giese /9/, it can be shown that, if the rate constant for the
addition of cyclohexyl to styrene is set equal to 1, the relative
rate constant for the addition to PhMAH is 2900. This is four times
higher than the corresponding rate constant for MAH (730), which
itself again is about twice as reactive to cyclohexyl as is FMAH.
These results are in excellent agreement with our co- and terpolymerization data, if we assume, that a propagating radical ending on a
styrene unit behaves in principle similar to the cyclohexyl radical.
We, therefore, think that the alternating copolymerization of MAH
and its derivatives with styrene may be explained without taking into
account the participation of EDA complexes just by the fact that the
rate constant e.g. for the addition of a PhMAH molecule to a propagating chain ending on a styrene radical is so much higher than the
homopropagation step of styrene that it will take place almost
exclusively.

        From our copolymerization data we tried to estimate the Q- and
e-values of the MAH-derivatives according to the Alfrey-Price scheme.
However, since the reactivity ratio $r_{XMAH}$ is approximately zero, the
uncertainty  especially of the Q-values is large. If the differences
in the e-values between MAH and the corresponding derivatives, $\Delta e$,

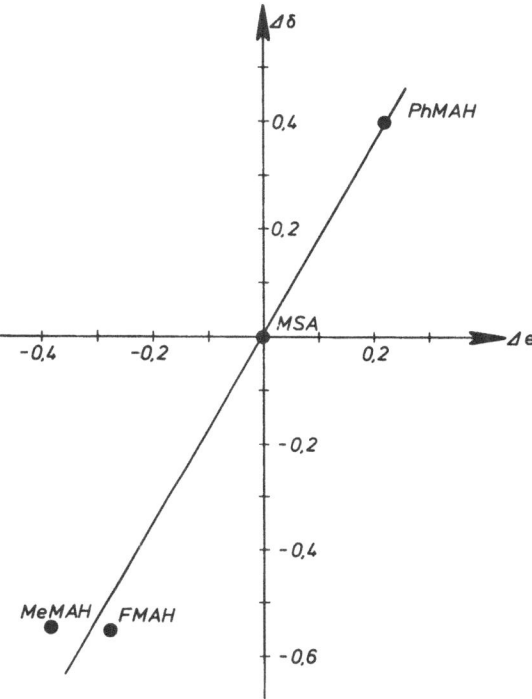

Fig. 7.   Plot of differences in e-values vs. differences in chemical
          shifts of the vinylic protons for various MAH derivatives


obtained from the copolymerization with styrene are plotted versus
the differences in the chemical shifts $\Delta\delta$ of the vinylic protons
obtained from [1]H-NMR measurements, a linear relationship is achieved
(fig. 7). Since both, e-value and chemical shift can be considered
as a criterion of the acceptor strength of the monomer, at least the
e-values of the substituted maleic anhydrides seem to be satisfacto-
rily describable by the Alfrey-Price formalism.

2-Phenylmaleimide

     Maleimide and maleic anhydride behave very differently in poly-
merization reactions. Up to date, of all maleimide derivatives substituted at the double bond only citraconimides have been polymerized
to low molecular weight products /10/.

     For comparison to the polymerization behavior of MSI on one
hand and to PhMAH on the other, we synthesized 2-phenylmaleimide
(PhMI) according to a patent /11/ by treating PhMAH with a solution
of ammonium acetate in dioxane. The product was carefully purified

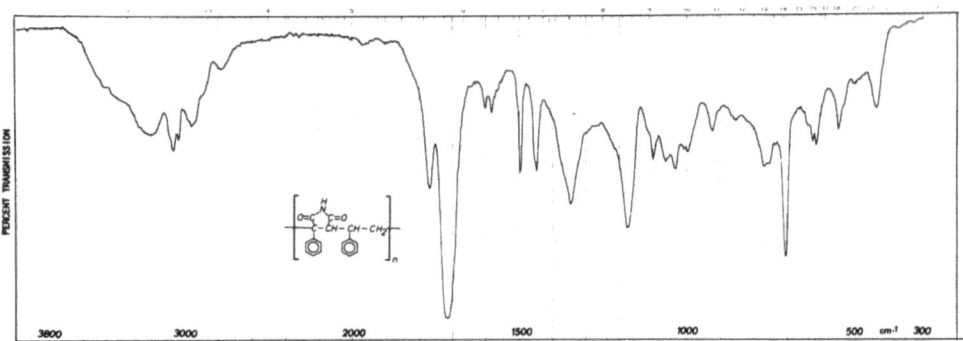

Fig. 8.   IR-spectrum of poly-2-phenylmaleimide

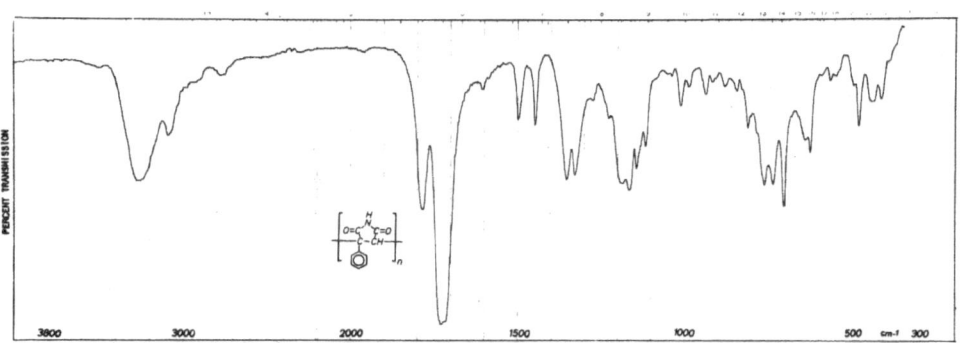

Fig. 9.   IR-spectrum of poly(2-phenylmaleimide-co-styrene)

by chromatography using $Al_2O_3$.

    While homopolymerization of PhMI could not be achieved in
solution, polymerization in the melt, i.e. above 436 K, was successful. The reaction is fast; about 6o% conversion were obtained after
4o min reaction time using γ-rays with a dose rate of
$1,5 \times 10^3$ $Jkg^{-1}h^{-1}$ at T = 443 K. At higher conversion the rate
decreases strongly due to the high viscosity of the medium and
approaches zero at about 70% conversion. The IR spectrum (fig. 8)
clearly shows that polymerization proceeds via the vinylic double
bond under retention of ring structure. The polymer is hard and
brittle, the molecular weight is low ($P_n$ = 10 - 15).

    Copolymerization of PhMI was carried out with styrene (fig. 9)
and isoprene and compared to the corresponding copolymerization
experiments with PhMAH. It can be seen from fig. 10 that copolymerization with styrene results in a mainly alternating product, while in

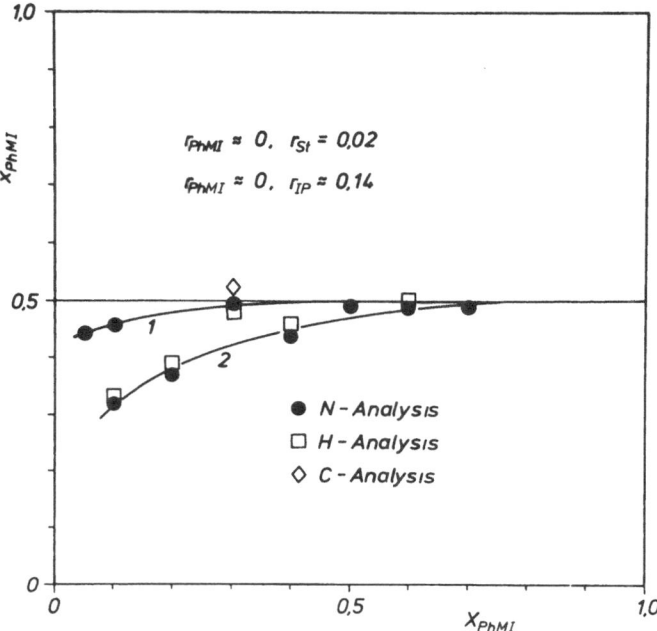

Fig. 10.   Copolymerization of PhMI with styrene (1) and isoprene (2)
solvent: THF;   T = 345 K; dose rate: $3,0 \times 10^2$ $Jkg^{-1}h^{-1}$

the case of isoprene longer isoprene sequences are formed for monomer
mixtures containing more than 40 mole% isoprene. In both systems the
reaction rate is much smaller for PhMI than for the corresponding
reactions with PhMAH. The copolymers are soluble in THF, dioxane and
DMF; the PhMI/styrene copolymer shows the high softening temperature
of about 593 K.

## Polymer Analogous Reactions

While hydrolysis and esterification are easy in the case of the
MAH/styrene copolymers, long reaction times or high temperature are
necessary for the PhMAH/styrene copolymers due to the bulky phenyl
groups. Almost no reaction takes place if e.g. $H_2O$ or KOH is replaced
by the more voluminous MeOH. Introduction of an electron withdrawing
substituent like fluorine in FMAH drastically increases the rate of
hydrolysis and of esterification. The completely hydrolyzed fluoromaleic acid/styrene copolymers are soluble in water and alcohols.
Conversion of XMAH/styrene copolymers to the corresponding imide/
styrene copolymers proceeds successfully and without side reactions
by treatment with a solution of ammonium acetate in dioxane. While
after a reaction time of 50 h, 94% of the anhydride groups in poly-
MAH and 85% of those in a MAH/styrene copolymer were converted to

imide groups only about 15% of the anhydride groups had reacted in the case of the PhMAH/styrene copolymer. Partial introduction of imide groups results in higher softening temperatures. Surprisingly, even the sterically hindered PhMAH/styrene copolymer can easily by converted to more than 95% to the corresponding phenylhydrazide/ styrene copolymer when treated with an aqueous solution of 50 mole% hydrazine and catalytically effective traces of acetic acid /12/. No side reaction is detectable by IR and elemental analysis; the product has a softening point of nearly 590 K, which is about 60 degree higher than that of the PhMAH/styrene copolymer.

ACKNOWLEDGEMENT

    We gratefully acknowledge the financial support of this work by the Bundesministerium für Forschung und Technologie. We are indebted to Dr. B. Giese, Darmstadt, for the determination of the relative rate constants for the addition of cyclohexyl radicals to the MAH derivatives.

REFERENCES

1.  R. K. Hill, A Convenient Preparation of Arylmaleic Anhydrides,
    J. Org. Chem. 26:4745 (1961).
2.  J. W. Bozzelli, K. S. Dennis (Dow Chemical Comp.), High Molecular
    Weight Polymers of Phenylmaleic Anhydride, US-Pat. 4,147,852
    (1979).
3.  M. S. Raasch, R. E. Miegel, J. E. Castle, Mono- and Difluoro-
    butenedioid Acids, J. Amer. Soc. 81:2678 (1958).
4.  E. C. Chapin, G. E. Ham, C. L. Mills, Copolymerization.
    VII. Relative Rates of Addition of Various Monomers in Copoly-
    merization, J. Polym. Sci. 4:597 (1949).
5.  E. Tsuchida, T. Tomono, and H. Sano, Solvent Effects on the
    Alternating Copolymerization Systems, Makromol. Chem. 151:245
    (1972).
6.  R. B. Seymour and D. P. Garner, Relationship of Temperature to
    Composition of α-Methylstyrene and Maleic Anhydride, Polymer
    17:21 (1976).
7.  R. G. Farmer, D. J. T. Hill, and J. H. O'Donnell, Study of the
    Role of Charge-Transfer Complexes in Some Bulk-Phase Free-
    Radical Polymerizations, J. Macromol. Sci.-Chem. A 14 (1):51
    (1980).
8.  K. Dodgson and R. Ebdon, On the Role of Monomer-Monomer Donor-
    Acceptor Complexes in the Free-Radical Copolymerization of
    Styrene and Maleic Anhydride, Eur. Polym. J. 13:791 (1977).
9.  B. Giese and J. Meister, Die Addition von Kohlenwasserstoffen
    an Olefine - Eine neue synthetische Methode, Chem Ber.
    110:2588 (1977).

B. Giese and J. Meixner, Korrelation zwischen Radikalreaktivi-
täten und Copolymerisationsparametern, Angew. Chem. 92,
3:215 (1980).

10.  T. V. Sheremeteva, G. N. Larina, V. A. Gusinskaye, Preparation
of High Molecular Weight Compounds on the Basis of Cyclic
Imides and Diimides of Dicarboxylic Acids, J. Polym. Sci.
C 16:1631 (1967).

11.  K. Ichimura, H. Ochi, α-Arylmaleimide Derivatives, Chem.
Abstr. 83:9559 k (1975).

12.  H. Naarmann, M. Patsch, H. Eilingsfeld, and M. Marse, Heat-
hardenable Coating Material containing a 1,2-Diaza-3,6-
Cyclohexanedione Compound, DPA 1915576 (1970).

# PHOTOTHERMAL DEGRADATION OF ETHYLENE/VINYLACETATE COPOLYMER

R. H. Liang, S. Chung, A. Clayton, S. Di Stefano,
K. Oda, S. D. Hong, and A. Gupta

Jet Propulsion Laboratory
California Institute of Technology
4800 Oak Grove Drive
Pasadena, California   91109

## Abstract

Ethylene/vinyl acetate copolymer (EVA), which has desirable elastomeric properties, low material cost and easy processability, is being considered as a candidate encapsulation material for photovoltaic modules. However, without protection from UV irradiation, EVA is expected to undergo photodegradation similar to that encountered in polyethylene and polyvinylacetate. We have carried out photothermal degradation studies of a "stabilized" formulation of EVA in the temperature range of 25°C to 105°C under different oxygen environments. At low temperature (25°C), slow photooxidation occurred via electronic energy transfer involving the UV absorber incorporated in the polymer. But no change in physical properties of the bulk polymer can be detected up to 1500 hours of irradiation. At elevated temperature, leaching/evaporation of additives took place leading ultimately to chemical crosslinking of the copolymer and formation of volatile photoproducts such as acetic acid.

## Introduction

Photovoltaic modules need protective encapsulation systems in order to ensure service life of twenty years or more. The heart of the encapsulation is the pottant, the elastomeric material which protects the solar cell against mechanical impact. Ethylene/vinyl acetate copolymer, (EVA), having the required elastomeric properties, low cost, and easy processability is considered a leading candidate to be used for encapsulation of solar cells.[1] However, like most of the low cost elastomers without additives, EVA undergoes relatively rapid degradation when ex-

267

posed to solar ultraviolet radiation. In addition, it was observed that during outdoor deployment, hot spots (up to 150°C) are
developed in the PV modules. These hot spots can cause accelerated degradation leading to premature failure of the encapsulant.
We have carried out studies of the photothermal degradation of a
"formulated" EVA composite system in the temperature range of 25°C
to 105°C under three different oxygen environments: in open air,
with limited access to $O_2$, and in a dark closed stagnant ov-
en.[2,3] The objectives are to characterize the principal degra-
dative products, and to identify modes of degradation leading to
an eventual development of a perfomance prediction model. The
composition of the EVA copolymer system used is listed in Table
1.[2]

Table 1.  Formation of EVA A9918

| COMPOUND | TRADE NAME | FUNCTION |
|---|---|---|
| ETHYLENE/VINYL ACETATE | EVA | ELASTOMER |
| 2-HYDROXY-4-n-OCTYL BENZOPHENONE | CYASORB UV 531 | U.V. ABSORBER |
| PHENYLPHOSPHITE ESTERS | NAUGARD-P | ANTI OXIDANT |
| BIS (2,2,6,6-TETRAMETHYL-PIPERIDINYL-4) SEBACATE | TINUVIN 770 | U.V. STABILIZER |
| 2,5-DIMETHYL-2,5-BIS-(t-BUTYL PEROXY HEXANE) | LUPERSOL | CROSS-LINKING AGENT |

Experimental

An experimental set up, illustrated in Figure 1, for studying
the degradation of EVA films under various conditions of temperature, UV light and $O_2$ levels was designed. The set up consists of
a removable 450 watt medium pressure Hg lamp equivalent to 6-10
suns (295-380 nm) enclosed in a quartz water cooling jacket. The
filtering effect of the atmosphere is taken into account by the
attachment of a 1/8" thick pyrex ring around the cooling jacket in
the vicinity of the sample placement. An annular transparent pyrex oil bath is then fitted around the ring. EVA films (3" x 1"
in size) were placed directly on the inner surface of the oil
bath, parallel to the pyrex ring but not in direct contact with
it, allowing full access of oxygen. The oil bath was thermostatically controlled by electrodes connected to a variable transfor-
mer. This enabled controlled accelerated testing of films at
70°C, 85°C, and 105°C as well as at room temperature (25°C). To
simulate limited oxygen access, a segment of a pyrex cylinder the
size of the inner circumference of the annular oil bath was placed

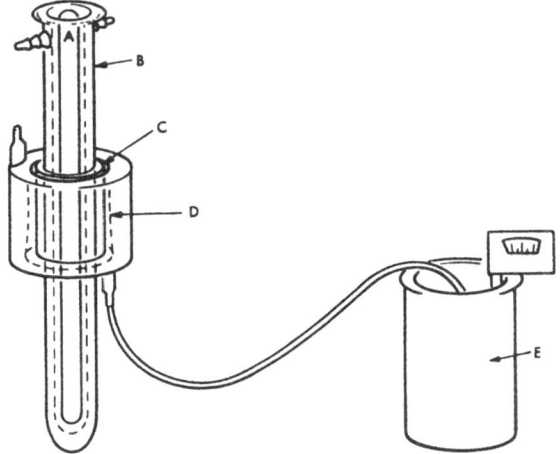

A = 450 WATT MEDIUM PRESSURE HG LAMP

B = PYREX $H_2O$ COOLING JACKET

C = 1/8" THICK PYREX RING

D = OIL JACKET

E = THERMOSTATICALLY CONTROLLER HOT OIL BATH

Figure 1. Photothermal Radiation Apparatus

over the sample film.

EVA films were exposed to pyrex filtered irradiation (290-400 nm). A portion of the irradiated samples were used to measure tensile modulus. Another portion of the samples were soxhlet extracted in hot $CH_2Cl_2$ for 168 hours. The sol was separated from the gel and UV-VIS absorption spectra of the sol were recorded on a Cary 219 spectrometer. The extract was then evaporated to dryness under nitrogen and the residues were dissolved in enough chloroform to make a 0.1% W/V solution. FT-IR spectra of the ex- tracted residues were obtained by evaporating a known amount of chloroform solution on KBr plates. The chloroform solution was also used for GPC analysis. The low molecular weight components (e.g., the residual curing agent and stabilizers) were analyzed using two capillary columns packed with μ-styragel (2x100A°). The high molecular weight analysis was carried out by using 4 μ- styragel columns ($10^5$, $10^4$, $10^3$, 500A°). Swelling experiments were also carried out on the gel in order to obtain the chemical crosslinking density.

## Results and Discussions

Photodegradation of EVA films was studied at room temperature

(25°C) in order to serve as base line for tests conducted at higher temperatures.  Fig. 2 and 3 show the changes in additive concentration as a function of room temperature exposure period.
Whereas there is no change in the UV absorber and anti-oxidant
concentration (Fig. 2), there was an efficient removal of residual

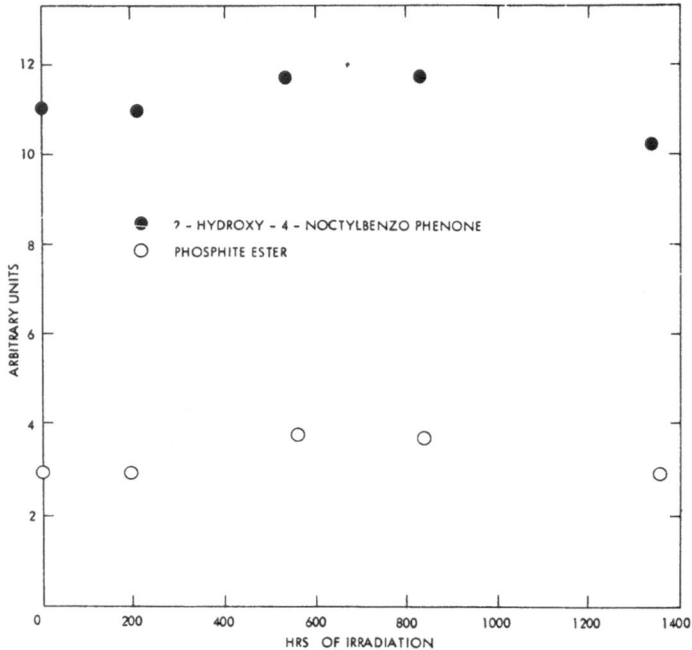

Figure 2. Concentration of UV Absorber and Antioxidant of EVA
          (A9918) as a Function of Photoaging as Detected by HPLC

curing agent (Fig. 3).  Fig. 4 illustrates the normalized difference FT-IR intensities.  There is a slow increase in the intensity
of the hydroxyl absorption at 3360 $cm^{-1}$.  Since most (90%) of the
light flux in the 290-360 nm region is absorbed by the UV absorber
incorporated in the polymer, an electronic energy transfer needs
to be postulated in this system which would presumably transfer
the electronic energy to the peroxide causing it to decompose and
form hydroxyl groups.  Table 2 shows the weight loss and the molecular weight distribution of the uncrosslinked polymers.  No significant change molecular weight distribution nor percent weight
of extractable were found.

     Control experiments carried out by heating EVA films in a
stagnant oven    (absence of irradiation), have also been carried
out to investigate the thermal degradation of EVA.  Figures 4 and
5 illustrate transmission change as a function of thermal and
photothermal aging time at 400 nm and 360 nm respectively.  There

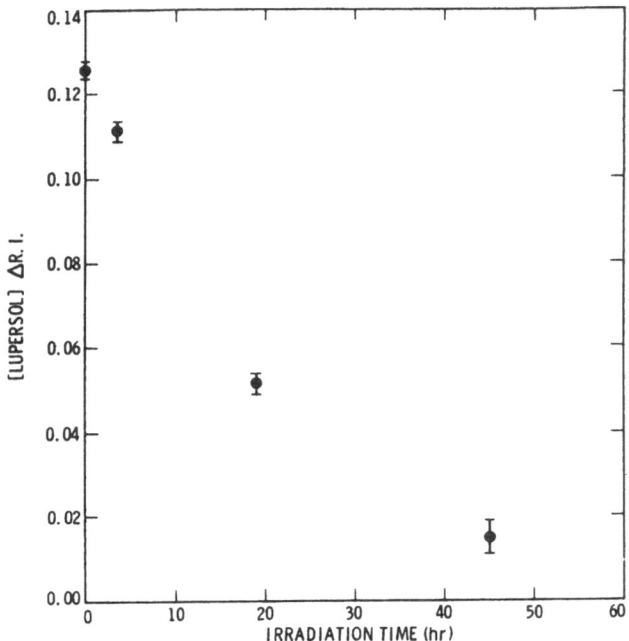

Figure 3.   Concentration of Residual Curing Agent of EVA (A9918)
            as a Function of Photoaging at 30°C as Detected by HPLC

is a definite increase in absorption at wavelength longer than 400
nm for the dark over-aged samples, which causes the samples to appear yellow.   Transmission at 360 nm and shorter λ also shows a
distinct increase which is caused by a gradual loss of the UV absorber.   Consideration of the data at 360 nm and 500 nm demonstrate that the yellowing effect shows up best at 450 nm, since
there is no compensating increase of transmission due to the loss
of the UV absorber at this wavelength.   At 360 nm however,

Table 2.   Percentage of Weight Loss Due to $CH_2Cl_2$ Extraction and
           Molecular Weight Distribution of the Extract of EVA
           Irradiated at 30°C

| IRRADIATION TIME (HRS) | % WT LOSS | $\overline{M}_n$ | $\overline{M}_w$ | $\overline{M}_w/\overline{M}_n$ |
|---|---|---|---|---|
| 0 | 21 | 35,000 | 101,000 | 2.91 |
| 168 | 23 | 38,000 | 114,000 | 3.01 |
| 524 | 23 | 34,000 | 112,000 | 3.26 |
| 836 | 22 | 41,000 | 116,000 | 2.84 |
| 1078 | 22 | 35,000 | 106,000 | 3.00 |
| 1388 | 20 | 38,000 | 108,000 | 2.88 |

the increase in absorption due to yellowing is more than compen-
sated for by the gain in transmission due to the loss of UV absorber, and hence there is a net gain in transmission at 360 nm.

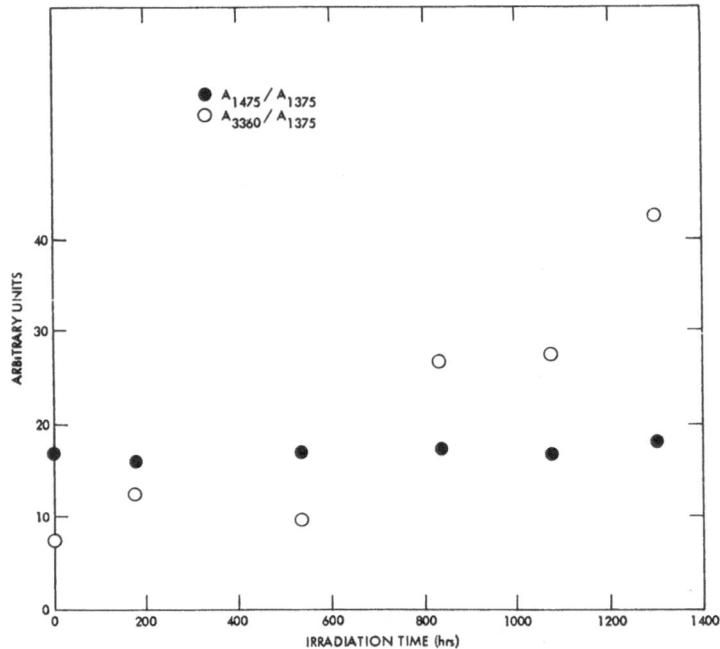

Figure 4.   Normalized FT-IR Intensities of EVA (A9918) as a Function of Photoaging at 30°C

        Transmission changes of photothermally aged samples were
complicated and reflected contribution by three processes:  (1)
loss of additives, (2) build up of polymer photo-products and (3)
bleaching of these photooxidation products.  Loss of additives
such as the UV absorber causes a gain of transmission at 360 nm
without causing any change in transmission at 400 nm. Build up of
photooxidative products initially causes yellowing, i.e., increase
in absorption at 400 nm and a greater rate of increase at 360 nm.
This increase in absorption then causes accelerated photolysis
which in turn leads to bleaching of these absorbing species.  The
result is attainment of a photostationary equilibrium in transmission at 400 nm (Figure 5).  Absorption at 360 nm is clearly
dominated by the contribution from the UV absorbers and the loss
of UV absorber therefore results in gradual increase in transmission at 360 nm.  Using the absorption data, we can estimate the
photoproducts formed via photothermal degradation actually absorb
eight times as much light as 360 nm than they do at 400 nm (i.e.,
$\xi_{360} = 8\xi_{400}$), but this increase in absorption is reversed by the
decrease in absorption of the UV absorber.  Thus transmission also

attains a photostationary equilibrium at 360 nm (Figure 6).
Quantitatively, loss of the UV absorber can be estimated from
Figure 5 and Figure 6 and the relationship $\xi_{360} = 8 \times \xi_{400}$ for the
photothermal degraded products.

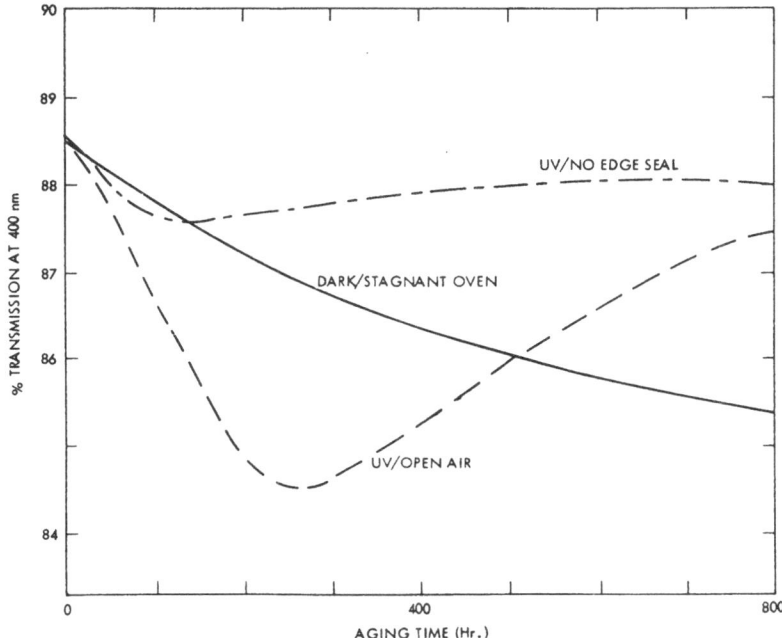

Figure 5.   Change in % Transmission at 400 nm of EVA Film (A9918)
            as a Function of Photothermal Aging at 105°C and 6 Suns

It is important to note that while EVA readily turns yellow in a
dark oven, loss of optical transmission is not a failure mode for
EVA under photothermal aging.

Our room temperature studies of EVA identified rapid destruction of peroxide and a slow formation of hydroxyl group as the
principal degradation modes at 25°C. We have found infrared spec-
troscopic evidence of acetic acid being formed after 800 hours of
aging (3.2 years of outdoor exposure) and a faster rate of formation of hydroxyl group both at 105°C after 800 hours. The sealed
samples (UV/no edge seal) had significantly higher concentration
of acetic acid than the oven aged samples, while very little or no
acetic acid can be detected in the open (UV/ambient oxygen) (Figure 7). In a stagnant oven a certain degree of evaporation can be
achieved, while in a sealed sample (UV/no edge seal) a significant
amount of acetic acid is retained.

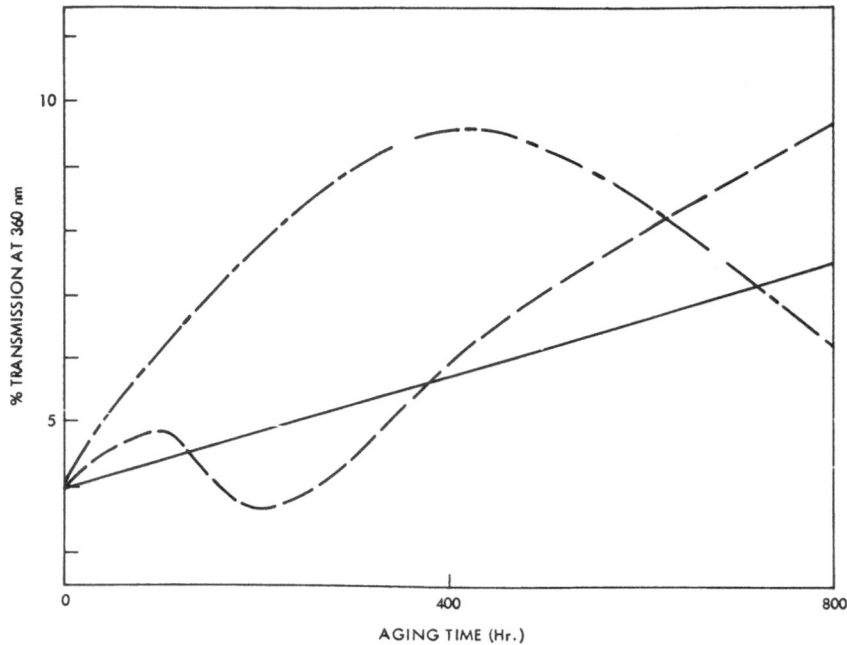

Figure 6.   Change in % Transmission at 360 nm of EVA Film (A9918)
            as a Function of Photothermal aging at 105°C and 6 Suns

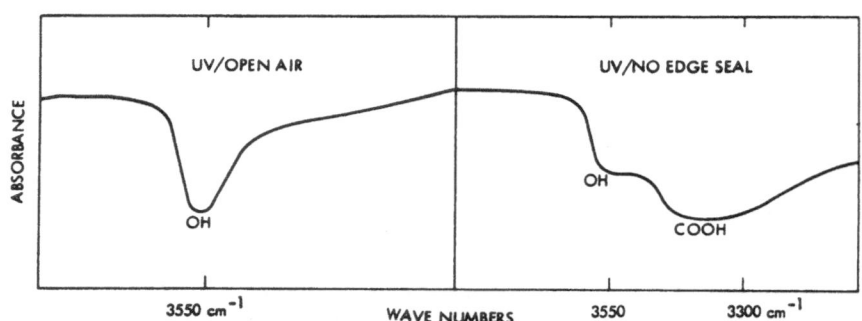

Figure 7.   Difference FT-IR Spectra of EVA Film (A9918) Photother-
            mally Aged at 105°C and 6 Suns in a) UV/Ambient Oxygen
            Sample; b) UV/No Edge Seal Sample

        Weight loss data of EVA as a function of photothermal aging
have been collected at 70°C, 85°C, and 105°C.  They all exhibited
similar profiles as a function of aging time.  At 70°C, gradual
weight loss up to 0.5% after 500 hours of aging, which is
equivalent to 2 years of outdoor exposure, was observed.  At 85°C,

weight loss up to 1% was observed after 800 hours of aging (3.2 years outdoor). Figure 8 illustrates the weight loss profiles of EVA samples aged at 105°C.

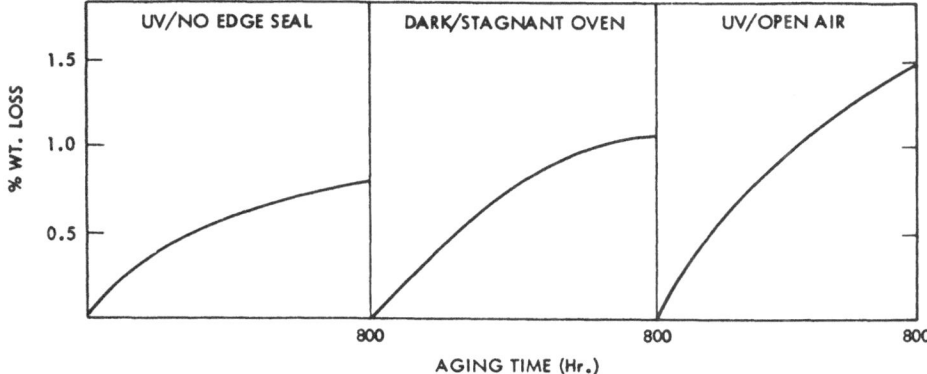

Figure 8.  Weight Losses of EVA Films (A9918) as a Function of Photothermal Aging at 105°C and 6 Suns

Together with the FT-IR results (previous paragraph) these weight loss data can be interpreted in terms of the following three sequential processes:  (1) loss of additives, (2) oxygen uptake through photothermal oxidation leading to formation of volatile products (i.e., acetic acid) and non-volatile products (i.e., hydroxyls) and (3) evaporation of the volatile products. Weight loss is a result of process (1) and (3), while process (2) leads to weight gain.

Initially all three samples experienced a linear weight loss region due to loss of additives.  Subsequently they all leveled off to a different percentage of weight loss.  Figure 9 is used to interpret the data.  In the UV/no edge seal sample (Figure 9a), where process (3) is not prominent, the difference between the actual weight loss and the extrapolated weight loss due to loss of additives AB is the result of total oxygen uptake through oxidation.  In the dark/oven sample (Figure 9), the additional weight loss BC must be due to partial evaporation of the acetic acid in a stagnant oven.  In the open (UV/-ambient UV) sample, BD represented the total amount of acetic acid (or volatile products) formed and evaporated and AD represented the amount of oxygen uptake through oxidation to form the non-volatile products.  Additional information can be obtained from these weight loss data measured at different temperatures.  Rate of loss of additives at 105°C can be estimated from the initial slope of weight loss in Figure 8.  When this rate is compared with that obtained at 85°C,

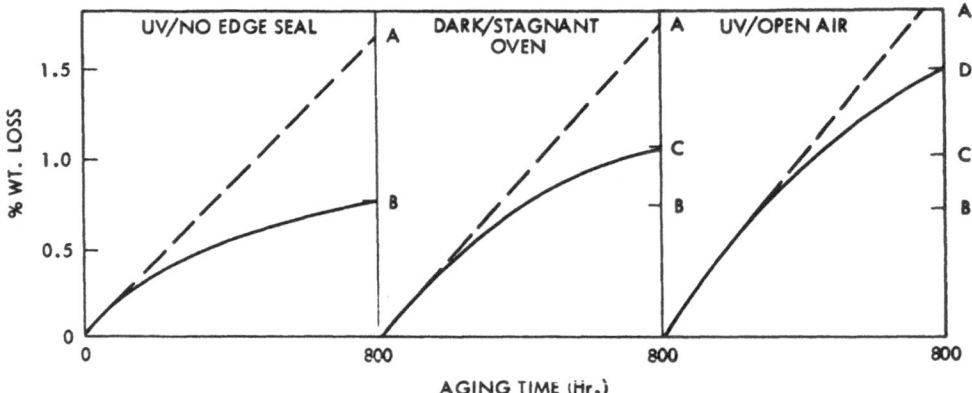

Figure 9.   Interpretation of the Weight Loss Profiles of EVA Films
            (A9918) as a Function of Photothermal Aging at 105°C
            and 6 Suns

energy of activation of the weight loss process is calculated to
be 7Kcal/mole, a typical value for physical processes.
        Stress-strain experiments were carried out before and after
aging in order to determine the change in tensile modulus due to
photothermal aging and are illustrated in Figures 10 and 11.   Table 3 shows the proportion results of the molecular weight analysis, the proportion of the extractable and crosslinking density as
a function of photothermal aging time.   Apparently EVA undergoes
chemical crosslinking when it is subjected to photothermal aging
which result in the decrease in the proportion of the extractables
and in apparent modulus, and in an increase in the chemical cross-
linking density.

Table 3.   Percentage of Extractable, Molecular Weight Distribu-
           tion ($M_w$) of the Extract and Crosslinking Density of
           EVA as a Function of Photothermal Aging

| TEMPERATURE °C | TIME OF TEST (HOURS) | TEST CONDITIONS | % EXTRACTABLE | $\overline{M}_w$ | CROSSLINKING DENSITY (mole/cm³) |
|---|---|---|---|---|---|
| CONTROL | | AS RECEIVED | 48 | 206,000 | $1.12 \times 10^{-6}$ |
| 85 | 800 | OVEN | 35 | 168,000 | $2.29 \times 10^{-6}$ |
| | | UV/AMBIENT AIR | 33 | 118,000 | $4.32 \times 10^{-6}$ |
| | | UV/NO EDGE SEAL | 29 | 75,000 | $7.80 \times 10^{-6}$ |
| 105 | 800 | OVEN | 37 | 174,000 | $1.33 \times 10^{-6}$ |
| | | UV/AMBIENT | 33 | 91,000 | $5.86 \times 10^{-6}$ |
| | | UV/NO EDGE SEAL | 34 | 44,000 | $10.10 \times 10^{-6}$ |

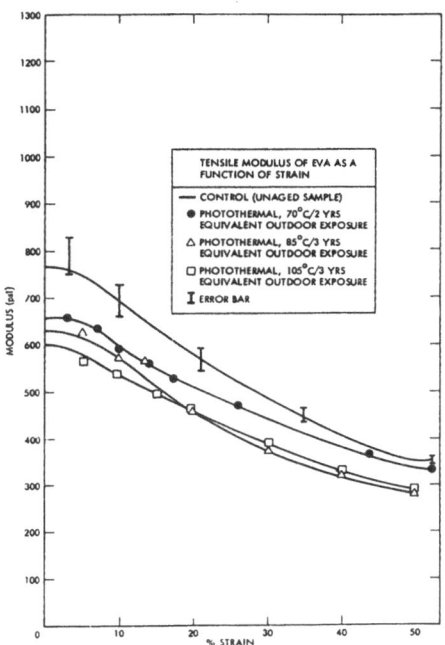

Figure 10. Stress-Strain Curves of EVA Films (A9918) as a Function of Photothermal Aging in Air at 6 Suns and 105°C

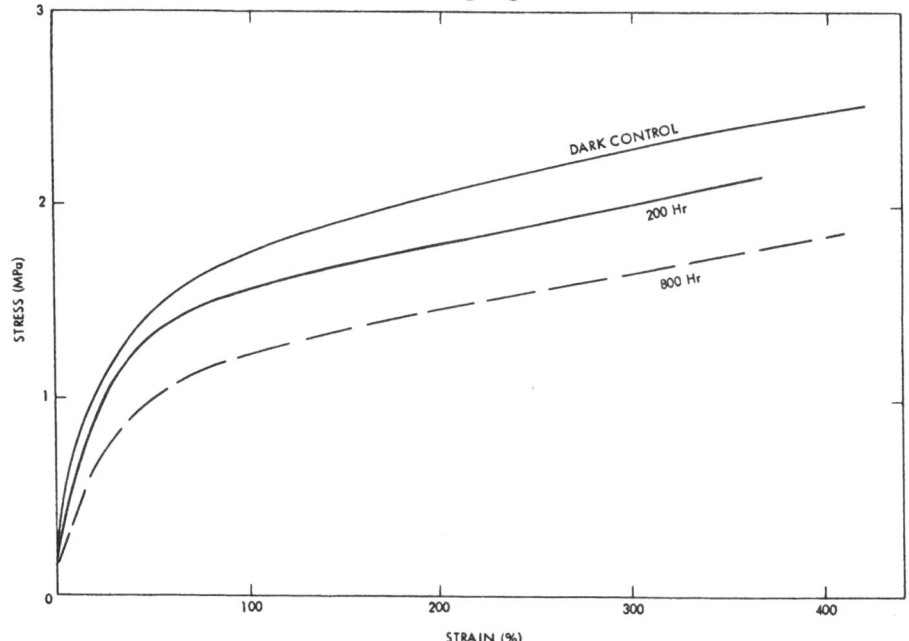

Figure 11. Tensile Modulus of EVA Films (A9918) as a Function of Strain

References

1.  Carroll, W. and Coulbert, C., Cuddihy, E., Gupta, A., and Li-
    ang, R., "Photovoltaic Module Encapsulation Design and Materi-
    al Selection: Volume I," JPL Internal Document No. 5101-177,
    Jet Propulsion Laboratory, Pasadena, CA, November 1, 1981.

2.  Cuddihy, Edward F., "Encapsulation Materials Status to Decem-
    ber 1979," JPL Internal Document No. 5101-144, Jet Propulsion
    Laboratory, Pasadena, CA, January 15, 1980.

3.  Di Stefano, S., "Photocatalytic Degradation of a Cross-linked
    Ethylene Vinyl Acetate (EVA) Elastomer," Polymer Preprints,
    21, 178 (1980).

# SYNTHESES OF NITROGENOUS MATERIALS BY RF PLASMAS

T. L. Ward, O. Hinojosa and R. R. Benerito

Southern Regional Research Center
Agriculture Research Service
U.S. Department of Agriculture
New Orleans, Louisiana  70179

## INTRODUCTION

Radiofrequency (rf) plasmas of argon, nitrogen or air have
been used to make cotton more absorbent of either oil or water[1].
Free radical sites created on the substrate by the plasma picked up
oxygen from moisture or air to chemiluminesce and the electron spin
resonance (ESR) signal intensity was inversely related to the
chemiluminescence (CL).  The free radicals could initiate
polymerization some time after plasma irradiation was stopped.[2]
Irradiation of cotton with any of the aforementioned plasmas
resulted in addition of nitrogen to the cotton.  The nitrogen was
readily detected by electron spectroscopy for chemical analysis
(ESCA), but was not detected by multiple internal reflectance
spectroscopy (MIR).  Except for hydrogen produced by plasma effects
on cotton or derived from water adsorbed on the walls of the
reactor, there was none in those plasma systems.

Miller and Urey[3] reported formation of amino acids and other
organic materials by electric discharge in a mixture of methane,
ammonia, hydrogen and water.  Hollahan and Emanuel[4] produced protein-
like materials by subjecting a mixture of CO, $N_2$, and $H_2$ to an rf
field (13.56 MHz).  In both of these instances hydrogen cyanide and
formaldehyde were shown to be precursors of the amino acids.

This is a report of the irradiation of cotton in rf plasmas
containing either mixtures of $N_2$ and $H_2$ or $NH_3$ gas.  Cotton was
located between the electrodes and downstream from them.  Modifica-
tions of the cotton, particularly the surface, by the various
plasmas were compared by application of the techniques of ESCA, MIR

and electron microscopy. Other tests such as those based on TLC, UV absorption, ESR and amino acid analysis provided additional information. For comparison with cotton, other sources of carbon in different states of oxidation were irradiated between the electrodes.

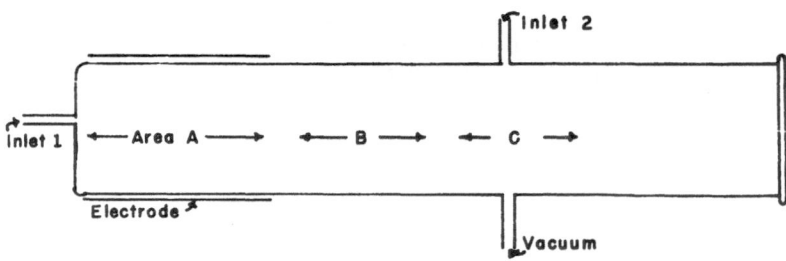

Fig. 1. Plasma reactor showing location of electrodes, sample locations (A,B,C), gas inlets (1,2) and outlet to vacuum.

EXPERIMENTAL

The 0-100 W, 13.56 MHz rf generator, plasma reactor, and general operating procedures were described previously.[5] After irradiation, samples were kept in an argon atmosphere for all loading and transfer operations. Figure 1 shows the three positions in the reactor (A, between the electrodes; B, downstream adjacent to the electrodes; C, further downstream) which were selected for sample location. All locations were within the colored area of plasma. The selected plasma gas or mixture of gases entered the reactor between the electrodes at inlet 1 with flow rates adjusted so that the total flow rate was 0.1 standard $cm^3$/s. Unless otherwise indicated, samples were irradiated at 53.32 Pa (400 mTorr) and 40 W continuous power. The cotton was 80 X 80 woven printcloth that had been desized, scoured, and bleached and cut into 1.5 X 4.0 cm pieces. Samples were supported on the tips of glass prongs so that virtually all sample surface was exposed to the plasma. Methanol was analytical grade, carbon was practical grade decolorizing powder and the formaldehyde was generated by heating paraformaldehyde in a separate glass flask. The other gases were commercial-grade cylinder quality.

Samples were examined by ESCA, MIR, ESR, SEM and by microscopic methods based on layer expansion and cuene solubility as described previously.[6]  The UV and amino acid analyses were made on a 300 µg/ml water solution of the plasma product.  The amino acid analysis was by GLC technique of Kaiser et al.[7]

RESULTS AND DISCUSSION

MIR spectra obtained on samples irradiated at three locations in the reactor (A, B, C) in plasmas of $N_2$, $NH_3$, or mixtures of $N_2$ and $H_2$ are shown in Figure 2. An untreated control is included for comparison. The most obvious difference between spectra of either the control or samples treated in pure nitrogen plasma and those treated in either ammonia or the mixtures of hydrogen and nitrogen is the band of 4.7 µm indicative of N to C triple bonds. In the case of samples irradiated in mixtures of $H_2$ and $N_2$, this band is most prominent for samples located just downstream from the electrodes (B). With the 1:1 ratio of hydrogen to nitrogen, it was smaller for samples further downstream (C), and did not appear at all for samples located between the electrodes (A). While the relationship of intensity of the peak to location in the reactor is the same for $NH_3$ plasma, the intensity of the band for every location is much less than for the $N_2$-$H_2$ plasmas.

Additional changes are noted at the 3.1 µm area associated with CH or NH stretching for samples in all plasmas used except that of pure nitrogen. The area from about 5.8 to 7 µm was intensified for all of the irradiated samples. This was interpreted as added carbonyl groups and amide groups showing NH bending vibrations.

Figure 3 shows the changes in MIR spectra when two different power levels (20W and 40W) were used in irradiating samples for the same duration at the three locations in a plasma of equal parts of $N_2$ and $H_2$. The amount of nitrogen added as shown by the MIR band at 4.7 µm was greater for the higher power and greatest at location B. This indicates that for deposition of nitrogenous material, the power level should be sufficiently high to maintain the plasma glow about the samples and that the sample should be located just downstream from the electrodes.

MIR spectra for samples irradiated for varying periods of time with other parameters maintained constant are presented in Figure 4. The MIR bands previously noted were intensified with longer

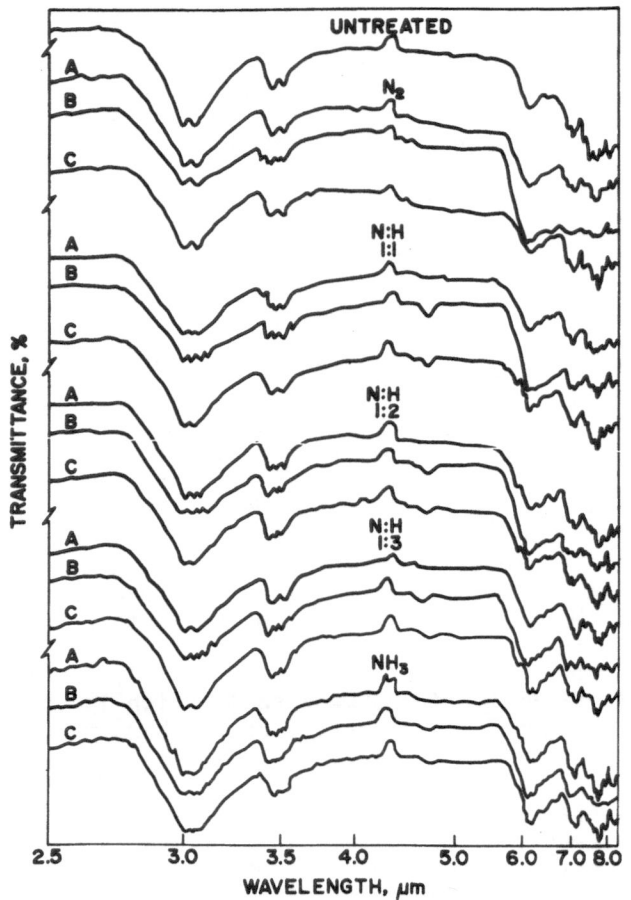

Fig. 2.  Multiple internal reflectance infared spectra of cotton
         irradiated in rf plasmas, 7200s, 40W, 53.32Pa at location
         A, B, or C.

irradiation times, although this was not as noticeable for samples
located between the electrodes.  The advantage of the location
adjacent to the electrodes is also evidenced by the intensity of
the peaks at 4.7 and 6 μm.  When periods longer than 7200 s (2h)
are used, etching of surfaces can be a problem for samples located

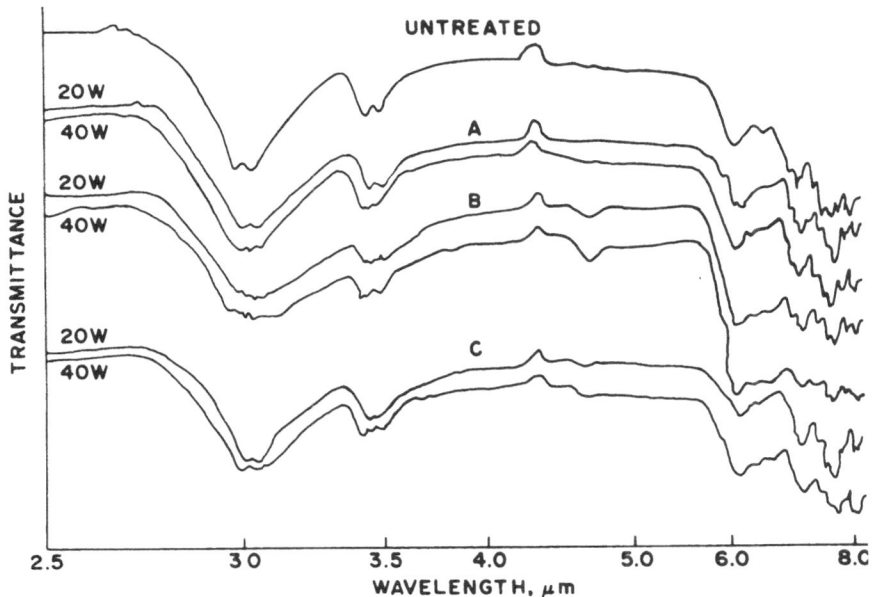

Fig. 3.  Multiple internal reflectance in infared spectra of cotton
         irradiated in rf plasma of 1:1 mixture of $N_2$ and $H_2$,
         7200s, 53.32 Pa, 20 or 40W at location A, B, or C.

between the electrodes.   This was especially true when $NH_3$ plasma
was used.

     Figure 5 shows that although plasma flowed over both the upper
and lower surfaces of the sample suspended horizontally on glass
teats, MIR bands associated with irradiation were modified more
extensively on the upper than on the lower surface.

     Carbon powder, carbon monoxide, formaldehyde, methanol and
methane as well as cotton were placed between the electrodes during
irradiations of a cotton substrate in a plasma of equal parts of $N_2$
and $H_2$.  MIR spectra of the irradiated cotton substrates located
downstream adjacent to the electrodes are shown in Figure 6.

     When methane was the carbon source the band at 4.7 μm in the
MIR spectrum was similar to that when cotton served as the source
of carbon.   However, the spectrum of the product formed with
methane differed in other areas and the product itself was visibly

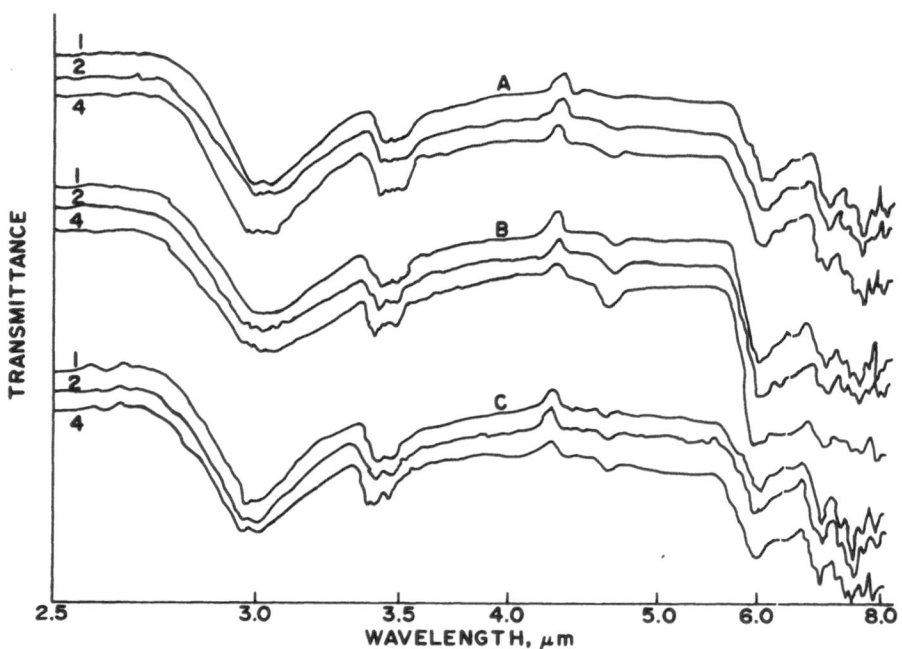

Fig. 4.   Multiple internal reflectance infrared spectra of cotton
          irradiated in rf plasma of 1:1 mixture of $N_2$ and $H_2$, 53.32
          Pa, 40W at location A, B, or C for: 1 (3600s = 1h), 2
          (7200s = 2h), 4 (14,400s = 4h).

different from that produced when cotton was between the electrodes.  All of the carbon compounds other than methane and cotton
had similar spectra and gave no product like the one produced when
cotton was used.  CO did give an increased absorption at 5.8 μm but
no visible product.

Figure 7 presents the ESCA spectra for the $O_{1s}$, $N_{2s}$, and $C_{1s}$
electrons in the surfaces of cotton samples after irradiations in
plasma of equal parts of $N_2$ and $H_2$ while located adjacent to the
electrodes but with different sources of carbon between the electrodes.  The obvious difference is in the carbon spectrum with
cotton between the electrodes.  This is the only carbon spectrum
with a distinct shoulder at higher $E_{BE}$ of 289 eV.  All other peaks
for $C_{1s}$, $N_{1s}$, or $O_{1s}$ electrons are singular and symmetrical.
Relative intensities of the $C_{1s}$ to $N_{1s}$ peaks indicate that deposition of nitrogen downstream is greatest when cotton is between the
electrodes.  These ESCA spectra suggest that the activated carbon
species responsible for formation of the nitrogenous product
derived from cotton differs from that derived from the other carbon
materials.

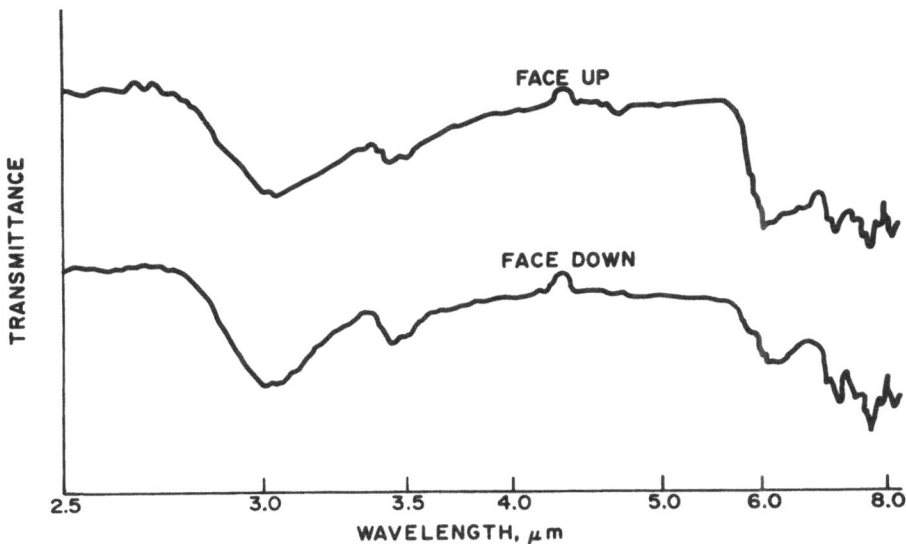

Fig. 5.   Multiple internal reflectance infrared spectra of upper
          and lower surfaces of cotton fabric irradiated in rf
          plasma of 1:1 mixture of $N_2$ and $H_2$, 3600s, 53.52 Pa, 40W,
          location B.

Table 1 gives peak areas and ratios of atoms from ESCA data of
cottons irradiated in the different plasmas of nitrogen and
hydrogen or ammonia.  Nitrogen was added to cotton from every
plasma tested.  When mixtures of molecular hydrogen and nitrogen
were used, considerably less nitrogen was added to samples located
between the electrodes as compared to those downstream.  Amounts
added at the three locations from pure nitrogen or ammonia plasmas
were equivalent.  Considerably less oxygen was added to samples
located downstream (B,C) when mixtures of $H_2$ and $N_2$ were used.
With ammonia plasma there is little difference due to location in
either peak areas or ratios of atoms.  With samples located
downstream (B,C) the amount of nitrogen added was decreased as the
proportion of hydrogen in the mixture was increased.

The binding energy ($E_{BE}$) for $N_{1s}$ electrons for all the
irradiated samples was 63.87 aJ (398.7 eV) and in a range associated
with either RC≡N or $RCONH_2$ structure.[8]  A nitro group would have a

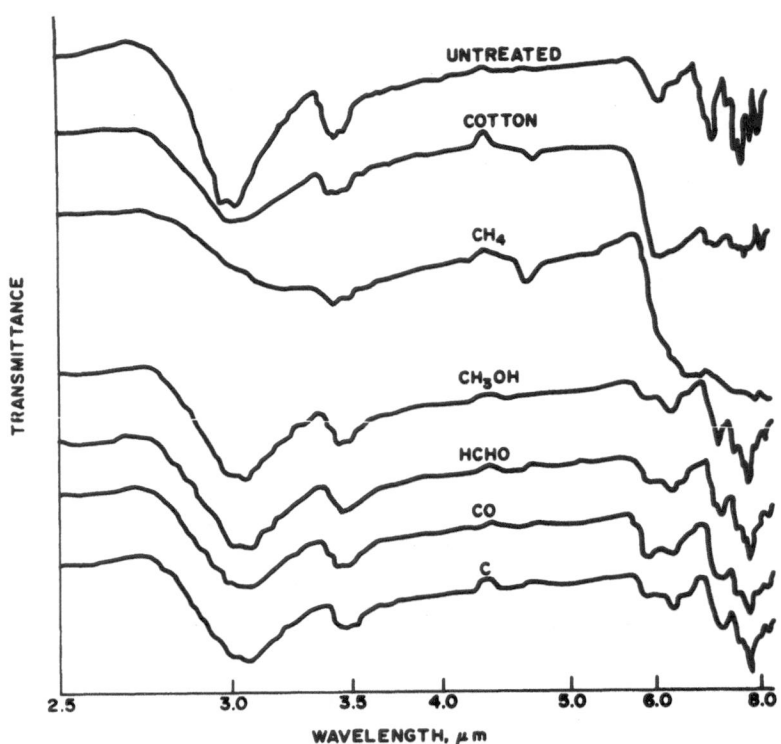

Fig. 6.   Multiple internal reflectance infrared spectra of cotton
located at position B with 6 different sources of carbon
located between the electrodes during irradiation in 1:1
mixture of $N_2$ and $H_2$, 3600s, 53.52 Pa, 40W.

noticeably higher $E_{EB}$. The $C_{1s}$ peak for samples irradiated in all
plasmas except pure nitrogen exhibit a new shoulder at a higher
binding energy in the range associated with carbonyl groups.[1]

Nitrogen does add from a plasma of pure nitrogen and yet no
solid material is formed. The mixture of nitrogen and hydrogen or
pure ammonia gas as plasmas resulted in formation of a solid
material. The mixture of nitrogen and hydrogen resulted in a
coating on the surface of the fibers while ammonia plasma produced
a new material primarily in the outer walls of the fibers (Figure
8).

Table 2 shows that cotton fabric treated in plasmas of ammonia
while located between or adjacent to the electrodes exhibited about
a 20% increase in resistance to wrinkling in the dry state. There
was no increased resistance to wrinkling in the wet state

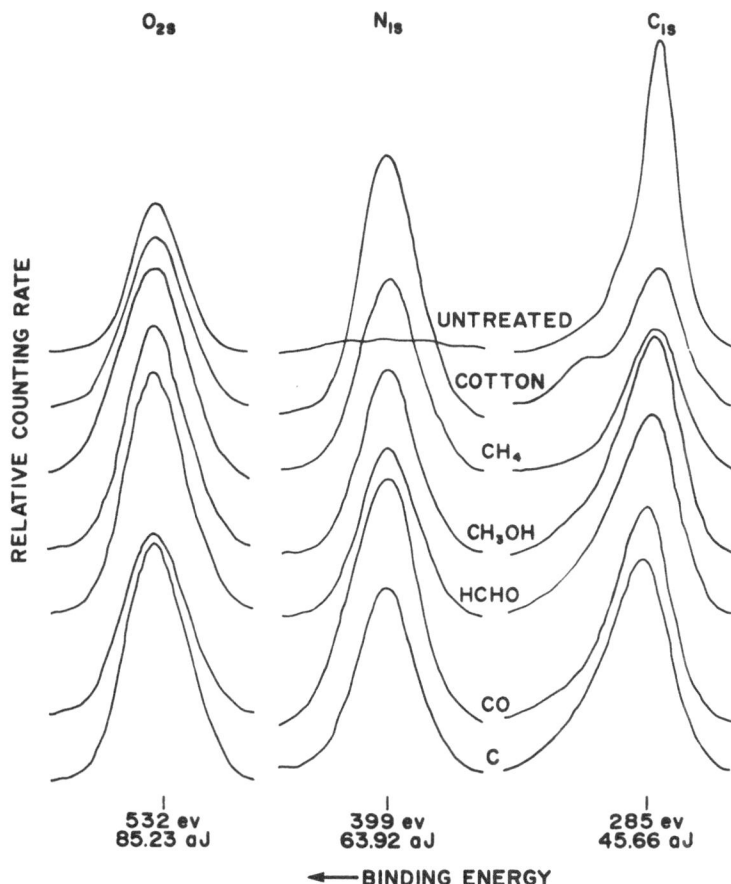

Fig. 7.   ESCA spectra of $C_{1s}$, $N_{1s}$ and $O_{1s}$ electrons in surface of
          cotton located at B with 6 different sources of carbon
          between electrodes during irradiation in 1:1 mixture of $N_2$
          and $H_2$, 3600, 53.52 Pa, 40W.


regardless of location in the reactor.  Cotton fabric treated in
plasmas of mixtures of nitrogen and hydrogen showed no change in
wet or dry recovery from wrinkling.  This may be because the new
nitrogenous material formed in those mixtures is on, rather than
in, the fiber surface.

    The methacrylate layer-expansion technique uses layering
caused by the polymerization of methacrylate that has penetrated
interfibriallar area to indicate the location and amount of
polymerization in the fiber.  Fibers of untreated cotton layer

Table 1.  ESCA peak area and ratios of atoms calculated from peak
          areas of cotton irradiated in RF plasmas, 7200s, 40w
          53.32Pa.

| Plasma Gas | Reactor Zone * | Peak area, arbitrary units | | | Ratio of atoms | | |
|---|---|---|---|---|---|---|---|
| | | $N_{1s}$ | $C_{1s}$ | $O_{1s}$ | N:C | O:N | O:C |
| $N_2$ | A | 277 | 1117 | 1539 | .25 | 5.56 | 1.38 |
| | B | 320 | 1056 | 854 | .30 | 2.67 | .81 |
| | C | 318 | 1274 | 1235 | .25 | 3.88 | .97 |
| $N_2:H_2$ | | | | | | | |
| | A | 178 | 883 | 1126 | .20 | 6.32 | 1.28 |
| | B | 382 | 1002 | 778 | .38 | 2.04 | .78 |
| | C | 426 | 1059 | 619 | .40 | 1.45 | .58 |
| 1:2 | A | 149 | 1168 | 1299 | .13 | 8.72 | 1.11 |
| | B | 374 | 1146 | 674 | .33 | 1.80 | .59 |
| | C | 353 | 1299 | 738 | .27 | 2.10 | .57 |
| 1:3 | A | 164 | 1222 | 1331 | .13 | 8.12 | 1.09 |
| | B | 322 | 1168 | 794 | .28 | 2.47 | .68 |
| | C | 306 | 1336 | 590 | .22 | 1.93 | .43 |
| $NH_3$ | | | | | | | |
| | A | 214 | 1133 | 1146 | .19 | 6.76 | 1.28 |
| | B | 246 | 1062 | 1482 | .23 | 6.02 | 1.40 |
| | C | 226 | 1242 | 1402 | .18 | 6.20 | 1.13 |

* A, between electrodes; B, downstream adjacent to electrodes;
  C, downstream after B.

in an even manner throughout the cross section of the fiber.  Cross-
linking of cellulose chains and/or polymer deposits in the fiber
can prevent layering.  Fibers subjected to plamas of nitrogen and
hydrogen expanded as do fibers of untreated cotton.  Only cotton
fibers that had been subjected to ammonia plasma had area that
resisted layer expansion.  Electron micrographs illustrating
resistance to layer expansion are shown in Figure 9.  More
resistance to layering is exhibited by the samples located either
between or adjacent to the electrodes.

     Cross sections of fibers from a cotton after ammonia plasma
treatment and from an untreated control were examined by techniques

Fig. 8.   Scanning electron micrographs of cotton fiber (100X, 5
          KV), X (untreated control), Y (irradiated in 1:1 mixture
          of $N_2$ and $H_2$, 7200s, 40W, 53.52 Pa), Z (irradiated $NH_3$,
          7200s, 40W, 53.52 Pa).

Table 2.   Effect of ammonia plasma upon weight and wrinkle
           recovery of cotton printcloth[1]

| Location in reactor[2] | A | B | C | Untreated |
|---|---|---|---|---|
| Weight loss, % | 12.27 | 3.99 | 3.90 | ---- |
| Wrinkle recovery (W + F) m rad | | | | |
| Conditioned | 3368 | 3438 | 2932 | 2880 |
| Wet | 2729 | 3001 | 2984 | 3158 |

[1]Treated 3600s, 150 μm Hg, 40W.
[2]A, between electrodes; B, downstream next to A; C, downstream
   after B.

of transmission electron microscopy.  Electron micrographs of cross
sections after soakings in 0.5 M cupriethylenediamine hydroxide
(cuene) are presented in Figure 10.  Untreated cotton dissolves in

LAYER EXPANSION

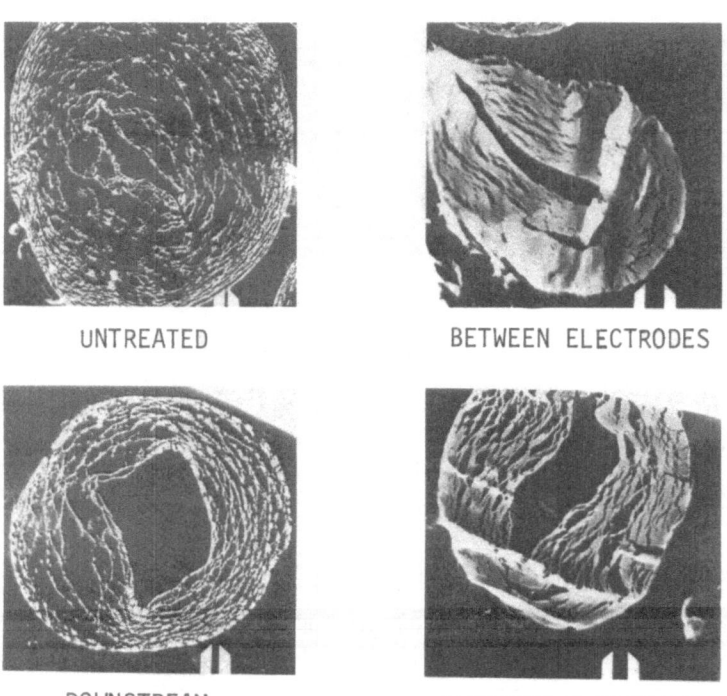

UNTREATED                    BETWEEN ELECTRODES

DOWNSTREAM                   ADJACENT ELECTRODES

Fig. 9.  Transmission electron micrographs of cross sections of
         fibers from untreated and ammonia plasma treated cotton
         printcloth subjected to methacrylate layering
         technique.


cuene while crosslinked fibers do not.  The surface of the ammonia
plasma-treated fiber resisted cuene.  The layering technique showed
reaction more evenly distributed throughout the cross section.  It
may be that there are either two different reaction products or two
different degrees of reaction when cottons are subjected to $NH_3$
plasmas.

    In our previous studies irradiation of any polysaccharide in
rf plasma resulted in a strong ESR signal indicating free
radicals[1].  The free radicals have been associated with excited
carbonyls which produce chemiluminescence (CL).  The two are
reciprocally related, that is, as ESR decays CL is produced.  The

CUE SOLUBILITY

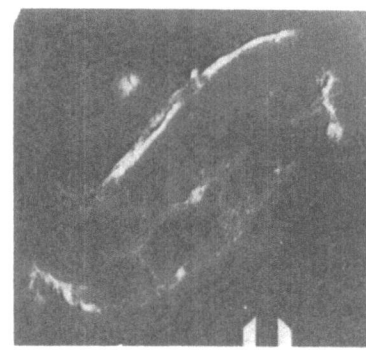

UNTREATED                                    ADJACENT ELECTRODES

Fig. 10.  Transmission electron micrographs of cross sections of
          fibers cross sections soaked in 0.5 M
          cupriethylenediamine hydroxide (cuene) for 1800s.

cotton irradiated in any of the plasmas tested has strong ESR
signals.  The signals decayed with time; slowly in a dry argon
atmosphere, and rapidly in moisture or air.  All gases as plasmas
produced similar spectra having the same (2.0033) g-factor which
exceeds that of the carbon standard (2.0028).  Norman and Pritchett
correlated g-factors with structure of certain aliphatic radicals,
showing that the g-factors of carbon radicals increase as attached
groups change from saturated to carbonyl.[9]  The increase in the $E_{BE}$
of the $C_{1s}$ electrons as determined by ESCA and the increase of the
IR band at 6 μm of samples after plasma irradiations indicated the
presence of carbonyl groups.

    Irradiation of cotton in all the plasmas resulted in CL when
the sample was exposed to oxygen in moisture or air.  Only 1/3 as
much CL was exhibited by cotton irradiated in ammonia plasma as by
cotton irradiated in the other plasmas even though the ESR signals
were of similar intensities.  The CL decays more slowly for ammonia
treated cotton.  This may mean that these free radicals sites are
more protected from moisture or air by the water insoluble material
in the fiber surface.

Table 3 gives the CL both immediately after irradiation in an equimolar nitrogen-hydrogen plasma and after exposure to air.  Also included are the ESR signal intensities.  The signals were obtained from a sample located just downstream from the electrodes and irradiations were conducted with various sources of carbon between the electrodes.  Either cotton or methane between the electrodes resulted in far more CL than did any of the other sources of carbon. Methane had the least ESR signal of any carbon source.  It would seem than when methane is used, the CL must be generated via a different mechanism that when cotton was between the electrodes.  The data indicate that there is no direct relationship between oxidation state of the carbon source between the electrodes and either the ESR or CL intensities.

The nitrogenous material produced by the ammonia plasma is not removed from the treated cotton by water.  The material produced by the mixture of hydrogen and nitrogen is soluble in water.  With cotton between the electrodes, a plasma of equal parts of nitrogen and hydrogen coated a glass plate located downstream at "B".  The coating was examined by MIR and gave the spectrum in Figure 11.  The spectrum of the coating looks similar to that of a cotton located in the same place in the reactor.  There is an accentuation of the bands at 4.7 and 5.8 μm.  The coated material on the glass plate was soluble in water, acetic acid, methanol, phenol, DMF, and DMSO. It was relatively insoluble in carbon tetrachloride, benzene, cyclohexane, and methyl ethyl ketone.

Table 3.   ESR and CL intensities of cotton with different sources of carbon between reactor electrodes.[*]

| Carbon source | CL (counts/60s) | | ESR intensity (arbitrary units) |
|---|---|---|---|
| | Immediate | After air | |
| Methane | 870,000 | 9,970,000 | 5.1 |
| Methanol | 25,000 | 185,000 | 6.2 |
| Formaldehyde | 18,000 | 190,000 | 8.3 |
| Carbon monoxide | 40,000 | 160,000 | 13.5 |
| Carbon powder | 15,000 | 135,000 | 7.6 |

[*]14,400s, 40w, 53.32Pa, ESR and Cl measured on cotton located adjacent to electrodes in a 1:1 $N_2$ and $H_2$ plasma.

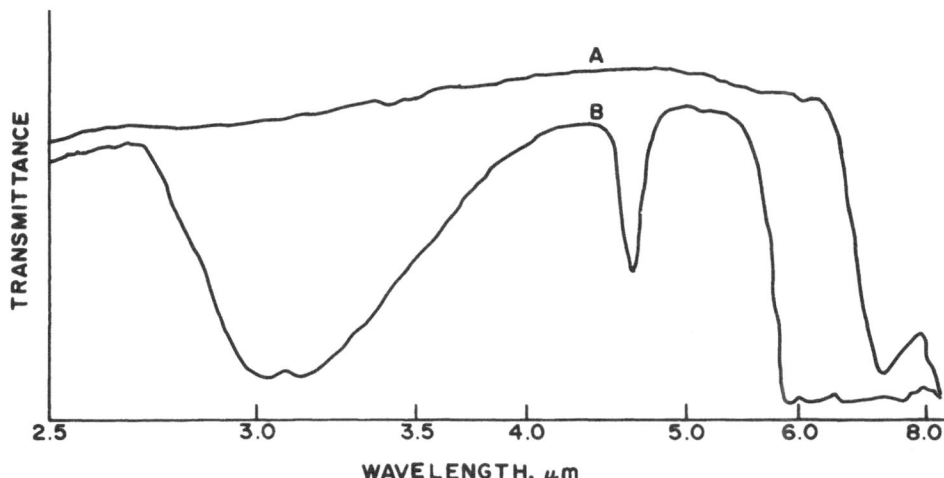

Fig. 11. Multiple internal reflectance infrared spectra of: A
(glass collector plate) B. (coated glass plate).
Plate located adjacent to electrodes during irradiation
in 1:1 mixture of $N_2$ and $H_2$, cotton between electrodes,
40W, 53.52 Pa, 14,4000s.

The coating of nitrogenous material was removed from the glass
collector plate by washing with doubly deionized water. The solution was filtered and analysis showed it to contain 300 µg/ml of
material. Filtration removed a small amount of material that was
found to be unreacted cellulosic material.

The solution was examined by ultraviolet spectroscopy for
absorption between 200 and 360 nm and no bands were observed.
Certain protein materials as well as aromatic or conjugated structures give bands in this area. Although about 20 amino acids are
found in proteins, only phenylalanine (282 nm), tyrosine (303 nm),
and tryptophan (350 nm) absorb and emit in the range scanned.[10]
The CL emitted by cotton when the coating is on it, rather than on
the glass, indicates that the CL is due to the activated cotton
substrate rather than to the coating.

Figure 12 is an ascending thin layer chromatogram of the
water soluble coating formed on glass in a plasma of equimolar
nitrogen and hydrogen with cotton between the electrodes. It was
on a precoated Gil-G-25 TLC plate from Brinkman Instrument Co.* and
with a 7:3 propanol: water eluate. The TCL color was developed
with ninhydrin spray which is a colorometric indicator of amino
acids. The four to five color bands indicate that the coating may
contain amino acids.

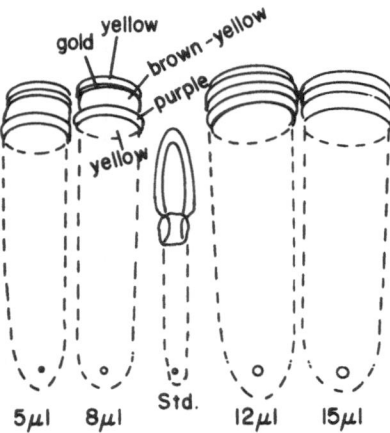

Fig. 12.   Thin layer chromatogram of coating formed on glass by 1:1
          mixture of $N_2$ and $H_2$ as plasma, cotton between
          electrodes, 40W, 14,400s, 53.52 Pa, Eluent (7:3
          propanol:water), minhydrin spray.

Table 4 lists the 17 amino acids found in the water soluble
coating produced on the glassy plate by the equimolar nitrogen-
hydrogen plasma with cotton between the electrodes.  The total
µg/ml of 284.9 represents 94.97% of the 300 µg/ml that was in the
solution.  Glycine is the predominant amino acid with aspartic
acid, alanine, serine, glutamic acid, and phenylalanine present at
the 5% or greater level.  Referring back to the UV scan, materials
that contain phenylalanine with tyrosine but no tryptophan emit at
303 nm the same as tyrosine alone.[10]  This together with the low
level of tryosine in the coating may explain the lack of an
absorption band in the UV scan.

*Names of companies or commercial products are given solely for the
 purpose of providing specific information; their mention does not
 imply recommendation or endorsement by the U.S. Department of
 Agriculture over others not mentioned.

Table 4.   Analysis of coating on glass plate.[*]

| Amino Acid | μg/ml | Weight % |
|---|---|---|
| Alanine | 22.2 | 7.79 |
| Valine | 3.4 | 1.19 |
| Glycine | 122.1 | 42.86 |
| Isoleucine | 4.5 | 1.58 |
| Leucine | 6.0 | 2.11 |
| Proline | 3.8 | 1.33 |
| Threonine | 4.0 | 1.40 |
| Serine | 20.3 | 7.13 |
| Methionine | 4.4 | 1.54 |
| Hydroxyproline | 2.4 | .84 |
| Phenylalanine | 15.5 | 5.44 |
| Aspartic acid | 30.0 | 10.53 |
| Glutamic acid | 26.1 | 9.16 |
| Tyrosine | 3.1 | 1.09 |
| Lysine | 6.1 | 2.14 |
| Histidine | 7.1 | 2.49 |
| Arginine | 3.9 | 1.37 |
| Cystine | | |
| Tryptophan | | |
| Totals | 284.9[**] | 99.99 |

[*]1:1 nitrogen-hydrogen plasma, plate adjacent
   to electrodes, cotton between electrodes.
   40W, 14,400s, 53.32Pa.
[**]Test performed on solution of 300 μg/ml.

CONCLUSIONS

    Irradiation of cotton in rf plamas of mixtures of hydrogen and
nitrogen resulted in formation of a nitrogenous material that was
coated on either cotton fabric or glass plates located downstream
from the electrodes.  The coating was water soluble and separated
into 5 fractions in a thin layer chromatographic test.  Amino acid
analysis showed that 17 different amino acids constituted almost
95% of the coating material.

    Irradiation of cotton in rf plasma of ammonia resulted in
formation of a1nitrogenous material in the surface of the fibers.
This material was insoluble in cupriethylenediamine hydroxide
solution which dissolves untreated cellulose.  The treated fibers
resembled those of a conventionally crosslinked cotton.  The cotton

fabric treated with $NH_3$ plasma showed a 20% improvement in resistance to wrinkling in the dry state. There was no improvement for the wet state.

Nitrogen adds from any of the plasma gases but the nitrogenous solid material was formed only when cotton was located between the electrodes of the reactor. Other sources of carbon in different states of oxidation were tried between the electrodes and none of them resulted in the same product as was obtained with cotton. It is believed that the cotton serves as a source of carbonaceous moieties which together with activated species of N, O and/or H in the plasma can form amino or amide structures.

Activated sites formed on cotton by irradiation in any of the plasmas were scavenged by oxygen from moisture or air and emitted chemiluminescence. The nitrogenous coating from plasmas of mixtures of nitrogen and hydrogen did not impede reaction of the activated sites with oxygen. The material produced by ammonia plasma did reduce formation of chemiluminescence. This may be due to shielding by the newly formed material.

ACKNOWLEDGMENTS

The authors thank Donald Soignet for ESCA, Edith Conkerton for TLC and Wilton Goynes and Jarrel Carra for microscopy.

REFERENCES

1.  T. L. Ward, H. Z. Jung, O. Hinojosa and R. R. Benerito, J. Surface Sci. 76, 257 (1978).
2.  T. L. Ward, H. Z. Jung, O. Hinojosa and R. R. Benerito, J. Appl. Poly. Sci. 23, 1987 (1979).
3.  S. L. Miller and H. C. Urey, Science, 130, 245 (1959).
4.  J. R. Hollahan and C. F. Emanuel, Biochim. Biophys. Acta, 208, 317 (1970).
5.  H. Z. Jung, T. L. Ward and R. R. Benerito, Text. Res. J. 47, 217 (1977).
6.  R. T. O'Conner, Instrumental Analysis of Cotton Cellulose and Modified Cotton Cellulose, New York, Marcel Dekker, Inc. pp. 43, 215, 233, 441 (1972).
7.  F. E. Kaiser, C. W. Ghrke, R. W. Zumwalt and K. C. Kero, J. Chromatog. 94, 113 (1974).
8.  Siegbahn, Kai, et al., ESCA; Atomic, Molecular, and Solid State Structure Studied by Means of Electron Spectroscopy, Nova Acta Regiae Societies Scientiarum Upsaliensia, Upsala, Ser. IV, Vol. 20, pp. 108-118 (1967).

9.  R. O. C. Norman and R. J. Pritchett, Chem. Ind. 50, 2040
    (1965).
10. E. J. Bowen, Luminescence in Chemistry, London, D. Van
    Nostrand Co., Ltd. p. 200 (1968).